Ethics, Health Policy and (Anti-) Aging: Mixed Blessings

ETHICS AND HEALTH POLICY

Series Editors

The Series Ethics and Health Policy disseminates international research in current ethical issues and moral dilemmas in health policy

Editorial Board

VOLUME 1

For further volumes:
http://www.springer.com/series/10228

Maartje Schermer • Wim Pinxten
Editors

Ethics, Health Policy and (Anti-) Aging: Mixed Blessings

Springer

Editors
Maartje Schermer
Erasmus MC University
Medical Center Rotterdam
Department of Medical Ethics
and Philosophy of Medicine
Rotterdam
The Netherlands

Wim Pinxten
Erasmus MC University
Medical Center Rotterdam
Department of Medical Ethics
and Philosophy of Medicine
Rotterdam
The Netherlands

ISSN 2212-1536 ISSN 2212-1544 (Electronic)
ISBN 978-94-007-3869-0 ISBN 978-94-007-3870-6 (eBook)
DOI 10.1007/978-94-007-3870-6
Springer Dordrecht Heidelberg London New York

Library of Congress Control Number: 2012952195

Springer is part of Springer Science+Business Media (www.springer.com)

Preface

This volume is the first in a new Springer series, Ethics and Health Policy. It reflects on the challenges that ethics and health policy face in aging societies. Against the background of an ever further increasing life expectancy, and taking the huge impact of aging populations on healthcare and health policy into account, it is hard to imagine a more appropriate subject to start this series with.

Although life in itself is highly valued by many, aging—both in its individual and in its societal dimensions—is of often experienced as a mixed blessing. In this volume, the tension between valuing longer lives and coping with the decline that currently comes with age is considered from a wide variety of views and scholarship on the ethical aspects of aging. It includes reflections on the public and personal views on (good) aging; biomedical attempts to understand and influence the aging process; and on (public) health policy in an aging society.

The volume is in large part the result of an international conference on Ethics and Aging, held in Amsterdam on 17 and 18 March 2011. The conference was organised by the department of Medical Ethics and Philosophy of Medicine of ErasmusMC University Medical Center, with kind support of the Dutch Organisation for Scientific Research (NWO). The success of the conference inspired the idea to bring all the perspectives and views expressed and discussed there together into a comprehensive volume.

We are very pleased that most of the speakers at the conference have elaborated their lectures into contributions to this volume and we are also grateful to those authors who have developed their short comments into full chapters. We hope that the multi-disciplinary, open and positive-critical atmosphere of inquiry that marked the conference is reflected in the present volume.

Some of the work in this volume, (i.e. the chapters by de Beaufort, Meulenberg, Pinxten, and Schermer) is the direct result of the project Aging: personalised genomics, empowerment, identity and medicalization, funded by the CSG Center for Society and the Life Sciences. We gratefully acknowledge their support.

We belief this volume will offer the reader an interesting and inspiring collection of perspectives on the mixed blessings of growing old; and we hope this will contribute to further productive discussions on ethics and health policy in aging societies.

Rotterdam, The Netherlands
June 2012

Maartje Schermer
Wim Pinxten

Contents

About the Authors

Inez de Beaufort is professor of Healthcare Ethics at ErasmusMC University Medical Centre Rotterdam and head of the department of Medical Ethics and Philosophy of Medicine. She is an honorary member of the Dutch Health Council, a member of committees of the Dutch organisation for Research in Healthcare, the Appraisal Committee for the Societal Aspects of the Healthcare Insurance (CVZ), the European Group on Ethics in Science and New Technologies as well as the Board of the International Association of Bioethics. She was the organiser of the 11th World Congress of Bioethics (Rotterdam 2012). She has been a member of four 'local' review boards, as well as a member of the National Dutch Council for Ethics and Medical Research, the CCMO. She has published on research, reproductive technologies, beauty and ethics, personal responsibility for health, ethics and obesity, and ethics in fiction and is interested in the way scientists are portrayed in fiction. She is also the joint Editor-in-Chief of the series Ethics and Health. Being old she is very interested in ethical issues regarding aging and longevity.

Laura Capitaine has a Master's degree in Philosophy (Ghent University) and an Erasmus Mundus Master of Bioethics (Leuven University, Case Western Reserve University, Università degli Studi di Padova). She is currently affiliated to the Bioethics Institute Ghent as a fellow of the Research Foundation Flanders. Laura's Ph.D. research project consists of an ethical analysis of the use of age as a criterion for decision making in healthcare. The main problems addressed include the use of age in the context of medically assisted reproduction and organ transplantation. In addition, the research also focuses on life extension and anti-aging.

Wim Dekkers is a medical doctor and philosopher by training. For many years he combined an academic position with part-time work as a general practitioner. At the moment he is associate professor Philosophy of Medicine and Principle Lecturer in the section of Ethics, Philosophy and History of Medicine of the Scientific Institute for Quality of Healthcare of Radboud University Nijmegen Medical Centre. He is mainly interested in philosophical and ethical aspects of palliative care, especially in conceptual and ethical questions relating to old age, vulnerability, dementia, mortality, death, and end-of-life-decisions.

Hans-Jörg Ehni studied Philosophy, Literature and Political Science in Stuttgart, Paris and Tübingen. He completed his Ph.D. in philosophy under supervision of Professor Dr. h.c. Otfried Höffe on the subject of moral evil. He is currently deputy director of the Institute for the Ethics and History of Medicine, University of Tübingen. His research is focused on ethics and aging, particularly on the ethics of new biomedical interventions into the aging process and on the ethics of biomedical research involving human subjects, particularly the ethics of placebo controls in clinical trials and international research cooperation.

John Harris is a British bioethicist and philosopher. He is the Lord Alliance Professor of Bioethics and Director of the Institute for Science, Ethics and Innovation at the University of Manchester. He was educated at the University of Kent and Balliol College, Oxford. He was one of the Founder Directors of the International Association of Bioethics and is a founder member of the Board of the journal Bioethics and a member of the Editorial Board of the Cambridge Quarterly of Healthcare Ethics. He is also the joint Editor-in-Chief of the Journal of Medical Ethics.

Govert den Hartogh has taught moral, Political and Legal Philosophy and Medical Ethics, as an assistant, associate and full professor, in the Department of Philosophy and the Faculties of Law and Medicine of the University of Amsterdam, from 1974 to 2008. As professor emeritus he still directs some research projects, including a project on indirect paternalism in the professional ethics of doctors. He has been a member of one of the Dutch regional review committees for euthanasia from 1998 to 2010, and is a member of the Dutch Health Council. His present research interests concern in particular philosophical issues concerning death and the right to life.

Marieke ten Have studied philosophy at the University of Amsterdam. Subsequently she performed a Ph.D. research about the ethical aspects of obesity prevention. This resulted in the thesis "Prevention of obesity: weighing ethical arguments". Marieke is currently employed at the Council for Public Health and Health Care and at the related Centre for Ethics and Health.

Göran Hermerén Professor and artist. He has been professor of Philosophy at several universities, and since 1991 professor of Medical Ethics at Lund University, Sweden. He was President of the European Group on Ethics in Science and New Technologies (EGE) in Brussels, and is currently chair of the Advisory board of the German Reference Centre for Ethics in the Life Sciences (DRZE) in Bonn and involved in several EU-funded research projects. He has numerous publications, mainly on ethical problems in medicine, besides authoring and editing books on research ethics, aesthetics and philosophy of art. He is also an artist, and has exhibited his abstract paintings and graphical works in Sweden and abroad. He lives in Lund, Sweden.

Cees Hertogh is professor of Geriatric Ethics at the department of Nursing Home Medicine and the EMGO-Institute of the Vrije Universiteit Medical Center, Amsterdam. He teaches methodology and ethics of care at the medical faculty and the specialist training course for elderly care physicians (GERION). He is also working as an elderly care physician at the Vivium Zorggroep, Naarderheem, Naarden. His research focuses on ethical issues in the care and treatment of vulnerable older people using empirical research methods (empirical ethics). He has gained expertise in care ethics, qualitative research in ethics, ethical issues in the use of assistive and surveillance technoloy, end-of-life-decision making and euthanasia in patients with dementia, advance care planning and (decisional) capacity.

Søren Holm is a prominent bioethicist and philosopher of medicine. He holds a chair in Bioethics at the Centre for Social Ethics and Policy, part of the School of Law at the University of Manchester in Great Britain and a part-time chair at the Centre for Medical Ethics, University of Oslo. Holm holds a master's degree in healthcare ethics from the University of Manchester and two doctoral degrees in medical ethics from the University of Copenhagen.

Bert Keizer is a physician and philosopher. His bestselling book about life in a nursing home, Dancing with Mr. D. (1994), has been translated into several languages. After writing a novel, Passing Festivities (1998), and an introduction to the philosophy of Ludwig Wittgenstein, he published a translation of letters by Emily Dickinson in 2008. Keizer is a columnist for the Dutch newspaper Trouw and medical journal Medisch Contact, and a regular contributor to The Threepenny Review in Berkeley.

Frits de Lange Professor of Ethics, Protestant Theological University, the Netherlands. Professor Extraordinary of Systematic Theology and Ecclesiology, University of Stellenbosch, South Africa. Among his publications: De mythe van het voltooide leven (The Myth of a Completed Life) 2007; De armoede van het zwitserlevengevoel (The Shabbiness of Third Age Hedonism) 2008; Waardigheid—voor wie oud wil worden (Dignity—for who wants to grow old) 2010; In andermans handen. Over flow en grenzen in de zorg. (In someone else's hands. On flow and limits in care) 2011. Further info: www.fritsdelange.com.

João Pedro de Magalhães graduated in Microbiology in 1999 from the Escola Superior de Biotecnologia in his hometown of Porto, Portugal, and then obtained his Ph.D. in 2004 from the University of Namur in Belgium, where he worked in the Aging and Stress Group led by Dr. Olivier Toussaint. Following a postdoc with genomics pioneer Prof. George Church at Harvard Medical School, in 2008 Dr. de Magalhaes joined the University of Liverpool as a Lecturer where he now leads the Integrative Genomics of Ageing Group (http://pcwww.liv.ac.uk/~aging/). The group's research broadly focuses on understanding the genetic, cellular, and molecular mechanisms of aging.

Frans Meulenberg studied Dutch literature and language (University of Amsterdam). Currently he works as researcher 'ethics and fiction' at the Department of Medical Ethics and Philosophy of Medicine at ErasmusMC University Medical Center, Rotterdam. As such, he focuses on the representation of ethical and philosophical dilemmas in novels and films. Besides that he is also a science-writer, publishing articles and books mainly on 'literature and medicine'.

Guido Pennings director of the Bioethics Institute Ghent, is professor of Ethics and Bioethics. He obtained his Ph.D. in Moral Sciences at the Free University Brussels. He is a member of the Ad Hoc Ethics Committee of the Centre for Reproductive Medicine of the Academic Hospital of the Free University Brussels, the Task Force on Ethics and Law of the European Society of Human Reproduction and Embryology (ESHRE), the National Advisory Committee for Bioethics, the Ethics Committee of the Services for the Blood of the Belgian Red Cross and of the Reflection Group for Biomedical Ethics of the Academic Hospital of the Free University Brussels.

Wim Pinxten is senior research fellow at Erasmus MC University Medical Center in Rotterdam. He has a background in religious studies (M.A.), law (B.A.), applied ethics (M.A.), and biomedical sciences (Ph.D.). As a researcher in medical ehtics and health law, he has been involved in research projects on the triage of elderly at Intensive Care Units (at the International Forum for Biophilosophy, Leuven), the ethical and legal aspects of pediatric clinical trials (at the University of Leuven), and the ethical aspects of aging (at the ErasmusMC, Rotterdam) His research interests are in the ethical, legal and social aspects of clinical research in (minor) human subjects, aging and senescence, global justice in healthcare, and rare diseases and orphan drugs.

Maartje Schermer is professor by special appointment of Philosophy of Medicine, at ErasmusMC University Medical Center in Rotterdam. She is a medical doctor and philosopher with a Ph.D. in Medical Ethcis from the University of Amsterdam (*The different faces of autonomy*, Kluwer 2002). She published on a wide variety of subjects, like 'the good life' in care for dementia patients, pragmatist ethics, ethics of new and emerging technologies, telecare, neuroethics and (neuro)enhancement. She is a member of the Dutch Health Council, chair of the Dutch society for Medicine and Philosophy, and editor of Filosofie en Praktijk (Philosophy and Practice). Current research interests concern the philosophy and ethics of human enhancement, neuromodulation, anti-aging and 'wishfulfilling medicine'.

Anders Schinkel received his Ph.D. from VU University Amsterdam with his dissertation Conscience and Conscientious Objections (Pallas Publications, Amsterdam, 2007). He has published articles in ethics and animal ethics, philosophy of history, philosophy of education, and other areas. He is assistant professor in Philosophy of Education at the Faculty of Psychology of Education of VU University, Amsterdam.

Sebastian C. Sethe has a background in law, bioethics and technology transfer and holds a Ph.D. in the jurisprudence of innovation governance (Sheffield). Much of his advisory work involves regenerative medicine and advanced therapies. Acting internationally, but with a particular focus on the UK and Germany, he has worked closely with various academic institutions and hospitals, recently with the St. Georges Medical School, the NorthEast England Stem Cell Institute and the Fraunhofer Institute for Cell Therapy and Immunology. He is also an Advisor with the pioneering London-based law firm Lawford Davies Denoon, specialising in personalised medicine.

Alies Struijs received her Ph.D. from VU University Amsterdam with her dissertation "Minorities policy and morality. Recognition of cultural identity in the perspective of liberal morality" Since 1999 she is senior advisor at the Council for Public Health and Health Care and coordinator of the related Centre for Ethics and Health. She is the joint Editor-in-Chief of the series Ethics and Health Policy.

Dorothea Touwen is assistant professor in Medical Ethics at the department of Medical Ethics and Health Law of the Leiden University Medical Center, Leiden, the Netherlands. Her research focuses on care for the elderly and the role of family. Topics include the decision making processes concerning the treatment of incompetent nursing home patients; the moral problems of informal caregivers in the care for elderly people with diminishing cognitive functioning living at home; the participation of elderly people with diminishing cognitive functioning in clinical research; and the assessment of decision making competence of elderly people by their family members.

John Vincent was trained as a Social Anthropologist at the University of Sussex. Following professional, political and voluntary association activity related to the condition of old people, he started academic research and publication in social gerontology in the 1980s. In the twenty-first century he has specialised in extending his critique of ageism to the anti-aging movement. His book publications include Inequality and Old Age (1995), and Old Age (2003). He retired from the position of Associate Professor in the Department of Sociology and Philosophy at the University of Exeter in 2009. He is a great-grandfather and beekeeper.

Part I
Setting the Scene: Views on Age and Aging

Chapter 1
Introduction

Maartje Schermer and Wim Pinxten

1.1 Envisioning Aging Societies

Western societies are aging. Over the past two centuries, it has become clear that life expectancy can increase significantly under the influence of human interventions in physical health, the environment (including socio-economical conditions), and personal lifestyle. For example, in the course of the twentieth century, there has been a gain in life expectancy of about 30 years in Western Europe, the USA, Canada, Australia, and New Zealand. Some countries, most renown Japan, have even experienced an even higher increase in life expectancy. (Christensen 2009) The increase in life expectancy and related demographic changes in affluent societies are ongoing, and can be expected to rise further in the future. In addition, Western societies can most likely be regarded as a sneak preview of what is to happen elsewhere in the world.

To date, the process of aging is surrounded by ambiguity. One the one hand, there appears to be a vast potential for 'healthy aging', suggesting that it is possible to live both longer and in better health. On the other hand, aging often introduces health problems, which confront individuals and society at large with important challenges. For example, coping with dependency and chronic ill health in the population of elderly will require significant material and human resources that are not easy to provide. So even though many may want to live long, relatively few may want to be old. Nonetheless, aging seems the only way to live long...

Our views of aging and elderly life are surrounded by a similar ambiguity. On the one hand, aging has the potential of disrupting our individual lives and even our entire social setting profoundly. Personally, many of us fear the decline coming with

M. Schermer (✉) · W. Pinxten
Department of Medical Ethics and Philosophy of Medicine,
ErasmusMC/University Medical Center Rotterdam, Rotterdam, The Netherlands
e-mail: m.schermer@erasmusmc.nl

W. Pinxten
e-mail: w.pinxten@erasmusmc.nl

M. Schermer, W. Pinxten (eds.), *Ethics, Health Policy and (Anti-) Aging: Mixed Blessings*,
Ethics and Health Policy 1, DOI 10.1007/978-94-007-3870-6_1,

old age, the dependence, loss of capacities, loss of social position and contacts. As a society, we tend to see the aging population as a burden: healthcare costs will rise, the numbers of retired people will lay a heavy claim on pension schemes, and there may not be enough younger people to care for all the elderly. On the other hand, aging is not univocally tormenting. There are many reasons to have sympathy for old people, as individuals and as a population. Certain virtues, such as experience, humility, generosity, or patience may encounter in elderly life the ideal setting to flourish. And even where virtue is hard to find, it should not be forgotten that the generation born before has built the affluent society we live in today, often through hard work and in difficult circumstances. Doesn't such a contribution deserve a good old age?

At present, aging thus seems to be a mixed blessing. *What then do we wish for the elderly? What do we wish for ourselves when we eventually retire and grow old? And what do we wish for the generations to come? In other words: what are our ideals regarding aging and old age?* For some, the ideal may be to age in good circumstances, within a social network and with adequate care for those who become frail, ill, or dependent. For others, the ideal may be to age in good health—taking the lucky few who live both long and healthy as the prototype of the aging individual. For them, healthy aging is becoming the new paradigm and a goal for healthcare, health policy, and research. And yet others may find it desirable to not age at all, but rather to stay young forever—breaking through the boundaries of our current healthspan and lifespan.

These divergent views of aging are highly relevant to ethics and policy, as different views of aging will generate different definitions of and approaches to the problems that an aging society generates. So we face the challenge of choosing what we wish for the elderly and deciding upon the kind of aging we want to facilitate and support. An important part of this volume is devoted to exploring the different views and ideals regarding age and aging, since these set the scene for discussing ethical and policy problems and for choosing directions.

1.2 Ethics and health policy

In this volume, we will limit ourselves to the ethical and policy issues that are related to health and healthcare. *What kind of aging society do we envision? How do we currently define the problems of an aging society? What opportunities to deal with these problems can be found in science and society? What to make of the promises of healthy aging for all, compression of morbidity and (indefinite?) extension of the human lifespan? How do we distribute our present, limited, resources fairly among cure and care, among basic research in biological aging, the development of clinical applications, and research in social and environmental factors? What responsibilities do we designate to the individual, the family, and society at large? And how do we cope with aging in its current form? Do we do justice to the elderly and provide them with the care they need and the respect and recognition they are due?* All these questions call for reflection within bioethics, and in service of health policy.

The focus of this volume explicitly exceeds biological and epidemiological perspectives on aging. It addresses ethical issues and policy concerns from a broad range of views on aging and on how aging affects the individual and society at large. This volume brings together diverse views from various disciplinary backgrounds—science, philosophy, medical ethics, bioethics, sociology and policymaking. Presuppositions on 'good' aging that form the backdrop to our thinking, acting and policymaking are brought to the surface and critically examined from various perspectives.

1.3 Overview of Contributions

The contributions to this volume have been arranged in four parts, discussing images and views on aging, care for the elderly in our present society, future prospects for prolongevity and policy issues respectively.

Part I, **Setting the Scene**, elaborates and discusses a variety of views on old age and aging, from scientific, biomedical, sociological, anthropological and ethical perspectives (Chap. 2–5), and also represents the perspective of a geriatrician and the elderly themselves (Chap. 6–7).

Wim Pinxten reports his findings from interviews with a number of internationally renowned Dutch scientists, who all work in different fields of aging research. *How do scientists look at the spectacular increase in life expectancy that took place throughout the last two centuries? What enables human beings to live longer than their ancestors did? Are there boundaries to further increases in life expectancy? And why does the increase in length of life come with important downsides?* In this chapter, it is explored how scientists have revisited aging against the background of recent demographic and scientific evolutions. It is also described what efforts scientists undertake to understand, predict, prevent, and treat aging and/or age-related conditions and diseases. Finally, the policy implications of the scientific approach of aging and age related problems are discussed.

John Vincent examines the anti-aging movement from a critical sociological perspective and indicates the role it plays in the contested nature of old age. He argues that the anti-aging movement identifies aging as merely an undesirable aspect of the external appearance or internal constitution of a person's body, which is capable of manipulation and remedy. He suggests some reasons for the rise of such views and argues this perspective is ageist and that old age should be celebrated—not eliminated. He challenges three criticisms of this anti-anti-aging position: that all 'life' is a self-evident good, that it is a rationalization for prolonging pain and suffering, and that it depends on an untenable traditional notion of human nature. He concludes that the ageism embedded in the cultural and scientific, knowledge-creating institutions has to be acknowledged and countered if a 'Third Age' renaissance for later life is to succeed.

Wim Dekkers asks whether we need a special anthropology of the aging person and if so, what this should look like. He answers the first part of the question affirmatively. He argues that everyone involved in the care for the elderly or in research in the field of aging holds an implicit anthropology which should be made explicit. Moreover, new developments in the field of biogerontology demand reflection on their impact on the human condition. Finally, attempts to find meaning in aging and to accept old age and death might function as a counterbalance against the utopian idea of eternal youth and the posthumanist idea of prolongevity and expanding the human life span. Regarding the second part of his question, Dekkers argues that a philosophical anthropology of aging should be informed by the results of qualitative research on how we experience aging, with particular attention to a phenomenology of the aging body. He dismisses the idea of an all-encompassing, culture- and time-independent anthropology, and calls for policymakers to seek the right balance between a "focus on vulnerability and the ability to look after oneself". He concludes that policymakers should not explicitly aim at prolongevity and lifespan extension, but instead focus on quality of life and quality of care for the elderly.

Søren Holm also discusses the anthropology of the aging person, more specifically the anthropology underlying work in bioethics—or lack thereof. He begins by observing that there is a voluminous bioethical literature on reproduction and the beginning of life, and an almost equally large literature on the very end of life, but the perhaps 20 or 30 years that many of us hoping to spend between retirement and death are very incompletely theorized in bioethics. Holm suggests that this is because the aging and old person falls outside of the standard, implicit anthropology of bioethics. He discusses the role that anthropologies play in bioethical thinking and excavates the implicit anthropology underlying much of English language bioethics. This holds that people are primarily interested in rational decisionmaking and fulfillment of preferences, and are otherwise non-gendered, a-temporal non-relational, and fully functional. He then shows how this 'standard anthropology' can help to explain why old age is under-theorized in bioethics, and primarily seen through a deficit lens. He calls for bioethicists to: "consider all the life stages in our thinking about anthropology and not to valorise one as the most important or most 'normal' life stage". Like Dekkers, he argues we should abandon the 'deficit model' of aging.

Bert Keizer, in a slightly different type of contribution, gives his personal reflections on old age, based on his decades long experience as a geriatrician. He explains why we do not like old people; and why we should better admit this—according to him—biological fact. He vividly describes the effects of aging on bodily functions, the alleged wisdom that comes with aging and the risk of 'running out of fashion'. He concludes with two dangers that are the result of our successes in prolonging life: the increased proportion of elderly that will suffer from Alzheimer's disease and the weariness of life that may come with (very) old age. He argues that the way we treat elderly, especially Alzheimer's patients, shows our collective dislike of the elderly. According to Keizer, policymakers should listen more carefully to the needs and demands of the elderly themselves, also regarding life-extension and the end of life.

In the next chapter, *Frans Meulenberg* and *Wim Pinxten* give an impression of these views of elderly people themselves. Based on a series of interviews with elderly people, they highlight some of the themes, questions, and dilemmas they raise. They describe the different strategies that these elderly use to ensure that their lives are not hollowed out by the deterioration that comes with age: *confirmation, integration* and *continuity*. Issues like discipline, daily routines, autonomy, relationships with other people as well as pets, and continuation of activities, as well as cherishing memories are all important elements in the stories of the elderly.

In **Part II, Taking Care**, the question of how we take care—and how we ought to take care—of the elderly in our (arguably ageist) society is central. *What do we owe the elderly? What is good care and is it distributed justly between the so-called third age and fourth age? How should we care for demented elderly, and how do we deal with death wishes of elderly people who considered their life 'completed'? And how can elderly themselves give meaning to aging and thus contribute to a good old age?*

Cees Hertogh starts from the observation that aging has two scripts: that of the third age-of vitality and successful aging-, and that of the fourth age, of frailty and deep old age. He sketches both faces of old age, and reflects on the future of old age, with a special focus on the concept op frailty, of which he tracks down the origins and transitions Hertogh is critical towards both 'scripts' of aging, since they both have ageist connotations, demanding of people that they "swim against the currents of senescence". Hertogh's thesis is that both the focus on successful ageing and the present approach to frailty represent a new form of ageism that works out as a negative incentive with regard to the care for the oldest old (cf. Keizer). Therefore, according to Hertogh, the fundamental policy question to be answered is "whether an ongoing investment in strategies to postpone or reduce frailty and to compress morbidity in the next generations is morally acceptable and worth striving for, as long as we don't equally invest in research aimed at an improved quality of life for the oldest old".

Dorothea Touwen discusses a specific ethical and policy issue in the care for demented patients. Because demented patients are often incompetent to decide, surrogate decision makers become involved in the process of deciding about treatment and care. These surrogates -often family members-, try to decide in the way the demented person himself would have decided. However there is a problem in interpreting expressions of demented patients concerning what they would like or what they desire, now that they are demented. Should their former way of life take precedence over what they seem to prefer now? Touwen addresses this problem of the relation between former wishes (whether or not written down in a living will) and the possible change of interests due to the change in circumstances caused by the process of dementia. She concludes with the policy suggestion that people should be encouraged appoint their own favored representative and to write advance directives not concerning what one wants to be decided, but who one wants to decide.

Govert den Hartogh addresses a complex topic, the death wishes of those elderly people who are 'weary of life'. Not infrequently very old people who don't have a life-threatening illness and do not suffer severely from physical symptoms have an explicit and permanent wish to die. Should the law permit physician-assisted suicide in such cases? Den Hartogh argues that the Dutch and Belgian euthanasia laws are right in stipulating that, for the assistance of doctors to be lawful, a voluntary and well-considered request is not enough. If this is paternalism, it is only of the indirect kind, and indirect paternalism is not subject to the same objections as direct paternalism. Both laws in addition require that the patient is suffering unbearably. According to den Hartogh this requirement can be satisfied in the case of the elderly, because even extreme suffering does not require physical symptoms. Finally he argues that if the requirement of unbearable suffering is not satisfied, it is not inhumane to withhold medical assistance, because very old people always have the alternative of stopping eating and drinking: "elderly people, have sufficient opportunities to end their own lives in a carefully prepared and humane way."

Frits de Lange takes a broader perspective on the question of generating meaning in old age. Good aging, according to De Lange, entails making the best of growing old, or, alternatively: making old age a good part of life, too. He argues that ethics can have both a constructive and a critical task in this endeavor. The constructive task of public ethics is to develop tentative visions of good aging. An inventory of ingredients of good aging can be made, according to de Lange, but pivotal is the question of meaning. Old life is worth living, as long as it makes sense to the individual person. An existential gerontology is urgently needed here. Besides being constructive, the ethical reflection on good aging also needs to be critical, and De Lange therefore concludes with questioning the implicit normativity of some master narratives and gerontological paradigms of 'normal' aging that guiding biopolitics in late modern societies.

Anders Schinkel, in his contribution on justice and the elderly, asks what we owe the elderly. While the philosophical debate about issues of justice related to the elderly tends to be restricted to a rather limited range of issues (e.g. health care rationing, the pension system), Schinkel argues that the underlying perspective of this debate should be widened considerably. His central claim is that a just society gives the elderly the recognition they are due—not (primarily) as a group, but as individuals. He argues that recognition is the primary act of justice, and subsequently discusses various issues related to the elderly—ageism, distributive justice, just care, and recognition in the family and society—from this perspective. Shinkel's perspective has firm policy implications. Concretely, the recognition perspective favours government support of autonomy, independence, and abilities and opportunities for social participation. However, rather than pointing to one particular policy, justice as recognition provides a test for any policy we do choose. And complementary to policy, elderly also have personal responsibilities, Schinkel points out.

Part III of the volume, Looking ahead, moves to a somewhat more speculative and futuristic perspective. It looks ahead to the ethical issues that are raised by the prospect of further biomedical success in fighting the physical problems, decay and diseases that come with aging; and possibly even stop or slow down aging itself. *Is the prospect of extreme longevity one to look forward to? Will everyone profit equally? Or will these new technologies turn out to be mixed blessings?*

Sebastian Sethe and *João Pedro de Magalhães* give an overview of the ethical issues that biogerontologists enquire in their research activities. As such, they focus on an insiders' perspective on ethical issues in biogerontology, rather than on the philosophical and social issues in curing aging. The authors remark that ethical challenges that have been addressed in other fields of research, require specific attention when it comes to aging research. For example, it must be taken into account that elderly are a vulnerable population that is prone to frailty, regenerates less quickly, and might raise complexities in the informed consent process. From the insiders' perspective, Sethe and de Magalhães consider how to communicate truthfully with non-specialists, as well as the scientific developments and potential technological advances. At a more philosophical level, they consider the motivation for conducting aging research, and the social impact that aging research may provoke, including issues related to costs, equity, overpopulation, stagnation, and stigmatization. As a conclusion, Sethe and de Magalhães suggest that protagonists in the field of biogerontology are subject to three ethical imperatives: (1) to represent the potential of aging research without hype but also without unwarranted political constraint; (2) to face the fact that aging causes suffering whereas the putative social drawbacks of controlling aging are speculative and contested; (3) to tackle the challenge of 'fighting aging' without fostering ageism.

John Harris argues that longevity is a rational good, since people want life and fear death. Life extension, however, simply happens by postponing death, and if death could be postponed indefinitely, immortality (not to be confused with invulnerability) logically sets in. In this sense, immortality would be a side effect of efficacious therapies to the life-threatening problems that elderly experience. Harris critically considers philosophical and ethical objections against life extension, including claims that that life extension would be unjust, pointless, self-defeating (because personal identity could not survive long periods of extended existence), or prohibitively expensive in terms in increased healthcare costs. He is not convinced that these arguments make out a strong case against life extension and immortality, and concludes that we should be slow to reject cures for terrible diseases, even if the price we have to pay for those cures is increasing life expectancy and even creating immortals.

Inez de Beaufort reflects on the relation between aging and appearance. Many aging people regret that their appearance changes over time, and would prefer to look younger. Others may find it logical to accept that age comes with changing looks. She emphasizes that appearance, sometimes strongly related to personal identity and lust for life, is morally relevant and important, and therefore worthwhile to consider seriously in the ethical debate. When considering the moral implications of rejuvenating enhancements, de Beaufort discusses three arguments in favour of rejuvenating enhancements: the 'proper season'-argument, the 'pressure from society'-argument, and the 'more important problems in the world'-argument. Challenging nature is not morally wrong, she concludes, and the proper season argument can be used against the elderly. Unless societal attitudes towards old people change, people will try to look younger. If they use a vitalizing approach—as opposed to a camouflaging one—then the appearance changes seem an extra benefit. Although there are, de Beaufort admits, more pressing problems in the world, one should not underestimate the burden of looking old.

Maartje Schermer addresses the ever-returning question of whether aging is a disease or not. She states this is important, because the way the relationship between aging and disease is framed influences ethical and policy debates about aging and anti-aging interventions. She proposes to use the conceptual triad of disease-illness-sickness to discuss the various perspectives and controversies around this question, including the issue of medicalization of aging. Schermer argues that we should not reduce aging to disease. The issue of whether we ought to accept the fact that our bodies age, or should try to intervene in the aging process, remains a fundamentally normative question, that returns within the different conceptualizations of health and disease, she demonstrates.

Hans-Jörg Ehni evaluates the prospect of biomedical interventions which might slow down, prevent or even reverse biological aging in relation to theories of just health care. The focus on extreme life extension in the current bioethical debates on enhancement has, according to Ehni, diverted the attention from the short-term outlook which biogerontologists provide on the possibilities of developing interventions into aging, and what realistic goals they might be able to achieve in the near future. A central question here is whether such new interventions into aging should be made widely available, and if this is likely to be possible or not. Against this background, Ehni discusses three alternative reactions to the prospect of limited access to interventions into aging: prohibition, enabling general access, and setting priorities in research. He argues that the third option is most promising, and that improving the current social context for a fair distribution of the benefits of new interventions into aging and setting priorities in research immediately (instead of just waiting to see what kind of impact they will have) might generate a 'longevity dividend' for all.

Finally, **Part IV**, Choosing Directions, takes a more explicit policy perspective. It deals with some concrete policy questions—whether biogerontology is a promising route for cost containment, and whether a greater emphasis on personal responsibility for healthy aging is desirable—and considers the various contributions to this volume from a policy point of view.

Alies Struijs and *Marieke ten Have* explore the ethical arguments related to demanding personal responsibility for lifestyle from the elderly in health policy. It is well-known today that the health problems that elderly experience may be linked to personal lifestyle choices. Should persons take responsibility for their choices and for the consequences of these choices, in a prospective or retrospective sense? Struijs and ten Have argue that responsibilization has been a trend in government health policy for decades. They explore two arguments in favour of responsibilization of the elderly in health policy: serving personal interest and relieving the pressure on collective means. In addition, they reflect on arguments against personal responsibility, more specifically, the 'blaming the victim'-argument, interference with autonomy and freedom of choice, difficult practical application, and respect for privacy. Finally, Struijs and ten Have discuss the idea of a regular health check for people over 45.

Laura Capitaine and *Guido Pennings* ask whether biogerontology is a promising route to cost containment in healthcare. One of the concerns regarding population aging is that it will cause healthcare costs to spiral out of control. Two well-known policy strategies for cost-containment are age-based rationing of healthcare and the privatization of Medicare. A recent proposal is to invest more research into the biology of aging. The idea is that such research will enable us to tackle all age-related diseases simultaneously, thereby creating an increased healthspan, reducing healthcare costs. Capitaine and Pennings evaluate the arguments for this proposed strategy by critically examining its fundamental presuppositions. They conclude that the problematic nature of these presuppositions raises serious questions concerning both the morality and the efficacy of biogerontology as a cost containment measure. Instead, they suggest a more promising route to cost containment in healthcare.

Göran Hermerén in the final chapter performs the formidable task of giving a critical overview of the policy implications of the various arguments, issues and points of view brought forward in the contributions to this volume. Drawing on his long experience in bioethics, political philosophy and international health policy, he proposes a framework to evaluate the various views on aging, identified problems and concerns, suggested solutions and policy implications expressed in this volume.

A number of questions that are directly relevant to health policy run through the whole volume and are addressed in multiple chapters and the four parts of the book. First, the question what the current state of affairs is in scientific, biomedical research on aging, and what the prospects for the future are, with regard to interventions in the aging process. This question is addressed in the chapters by Wim Pinxten, Sebastian Sethe and João Pedro de Magalhães, and Hans-Jörg Ehni.

A second question regards our image of aging and the elderly, and how we can encourage more positive images that do justice to the positive aspects of this phase of life. For public policy and health policy, it is very important to consider from which background image of aging and the elderly policy is being developed. Are the elderly mainly conceived of as a burden, or as a group that can make their own contribution to society? Are elderly people only seen as vulnerable and frail or only as 'vital senior citizens'; or, more realistically: as a very diverse group of people with diverse needs, wants and capacities. In some form or other, this issue is addressed in the chapters of John Vincent, Wim Dekkers, Søren Holm, Bert Keizer, Cees Hertogh, Frits de Lange and Maartje Schermer.

A third question concerns the issue of justice. How can policy contribute to a society in which justice is done to the elderly, and in which health among elderly is as fairly distributed as possible? Anders Schinkel, Hans-Jörg Ehni, Alies Struijs and Marieke ten Have, and Laura Capiteine and Guido Pennings address this question.

Fourth, what is good care for the vulnerable older age-groups? How can health policy contribute to good care? This is addressed in general terms in the chapters of Bert Keizer and Cees Hertogh, who warns against putting too much emphasis on the Third Age at the cost of forgetting about the Fourth.

The chapters of Dorothea Touwen and Govert den Hartogh address more specific issues in the care for the elderly, namely how to deal with former wishes of demented patients, and with death wishes of elderly.

Finally, a policy question looking ahead to the future is how to deal with increasing opportunities to intervene in aging and increase the (healthy) lifespan. Although at first sight, there seems little that would speak against this mission, some possible objections and caveats are discussed in chapters by John Harris, Inez de Beaufort, Hans-Jörg Ehni, and Laura Capitaine and Guido Pennings.

Reference

Christensen, K., G. Doblhammer, R. Rau, and J.W. Vaupel. 2009. Aging populations: The challenges ahead. *The Lancet* 374(9696):1196–1208.

Chapter 2
Built to Last?

Scientists' Views on the Definition, Prediction, Prevention, and Manipulation of Aging

Wim Pinxten

2.1 Introduction

As time goes by, individuals are deprived of many of the strengths, assets, and affluences they have edified earlier in life. Somewhat tragically, the loss that comes with age is not always graceful and may bereave human beings from some of their most intimate capacities, among which comprehension, appetite, sensitiveness, swiftness, and elegance. Such loss is likely to affect personal identity, independence, social integration, and wellbeing in many elderly individuals. Moreover, dealing with loss will often require a significant share of material and human resources. Whether we like it or not, *life is a losing game*.

Throughout history, the phenomenon of aging has been described and depicted in many ways. There are for example intriguing images of the ages of man, including the three ages of man painted by masters as Giorgione and Titian, or the seven ages of man in the famous monologue *All the world's a stage*. This part of Shakespeare's comedy *As you like it* portrays aging consistently with a commonsensical understanding of the life course, in which growth and decline are observed as a an obvious and natural process, an inescapable part of the human condition. Shakespeare's observation of the final life stage is not particularly flattering: '*Last scene of all, that ends this strange eventful history, is second childishness and mere oblivion, sans teeth, sans eyes, sans taste, sans everything*'. Luckily, it would be all too cynical to still represent old age as mere and inelegant loss, now that old age may present itself rather as a second youth than a second childishness. And also historically, negative depictions of aging have known their counterpart in images of longevity, rejuvenation, and immortality: Methuselah, the fountain of youth, and the elixir of life are just a few illustrations.

The historical accounts of aging, mortality, longevity, rejuvenation, and immortality have recently been challenged by a spectacular demographic evolution and new

W. Pinxten (✉)
Department of Medical Ethics and Philosophy of Medicine,
ErasmusMC/University Medical Center Rotterdam, Rotterdam, The Netherlands
e-mail: w.pinxten@erasmusmc.nl

M. Schermer, W. Pinxten (eds.), *Ethics, Health Policy and (Anti-) Aging: Mixed Blessings*, Ethics and Health Policy 1, DOI 10.1007/978-94-007-3870-6_2,

scientific insights in the process of aging. Against this background, images of senescence and old age are being redrafted in science and society, and the very process of aging—rather than age-related problems—has increasingly attracted scientific, ethical, and social interest.

In this chapter, it is explored how scientists define and approach aging in different fields of scientific enquiry. *What problems do they define? Which images of aging do they endorse? What are the objectives of their research? And what research outcomes do they envision?* Based on interviews with 11 leading scientists in different fields of biomedical enquiry, the background, objectives, drivers, results, progress, and way forward of research in the field of aging are explored.

Regardless the considerable variety in disciplines, research areas, and personalities, the views and opinions that were expressed during the interviews are surprisingly complementary. Therefore, it was possible to merge the information that was gathered in the interviews in a considerably coherent discourse.

The information in this chapter is mainly based on the information provided during the interviews. Although at many instances, scientists referred to published data (in general or to specific papers), these data were not explicitly or systematically taken into account in this chapter.

2.2 Aging Revisited

Life expectancy has increased spectacularly over the past two centuries: In entire populations, human beings live about twice as long as their ancestors did a few centuries ago. Over the past 160 years, life expectancy has increase with approximately 6 h for each day that went by (Oeppen and Vaupel 2002). *What factors can explain for this rather sudden increase in life expectancy? What human traits enable us to live longer than our ancestors did? Are there boundaries to further increases in life expectancy? And why does the increase in life length come with a downside?*

To answer these questions, we need to explore the main factors that are constitutive for the way in which we age, and the length of the lifespan we live: Genomics, environment, and lifestyle.

2.2.1 Genomics

Aging often comes with disease, and many diseases come with age. By consequence, it is relatively easy to designate a number of diseases and conditions as 'age related'. Such designation, however, can rather be an observation than a true understanding of the relationship between aging and disease. In order to explain *why* aging induces an increased mortality rather than *that* aging induces such increased mortality, insight in the biology of aging is required.

Over the past decades, new technologies have created very important opportunities to study the biology of aging, and such technologies continue to develop rapidly.

Genetic sequencing is exemplary in this respect: Since the discovery of the structure of DNA by Watson and Crick in 1953 and the first sequencing of an entire human genome in 2004, techniques for genetic sequencing have become much faster and cheaper. By consequence, the amount of genetic data that is generated is increasing exponentially, as scientists move from studying single genes, to exome sequencing (the mapping of approximately 30.000.000 base pairs), and currently towards full genome sequencing (*i.e.* 200-fold more). This has transformed the challenge of generating data into the challenge of processing and analysing data.

The analysis of genetic data can result in the discovery of defects in a single gene (e.g. a defect resulting in the generation of an inactive protein encoded by the gene) or in associations between genes and human traits, conditions, and diseases. Such associations are identified in large population studies, covering thousands of human subjects. For example, a study of genetic variance in height involved genomic data from 180.000 individuals.

Now that scientists are capable of gaining insights at the molecular level and sequencing the entire genome, what does this reveal about the process of aging? First, aging occurs in many organisms and species. Particularly organisms that reproduce sexually are vulnerable to aging, because there appears to be a trade-off between reproductive success and longevity in these organisms. From the viewpoint of evolutionary biology, selection in function of species survival lacks a rationale to promote population-wide longevity beyond a lifespan that is relevant for successful reproduction. This may explain why we experience so many age related problems now that our lifespan clearly exceeds the period that is relevant for conceiving and raising children (about 35–45 years, corresponding well with the life expectancy up to 1800). This may also explain that healthy aging appears to be a privilege of the lucky few to date. Second, if aging was not selected for, there is no purpose in aging and longevity from a biological point of view. Rather than being a functional biological process, aging seems to be a side effect of increased longevity due to man-made environmental changes. Among the interviewed scientists, there was even discussion whether aging is a biological process at all. Regardless the definition of aging as a biological process or not, there appears to be considerable consensus on what happens when we get older: The body looses part of its capacity to regenerate and to repair defects, and accumulates damage. Loss of regenerative capacity and accumulation of damage render the body more vulnerable to various pathologies, which may induce multimorbidity in elderly individuals and increase the likelihood of mortality. And also in absence of disease, the loss of regenerative capacity and accumulation of damage will affect the composition and functioning of the body and induce frailty. Third, genetics in itself cannot account for the changes in the length of our lives and the way in which we age. Given the fact that two centuries is a very short timespan in terms of natural selection, the spectacular increase in life expectancy in many societies cannot be explained by significant changes in the genome. Rather, such a sudden increase in longevity suggests the importance of other factors: The environment and personal lifestyle. This has implications for what genetic research can reveal. Given the millions of differences between the genomes of individual human beings, studies of the genome of long living subjects (for example, Mrs Andel, a

Dutch woman who was for some time the oldest person alive and disposed her body to science) are not replicable and will not enable scientists to unravel the 'secret' of longevity. Even if it is possible to associate genetic variance to longevity, such associations cannot give a clear indication on how long an individual will live, or whether someone will age in relatively good health or not. In addition, studies of the genome in twins revealed that, in general, the heritability of longevity is about 25 %, clearly suggesting that the length of our lives is influenced strongly by other factors than our genes. Fourth, aging is very heterogeneous. Individual beings grow and age at a different pace and in a different way. Also within one and the same individual, different tissues and organs appear to age at a different pace and in a different way. This renders it hard to define at what age aging sets in. As has already been pointed out, the personal genome only partly accounts for differences between aging individuals, since the heritability of longevity is rather limited. Some individuals born in long-living families, however, appear to have a genetic advantage for living long and in relatively good health. In these families, the heritability of longevity is higher than in the population at large: about 40 to 45 %. Fifth, aging has a high plasticity and is therefore open to manipulation. One the one hand, aging can be manipulated by changes in the environment and lifestyle. Caloric restriction is a well-known example in this respect: When the caloric intake of an organism (for example yeast, worms, mice, and monkeys) is restricted with 25–30 % in comparison to an organism that is fed ad libitum, longevity increases significantly. In addition, research in epigenetics demonstrates that environmental and lifestyle factors strongly influence the phenotypic expression of genes, from the very start of life onwards. For example, a large study of persons who suffered from prenatal malnutrition during gestation (born in Amsterdam between 1943 and 1947, in wartime conditions with food rations from only 400–800 kCal/day) revealed an increased incidence of diabetes, cardiovascular disease, hypercholesterolemia, obesity (in women), and many other health problems in this population.[1] This shows a considerable plasticity of aging and implies that, even if we are all unequal at the start, healthy aging needs not to be a privilege of the lucky few.

2.2.2 *Environment*

It has already been suggested that environmental changes are particularly relevant to the increase in life expectancy that occurred over the last two centuries. What changes in our environment have paved the way for humans to live so many years longer?

First, better hygiene has contributed to the significant reduction of the risk of infections and inflammations, and thereby eliminated a large number of preventable deaths. At the public level, important efforts have been made to provide people with clean (chlorinated) water and clean living environments (e.g. waste collection). At

[1] See: www.hongerwinter.nl

the personal level, personal hygiene has improved, for example due to the broad availability of soap and clean water. Second, the availability and quality of food have strongly improved. New techniques to produce, pack, and conserve food have rendered people less dependent on factors that are out of our control, such as favourable weather conditions. Apart from the availability and quality of food, also the quality of our diet has improved, by moving towards a more balanced diet in which all essential nutrients (e.g. proteins, fats, fibres, vitamins) are adequately represented. Third, our living conditions have been strongly improved by building better infrastructure in terms of for example housing, labour, transport, social contact, communication, and leisure. Fourth, biomedicine has been very successful in the timely diagnosis, prevention, and treatment of many conditions and diseases. Vaccines, antibiotics and other drugs, operative interventions, imaging techniques etc. have contributed significantly to better heath and longer lives.

Together, these four factors have changed the way in which we age, not only as individuals, but also as populations. Indeed, the increase in longevity that has taken place over the past two centuries can be observed in entire populations. Once again this illustrates that aging is certainly not a privilege of the lucky few. A classical, almost iconic example of longevity in entire populations is Japan, the country that hosts the highest rate of long living people worldwide. Such an example clearly shows that man-made interventions in the environment can induce large and fast increases in longevity in entire populations. The short time span in which the recent increase in life expectancy took place, precluded human organisms to adapt to these changes timely through selection via reproduction. By consequence, the body outlives the lifespan it was selected for and functions in an environment it was not selected for. This explains why longevity has not just increased, but increased longevity has caused numerous health related problems.

2.2.3 Lifestyle

Changing environments can work in favour of longer lives. But this is not always the case: Changing environments can also transform assets into perils. For example: If selection has rewarded the capacity to survive periods of famine, organisms become more vulnerable to obesity. And if selection has adapted organisms to a diet of food that is extracted from the near environment, switching to a diet of ready to consume food in which sugar is overrepresented renders organisms more vulnerable to diabetes. By consequence, there is a new killer in town: *lifestyle disease*.

In has become increasingly clear that in an aging and society where living conditions are largely man-made, one must be vigilant of man-made health problems, including lifestyle diseases. The situation in Okinawa, the Japanese province with the highest rate of centenarians, clearly illustrates this situation. While the centenarians in Okinawa can be regarded as a paradigmatic example of increased longevity, a considerable decrease in life expectancy can be observed in their grandchildren due to lifestyle changes. Another notorious example in this respect is Glasgow, where the local lifestyle has generated a longevity gap of about 14 years with Kensington, London.

While Okinawa and Glasgow illustrate that differences in lifestyle may generate differences in life expectancy of entire communities, the same is also true for individuals. Smoking, unhealthy diet, inactivity, and stress are major examples of lifestyle factors that negatively affect health and life expectancy of individuals. Also social factors, such as the educational level, creates longevity gaps of up to 7 years.

The importance of lifestyle can hardly be overestimated. One of the interviewees in this study emphasized that in absence of an appropriate lifestyle, medical interventions are not likely to be effective. It is thus not an option to search or wait for a magic bullet that allows us to endorse an unhealthy lifestyle and address all health problems with medical quick-fix interventions. The definition of a healthy lifestyle by the interviewed scientists does not reveal unexpected recommendations: Sufficient fruit and vegetables, no smoking, sufficient physical activity, and moderate wine consumption (2 glasses/day) (Khaw et al. 2008).

Together, genomics, environmental factors, and lifestyle suggest that aging is a highly complex phenomenon. Despite this complexity, there seems to be a substantial potential to explain, predict, prevent, and/or treat various problems that come with age. What actions, than, do scientists undertake to understand, predict, and manipulate the way in which young people become old, and old people experience the perils of old age.

2.3 Work in Progress

For many reasons, research in the field of aging is booming. On the one hand, people continue to live longer, many long-living individuals suffer from multiple (chronic) diseases, and societies at large are challenged to take care of an ever-growing number of frail elderly. Therefore, the urgency to deal with the health problems in the elderly is high, and research in various aspects of senescence and aging is most relevant. On the other hand, aging provides scientists in various field of enquiry with a solid framework to formulate research questions, interconnect data and disciplinary approaches, and interpret research outcomes. Regardless whether scientists focus on the study of disease, the genome, genetic expression, or social determinants of health and well-being, framing research in the context of aging (1) provides a relevant background against which research questions can be formulated and (2) integrates specific research in a broader, multidisciplinary, and socially highly relevant framework. Such framework sheds a new light on research findings: If we try to understand the complex process of aging in which genomics, environment, and lifestyle all play a key role, an incredibly large amount of research findings will be deemed relevant, even when such findings currently lack clear significance on their own, or lack potential to contribute to the development of clinical applications or patient care.

The fact that scientists enquire aging from multiple perspectives and disciplinary approaches, renders research in this domain very heterogeneous in its objectives, methods, and outcomes. At the same time—at least for what concerns the scientists who were interviewed in this study—these heterogeneous objectives, methods, and outcomes are rather complementary than contradictory.

According to the interviewed scientists, it is unlikely that aging as such will be eliminated. Even with an excellent maintenance of the body, a minimization of environmental challenges, and a maximisation of adequate bodily response to environmental challenges, aging will still be present in man. Like with a car, maintenance may help to keep it going much longer in a relatively good condition, but inevitably, it is bound to get exhausted at a certain point in time. In the end, we are not *built to last*.

If elimination is not the focus of research in the field of aging, what is? In their research, the interviewed scientists focus on (1) the biology of aging, (2) age-related diseases, (3) healthy aging, and (4) adequate care. These four domains are certainly not distinct fields of enquiry. However, for reasons of comprehensiveness and somewhat reductionist, they will be explored as four focal areas in this chapter.

2.3.1 Biology

Scientists may focus their endeavours on the biology of aging. The main aim of this type of research is to gain insight in the complex biological background of aging. How can aging be understood from the viewpoint of evolutionary biology? What is the genetic basis of senescence and longevity? What causes the degeneration of the body, resulting in loss of functionality and increased frailty? What is the role of specific cells, such as stem cells? What is the role of specific mechanisms, such as DNA repair? To gain an answer to this type of questions, genetic data are of key importance. It has already been pointed out that over the last decades, the sequencing of genes en entire genomes has become cheaper and faster. This enables to compare genetic data of thousands of individuals in large association studies, with the objective of revealing what genes are involved in aging and age related diseases. What results then, do such association studies generate? Most of all, association studies show that most age related conditions are complex, and involve many genes. For example, association studies identified about 95 genes related to hypercholesterolemia, and 80 genes related to osteoporosis. Linking genetic variance to particular age-related diseases is one thing, understanding the specific role of all the genes involved, however, is quite another. While a lot is known about some parts of the genome, other areas—so called gene deserts—have no known functional importance. This clearly impairs the explanatory powers of findings that associate genetic variance with particular age related diseases or conditions.

The fact that so many genes are involved in different kinds of biological processes has implications for the predictive power of genetic analyses. Can genetic markers help to predict health risks and/or diagnose diseases (already before the onset of pathologies) in a reliable way? For various diseases, including diabetes, cancer, and cardiovascular diseases, genetic markers have been identified. The predictive value of such markers however, seems questionable. For example, a study measuring successful markers for various diseases in a population of members from long-living families, indicated that the markers for these diseases were equally well represented in long living families than as in other populations. This clearly suggests that with the worst set of genetic markers, people may still grow very, very old.

The fact that many genes are involved in different kinds of biological processes also has firm implications for the manipulation of these processes. Given the high number of genes involved, and the (indirect) interaction of genes in genetic processes, manipulation is very complex. Many genetic processes have alternative pathways, and therefore tweaking particular genes will not suffice to do the trick. Rather, some sort of systems approach is required to get a grip on the process itself, and not just the functioning of a particular gene in this process.

Apart from the study of the human genome and the relationship between genetic variance and diseases and conditions, also animal research is very important. Animals have a shorter lifespan and can be nurtured in controlled conditions from conception to death, selected on genetic traits, bred with identical genomes, involved in experiments that would never be allowed in humans (for example experiments with stem cells or DNA repair), and sacrificed for the sake of science. Obviously, this opens up many research opportunities.

To date, genetic research in animals and humans has generated numerous findings that provide new insight in the genetic background of aging. However, for the bulk of these findings, translation into the clinical applications is not likely to take place in the near future. This seems not to be of major concern to the interviewed scientists. Rather than developing high-tech interventions in the genome to counter age related health problems (or to extend life further), their research seems to focus on a better understanding of genetic mechanisms in service of prevention, diagnosis, prediction, and the optimization of available therapies. As one researcher stated: "One could say: If it is my objective to help people age healthily, then I can stop doing my research in molecular biology. Because there is one thing that certainly will enable many more people to age healthily, and that is getting them—from middle age on—to have sufficient physical activity and a healthy diet. In fact, we already know that."

2.3.2 Disease

Regardless whether aging is considered a disease or not, the incidence of many diseases increases with age. By consequence, morbidity is higher in older subjects in comparison to their younger counterparts. *What renders elderly more vulnerable to disease? How can the risk to age-related diseases be reduced? When should prevention and treatment start? How can the onset of age-related diseases be postponed? How can imminent health threats be averted? How can morbidity be compressed further? How should multimorbidity in elderly be managed in healthcare? And what are the standards of good health in the elderly, when one recognizes that a certain degree of degeneration and/or ill-health is inevitable?* All these questions trigger research in the field of age-related diseases.

How do researchers envision the future impact and approach of age related diseases? Although life expectancy continues to increase, scientists pursue to decrease the timespan in which elderly suffer from various ailments. So research aims to contribute to a further rectangularization of the survival curve (Fries 1980) due to a further compression of morbidity. Several complementary and interrelated strategies are developed in this respect: (1) reducing of the risk to disease, (2) postponing the onset and/or slowing down the progress of diseases, and (3) reducing mortality from age-related diseases such as heart disease, cancer, and stroke.

First, whether an individual will start to suffer from age-related diseases is not just a matter of fate. Research at the intersection of genomics, environment, and lifestyle, enables scientists to search for different types of interventions that will reduce the risk to diseases. Here, it is all about a proactive and personalized approach to health and disease. Instead of awaiting pathologies and health complaints before entering the clinic, predictive tests would enable to tackle disease before pathologies develop and morbidity increases. To optimize primary, secondary, and tertiary prevention, it is enquired how health can be assessed and addressed in a timely fashion. Also a personalized approach is important in this respect: Since phenotype plasticity renders it possible to regulate the expression of genes to a certain extent (e.g. through environmental or lifestyle interventions), a personalized approach of health could aim at tweaking the genome in such a way that health problems are averted and risks are reduced.

Second, also when an individual will start to suffer from age-related diseases is not written in stone. Given the considerable plasticity of aging, age-related pathologies can be postponed and/or slowed down. For example, a healthy lifestyle will contribute to the good condition of tissues from early on in life, which is advantageous throughout the entire degeneration process. Those who are ahead at the start can cope with a larger degenerative losses before developing diseases. And those who invest in maintenance of their body can slow down the pace of the aging process. Or, for example applied to sarcopenia, individuals who developed more skeletal muscle mass at the peak of development, are able to cope with more loss of muscle mass before developing actual sarcopenia.

Third, when focusing on disease, also the main causes of mortality are targeted. New technologies can help to study how disease can be treated more effectively in aged bodies. For example, mouse skin cells have been reprogrammed to heart cells that are beating in vitro (Efe et al. 2011). Such cell cultures offer new opportunities to test the efficacy of drugs.

Combatting the main causes of mortality can also come with a downside. By addressing preventable causes of death, aging, senescence, and mortality will not be eliminated altogether. By consequence, the successful approach of one disease will thus not just discharge us from ill health, but also pave the way for other health problems. In other words, the successful approach of one disease may thus come at the cost of getting another. For example, now that some causes of preventable death are addressed adequately, dementia is on the increase. By consequence, rather than avoiding death, we may be choosing to die another day. Such choice may come at a high cost, since it may destabilize the overall frailty that enables human beings to degenerate slowly until they are completely worn out. By consequence, the end of life may become ill, rather than frail.

2.3.3 Health

Although one needs not to be healthy to be happy and good health, for its part, is no guarantee for happiness at all, health is highly valued. No wonder then that scientists search for strategies to age healthily.

How do scientists define a good health in elderly life? As has already been pointed out, in the views of the interviewed scientists, aging healthy is certainly not synonymous to the elimination of health problems in the elderly. Rather, the idea of good health in old age is inspired by models of healthy aging, such as members of long-living families, who live longer than average in a relatively good health.

If 'healthy agers' are the lucky few that can serve as a model for aging in good health, can something be done to enable the population at large to resemble these models? Research points out that this is certainly the case. For example, a the study of partners of members of long-living families, who obviously do not automatically share a similar set of genetic traits that favour healthy aging, can start to resemble their partners in the way in which they age as the outcome of a program of adequate lifestyle interventions, including healthy diet and physical activity. If initially the two partners have a different 'molecular profile' of aging, this profile is open to change, and changes can be measured within weeks, or at least three months from the start of such a program. The challenge ahead here, seems to tweak the genome is such a way that health and longevity are promoted. An overall objective here is to reduce the heterogeneity in the way in which individuals age, and to contribute to narrowing existing longevity gaps.

There is a broad consensus among the interviewed scientists that prevention and appropriate lifestyle is *the* way to gain in health in elderly life. It must be emphasized, however, that this observation needs not to work against basic research. Indeed, the importance of lifestyle does not suggest the priority of research in lifestyle interventions over basic research in any way. For example, if basic research can provide insight in the risks to diseases of individual persons, these findings can be used to support an adequate response to these risks.

What then, is constitutive for a lifestyle that supports longevity and healthy aging?

First, efforts can be made to discharge the human organism from harmful environmental challenges. For example, overloads of sugar, fat, or calories in general in the diet cause metabolic stress. Reduction of caloric intake and a balanced diet (e.g. the Mediterranean diet) will contribute to increased health and longevity. Also the reduction of other environmental challenges, such as smoking and stress is important in this respect. Second, the capacity of human organisms to respond to environmental challenges can be cultivated. Here, sufficient physical activity is the paradigmatic example of healthy lifestyle. More can be done to keep the body in good shape: Given phenotype plasticity, efforts can be made to keep the genome 'flexible', responsive to various environmental challenges. Now that man lives in a largely man-made environment, discharged from many ordeals (such as lack of household food due to a bad harvest), environmental challenges do no longer keep the genome flexible. Therefore, the overall resilience of the genome may be increased if it is challenged, for example in periods of fasting. Third, also socio-economical factors have a strong influence on longevity and healthy aging. Certain differences in social-economical conditions, such as level of education, still create considerable longevity gaps, that are strongly related to lifestyle. Also the importance of social networks was strongly emphasized in the interviews.

2.4 Care

> My mother lives in a home for the elderly. What happens there is all fine, but the emphasis is on care. My research is very distant from that. We try move closer, but it is very difficult to build bridges. In practice one is confronted with the urgency of care. It is good to have a nurse at the bedside. And this practice does not benefit from the wonderful genetic research that we are conducting.

Where frailty and disease cannot be prevented or adequately treated, care is required. *What does biomedicine has to offer to elderly in need of care? What care do elderly wish themselves? And how can care for the elderly be optimized?* These kind of questions are also addressed in aging research.

In a world growing old, the need for care increasingly enters the domain of healthcare. Addressing demographic aging in combination with a high incidence of chronic diseases and need for care in the population of elderly requires such a share of material and human resources that individual families will often find themselves incapable of handling this burden.

There are several perspectives on the provision of care. First, the need for care can be approached from the perspective of the available means to relieve healthcare problems. Here, it must be enquired what problems are considered to be health problems, and what is being done to address such health problems. For example, many patients tend to focus their call for care on specific health complaints for which they hope that a treatment will exist. Many doctors tend to respond to these calls with curative and preventive interventions, in the hope of generating a health gain. Focusing on available means to generate a health gain, however, may loose important aspects of care out of sight. Does treatment of individual disease truly generate health gain in elderly suffering from multiple morbidities? If yes, at what price health is improved? For example, does better health come with increased burden? And is such and approach favourable for the well-being of elderly?

Second, there is the perspective of the needs of the elderly. While elderly may search help for perceived health problems, they may have many more needs than those for which they seek help. Are elderly capable of framing their call for care in terms of the functionality they need to realize their personal objectives? And are healthcare professionals capable of limiting their curative interventions to what is necessary to support elderly in their functional needs, so that elderly are not exposed to unnecessary burdens? The answer to both questions appears be 'no'.

If the preventive and curative approach of individual diseases does not necessarily result in improved health and/or well-being, there is a need for coordination of care. And if well-being in the population of elderly does not necessarily require excellent health, priorities must be set for provision of care, certainly when it is acknowledged that societies do not dispose of abundant resources to provide all imaginable care for all citizens.

Also the challenge to determine what kind of care elderly need to be able to access sufficient sources of well-being, is addressed in the work of the interviewed scientists. The interventions here well exceed the domain of biomedical research, and also focuses on psycho-social aspects.

2.5 Discussion: Science, Ethics and Health Policy

Research in the field of aging is important. Demographic changes confront us with challenges. Elderly have problems. Science can contribute to address these challenges and problems. Science should address challenges and problems in aging societies from different disciplinary angles. Better health will improve the overall well-being and quality of life of the elderly. All these premises (and many more could be formulated) clearly show that research in the field of aging is not situated in a normative void. Consequently, health policy and medical ethics are called to address the ethical questions that emerge in science and society. *How long do we want to live? In what conditions do we want to become old? What aspects of aging are considered to be problematic? Should something be done about these perceived problems? What can be done about these problems? How do we choose among different scenarios that science and society envision? What do individuals and communities expect from science and society? What can society expect from individuals and communities? And how can priorities be set in a field as expansive and heterogeneous as human aging?* From the interviews with scientists, no clear answer can be given to these diverse and complex questions. Therefore, it seems particularly relevant to reflect on three major questions: First, what can science contribute to the realization of the elderly life that we envision as individuals and societies? Second, how can we not get lost in the complex, vast, multidisciplinary, and expansive research landscape? And third, how do we determine the appropriate way forward in a practice as complex and heterogeneous as aging research?

2.5.1 Begin with the End in Mind

Human life hosts many enterprises. To realize our objectives and to realize our selves, we need capabilities and opportunities. These capabilities and opportunities may be impaired by the degenerative loss that accumulates throughout the lifetime. Should we modify our objectives and moderate our desire for self-realization in function of the degenerative loss we experience? Should we combat the loss of capabilities and opportunities, so that the realization of our objectives and our selves is not compromised? Or should we opt for a middle way, as a balanced approach to aging? In other words: What do we want to get out of life, and how do we envision doing this?

To get a grip on what we want in any human enterprise we undertake, Stephen Covey advises in his famous *The 7 Habits of Highly Effective People* to begin with the end in mind. Particularly with regard to aging, this approach seems most relevant. *How do we want to come to our end? What will have made this end (un)worthy? What life will have preceded this end? And what would research and healthcare have contributed to the realization of such life?*

First, there are ample possibilities to provide the elderly with capacities and opportunities (including added life years) to realize lives that they envision. In less than two centuries, it has not only become possible to live much longer than our ancestors did, but also to remain in a relatively good physical and psycho-social condition. During the interviews, it was emphasized over and over again that environment and lifestyle have the most prominent role in this evolution, although also biomedicine is not unimportant. Second, notwithstanding the importance of lifestyle and environment, practice largely focuses on biomedical aspects. Given this focus, how do we envision biomedical research contributing to a better old age? And what challenges do ethics and policy face in this respect?

Focusing on the biomedical aspects of aging can have various outcomes. On the one hand, causes of preventable death can be eliminated successfully, as have done antibiotics, vaccines, statins, and many other successful interventions to avert imminent threats to our health and survival. This may prevent people from premature deaths, and contribute to the establishment of more equal opportunities for all to become old and realize ones aspirations. On the other hand, the elimination of one disease may come at the cost of getting another disease (e.g. less cardiovascular disease, more dementia), which we can try to eliminate in turn. However, at a certain point, frail individuals may no longer endure all efforts to avert threats to health and life. If we duly realize this, we should be vigilant of not reducing the capabilities and opportunities of the elderly by subjecting them to burdensome treatments. Draining all energy that is left to fight a losing battle, needs not to be noble. Nonetheless, also in the frail elderly, opportunities and capacities can be created. Doing so, however, will require thinking out of the box of biomedical paradigms. Summarized: Whether or not we will die hospitalized, after having consumed a disproportional part of our lifetime investment in healthcare during the last year of our life on burdensome treatments that did not save our life—as many people do today—is thus not just a matter of preference, coincidence or fate. It is a matter of choice.

Should we send the elderly to the frontline of medical innovation, battling to move the end of life to new frontiers? What would this bring? And what can we get out of other strategies? These questions seem most relevant. However, because there is so much that we do not know about aging, it is difficult to answer such questions in a detailed fashion. Nonetheless, regardless of what we do and don't know there seems to be a broad consensus on the fact that there is a lot that we can do about aging. In this respect, investing in a healthy lifestyle and environment is likely to yield longer lives and better health in entire populations, and refraining to do so, may equally well impair longevity and health in entire populations. So the main challenge ahead is to find a balanced approach to health in elderly life, in which biomedicine, lifestyle, and intervention play an appropriate role.

2.5.2 Staying Focused in the Bigger Picture

During the interviews, it became clear that aging is a powerful conceptual framework. For several reasons, it is worthwhile for scientists to think in terms of aging: Aging is everywhere. It is relevant to any human being from conception on. It is of high interest to individuals, families, communities, science, society, and policy, as it generates many problems that call for attention (which is also reflected in opportunities to obtain funding). It is relevant to science, as it enables to integrate miscellaneous findings into a bigger picture. And regardless whether one focuses on biology, disease, health, or care, many types of research can be easily reframed in the setting of aging.

Integrating various research findings into the bigger picture of aging research has an upside and a downside. On the one hand, such a bigger picture does more justice to findings that have no potential to a clinical translation (yet). They all can be interpreted as pieces of a big, enigmatic puzzle, that in the end will provide a 'big picture' that will not just show the relevance of these findings, but have the potential to transform the clinic as we know it. If we get a better grip on aging, including opportunities for early diagnosis, prevention, and personalized healthcare, we would no longer have to wait for pathologies and health complaints to occur. If such a big picture is in the making, and if this big picture will create the alleged new opportunities, a new, more proactive approach to healthcare is already in de pipeline.

On the other hand, there is also a downside. For sure, integrating many findings from various disciplinary approaches in one single framework opens up fantastic opportunities to improve our understanding of aging and to foster an adequate approach to the problems that elderly experience. However, in such a framework, it will be difficult to stay focused on what aging research is to pursue—if, at all, this is clearly determined, rather than for example just expanding itself as a field of scientific enquiry. And in a focused approach, we will face the challenge of balancing all different disciplinary approaches and points of attention. This is certainly challenging, as it is already clear today that it is difficult to balance between effort and investment between for example basic research, the clinic, prevention, psycho-social conditions, and hands-on care. How can it be prevented that one concern outcasts the other? And how can adequate priorities be set?

2.5.3 *Moving Forward on Solid Normative Grounds*

Science in the field of aging is moving forward steadily. How can we move forward on solid normative grounds? When reflecting on this question, it is useful to keep in mind *who* aging research is all about. In whose interest do we enquire and approach different aspects of aging? There is no univocal answer to this question.

First, aging research obviously concerns elderly, and in the interviews there was a large consensus that elderly have many and very heterogeneous needs. These needs clearly exceed the domain of biomedicine, and also include for example psycho-social and practical matters. *How can we respond to these needs optimally?* This is a challenging question in which a very important aspect may easily be overlooked: What do the elderly want themselves? In one interview, it was emphasized that we do know what we have on offer to help elderly, but that we don't know what is the demand. By consequence, we should be vigilant of not defining problems in function of the solutions that we have or envision. Rather, we should envision solutions that address the problems that elderly perceive themselves.

Second, aging is certainly not only a matter of the individual. In a world growing old, public policy is confronted with major challenges in providing just and adequate healthcare for the elderly. During the interviews, concerns about the costs of an aging society were frequently expressed. With more elderly and more that can be done in science and care, how can we set adequate priorities that serve the interests of the individual and society at large, without impairing scientific progress? This is certainly a difficult balance, since complexity is all over the place: Elderly are a very heterogeneous population with many but very different needs, and the scientific approach to aging is a very heterogeneous enterprise with many very different aspirations, objectives, methods, and outcomes.

Third, also science has an interest in aging research. In serving the interests of elderly and society at large, important opportunities can be created for scientific innovation and advancement.

2.6 Conclusion

Research in the field of aging is progressing steadily and calling for adequate ethical reflection and public policy. With many unresolved issues, the way forward seems a balanced and coordinated approach that truly serves the interests of elderly, society at large, and science. The remainder of this book provides input for reflection on such way forward.

Acknowledgments Grateful acknowledgement is made to all interviewed scientists:

Elaine Dzierzak, Professor of Developmental Biology, director of the Erasmus Stem Cell Institute, ErasmusMC, Rotterdam.

Jenny Gierveld, Professor Emeritus, Department of Sociology and Social Gerontology, Vrije Universiteit, Amsterdam & Honorary Fellow, Netherlands Interdisciplinary Demographic Institute (NIDI), The Hague.

Lisette de Groot, Professor of Nutrition and Aging, Division of Human Nutrition, Wageningen University.

Jan Hoeijmaekers, Professor Molecular Genetics, Department of Cell Biology and Genetics, ErasmusMC, Rotterdam.

Albert Hofman, Professor of Epidemiology, Department of Epidemiology, Erasmus MC, Rotterdam.

Michael Müller, professor of Nutrition, Metabolism and Genomics, Division of Human Nutrition, Wageningen University.

Frans van der Ouderaa, Director for Business Development, LUMC Leiden & Chief Scientific Officer, Leyden Academy on Vitality and Aging. Former Vice President Corporate Research at Unilever.

Joris Slaets, Professor of Geriatrics, Department of Internal Medicine, Universtity of Groningen.

Eline Slagboom, Professor of Molecular Epidemiology, LUMC Leiden.

André Uitterlinden, Professor of Complex Genetics, Erasmus MC Rotterdam.

Bas Zwaan, Professor of genetics, Head of the Laboratory of Genetics, Wageningen University.

References

Efe, J.A., S. Hilcove, J. Kim, H. Zhou, K. Ouyang, G. Wang, J. Chen, and S. Ding. 2011. Conversion of mouse fibroblasts into cardiomyocytes using a direct reprogramming strategy. *Nature Cell Biology* 13:215–222.

Fries, J.F. 1980. Aging, natural death, and the compression of morbidity. *New England Journal of Medicine* 303(3):130–5.

Khaw, K.T., N. Wareham, S. Bingham, A. Welch, R. Luben, and N. Day. 2008. Combined impact of health behaviours and mortality in men and women: The EPIC-Norfolk prospective population study, Plos Medicine 5(1):39–47.

Oeppen, J., and J.W. Vaupel. 2002. Broken limits to life expectancy. *Science* 296:1029–1031.

Chapter 3
The Anti-Aging Movement

Contemporary Cultures and the Social Construction of Old Age

John Vincent

3.1 Introduction

This chapter seeks to characterise the Anti-aging movement and indicate the role it plays in the contested nature of old age. I approach the topic from a cultural anthropological and critical gerontological perspective. My interest has been to investigate how old age is constructed across diverse cultures. The purpose of looking at the Anti-aging movement was to explore the contemporary western idea of old age by examining the contested cultural meanings of 'old age' manifest in their antipathy. The research on which this chapter is written in ethnographic is style and based on systematic observation of scientific and other anti-aging conferences, reading and analysis of journals, text-books, websites and other materials produced by a broadly defined Anti-aging movement. The original research is presented in previous publications; the intention here is to summarise that work and suggest some policy implications (Vincent 2003a, b, 2006a, b, 2007, 2008, 2009, 2011).

3.2 Social Movements

In all societies some use is made of social categories based on age. The use of age to classify people and attribute meaning and value in Western culture has a long and changing history (Minois 1989; Cole 1992; Katz 1996; Johnson and Thane 1998; Thane 2000; Macnicol 2006). Such change is manifest over the last sixty years, not least in social movements around the politics of identity. Broadly conceived, these social movements have sought to challenge received identities based on what anthropologists have in the past called 'natural categories'; notably sex, race, and age. These movements mobilise around a 'liberation agenda'; that is to say these

J. Vincent (✉)
University of Exeter, Exeter EX4 4RJ, Devon, UK
e-mail: JVincent@exeter.ac.uk

M. Schermer, W. Pinxten (eds.), *Ethics, Health Policy and (Anti-) Aging: Mixed Blessings*, Ethics and Health Policy 1, DOI 10.1007/978-94-007-3870-6_3,

social movements see themselves as liberating people from traditional stereotypes and enabling individuals to participate freely as full members of society (Touraine 1971; Weeks 1994; Bradley 1996; Isin and Wood 1999). However, there is a standard dilemma which manifests itself within the agendas of identity politics; namely, what priority should be afforded between the elimination of difference (making the cues invisible/ irrelevant), and the celebration of difference (valuing diversity and difference). Should we be colour blind or celebrate black as beautiful? Is femininity in dress and sexuality an oppressive personal style imposed by a male dominated society or something to be re-evaluated as having equal social value? This dilemma has been increasingly visible in debates on the significance of old age in the contemporary world.

There have been a number of social movements associated with the reappraisal of age-based social categories in the last thirty years. Two such developments are the focus of this chapter. They are the Third Age movement (Laslett 1989) and the Anti-aging movement. These movements present contrasting perspectives on the culturally devalued status of old age; the former seeks to celebrate old age, the latter to eliminate it. They have developed in the context of an aging population and of significant cultural and technological change. New discoveries in cell science, genetics and biochemistry offer the prospects of dramatically increased knowledge and control of the ways bodies change over the life course (Kirkwood 1999; Rose 1997; Bramstedt 2001; Benecke 2002; Fukuyama 2002, Post and Binstock 2004). The advances in the biology of aging should not be seen as simply causing cultural change, but part of the contemporary social and cultural milieu which not only responds to, but also frames scientific understanding (Kenyon et al. 1991; Dupré 1993; Gieryn 1999; Franklin and Lock 2003; Turner 2007).

3.3 The Diverse Phenomenon of Anti-Aging

The Anti-aging movement is not characterised by adherence to any particular knowledge system. The Anti-aging movement spans not only conventional medicine and orthodox science but holistic medicine; it is debated within diverse theologies and alternative knowledge systems (Mykytyn 2006a, b; Spindler 2008). The common theme which runs through all the manifestations of anti-aging is the identification of old age as a highly negative phenomenon of the body which requires action to defeat (Vincent 2006a, b). Indeed the anti-aging literature is replete with military images of the fight against old age (Vincent 2007). The position pays no heed to social age; as for example in senior citizen, church elder; or familial age, as in grandfather or great aunt. Similarly psychological aspects of aging such as maturity, fear of death, or wisdom are left out of the equation. Aging is seen entirely as a negative phenomenon of the body as an individualised biological entity.

I suggest the movement can be understood as having four linked foci differentiated by their location of the key identifying characteristic of old age. They are:

1. Old age as appearance. Old age is manifest on the surface of the body; this form of anti-aging seeks to eliminate the visual clues which mark people out as old. This approach is embedded in the aesthetics of a youthful body, not in the re-evaluation of older people's appearance. It is the subject of massive commercial activity, focussing in particular on women and an aesthetic which equates youth with health and beauty. Technical progress in these circumstances gives experts the ability to control and modify appearance. Examples might include the development of Botox (Ting and Freiman 2004; Cooke 2008) or the use of stem cell technology to rejuvenate hair colour (Nishimura et al. 2005). Hurd Clarke (2011), amongst others argues that such activity, while it might in some circumstances give pleasure, fundamentally reinforces highly negative attitudes towards older bodies.
2. Old age as a medical condition. Old age is seen as located within the organs of the body as a collection of symptoms. Anti-aging medicine seeks to cure these disease symptoms. Manifestations of old age are subject to diagnosis and cure and strategies are adopted to slow or to eliminate them. Technical progress in these circumstances is seen as eliminating the diseases of old age. Examples from research include the controversial use of male hormone replacement therapy by Ronald Klatz of the American Academy of Anti-Aging Medicine (http://www.worldhealth.net/whos-who/ronald-klatz/) or use of stem cells for heart repair by amongst others Bruno Gridelli (Triolo and Gridelli 2006). There is a key debate around the extent to which such strategies should aim to lengthen lifespan as opposed to merely increasing healthy life expectancy toward some 'natural' limit.
3. Old age as a biological process. Old age is identified as a biological process driven by the biochemistry of complex organic molecules, aging is located deep within the cell processes of all living things. This kind of anti-aging seeks to eliminate old age by techniques of bio-chemical control and optimum cell performance. Illustrative examples could include the work of the Geron Corporation on telomerase (http://www.geron.com/technology/telomerase/telomerase.aspx, Hall 2003), or that of James Carey genetically manipulating the lifespan of drosophila (Carey 2003). Constructing old age as cell function raises issues of a post-human future and a world stratified by biological powers including longevity (Bostrom 2008).
4. Old age as a prelude to death. Old age is understood as the period of life leading to death. Death is seen the ultimate failure of the body, aging is thus to be avoided by immortality. The strategy of this form of anti-aging is to devise techniques to postpone or avoid death through scientific study of rejuvenation. Serious money and biological expertise is being mobilised for such objectives. Aubrey de Grey's SENS (Scientifically Engineered Negligible Senescence) Foundation (http://www.sens.org/) and the Cryonics Institute (http://www.cryonics.org/) are two examples of institutions devoted to this end. Age as the prelude to death raises issues about the desirability of death, should we be seeking a good death rather than denying death any value. It also raises issues of the succession of generations.

3.4 Why has the Anti-Aging Movement Grown?

What gives the Anti-aging movement its salience in the contemporary world? Why has it become so important in how old age is defined and understood? We can draw on a number of developed sociological literatures to gain some insight into these questions. These include the sociology of medicine, of science, of the body, and of identity. Schermer Maartje in this volume (Chap. 16) has made excellent use of the sociology of medicine to examine claims about the understanding and control of aging in a medical context. I will look at the other three.

3.4.1 *Sociology of Science and the Focus on Life as Essentially Molecular*

The linked areas of the Sociology of Knowledge and the Sociology of Science provide historical and cultural contexts for claims about knowledge, including scientific and specifically biological knowledge, about human aging. This work can direct us to understanding how what counts as knowledge is contingent on social context (Mannheim 1952; Habermas 1987), how competing knowledge systems and scientific disciplines conduct 'boundary maintenance' (Barnes 1985; Barnes and Bloor 1996; Gieryn 1999), the trend towards reductionism manifest in scientific activity (Dupré 1993; Rose 1997), and the location of vital processes at increasingly microscopic levels (Rose 2001, 2007). A number of scholars, particularly in the sociology of science, have identified shifts in the cultural understanding of the essence of life (Franklin and Lock 2003; Lock and Farquhar 2007; Moreira and Palladino 2008). In the contemporary world the essences which are seen to animate life are the forces at play at the biochemical level in the cell. A related issue is the role of reductionism; in science in general and in scientists' understanding of life processes in particular. Scientists tend to look for explanations by seeking to divide problems into constituent elemental processes. This is an effective strategy but creates conceptual and communication issues when putting together a complete understanding of the whole, be that an organism an ecosystem or a process such as aging.

Rose's (2001, 2007) work examines the power struggles and cultural consequences of seeing life in this micro-molecular and fragmented way. He points out economic, professional, political and moral struggles over 'life itself'. Rose suggests five processes through which to understand changes to 'the politics of life itself'. They are molecularization, optimization, subjectification, somatic expertise, and the economics of vitality. The policing and control of life implies the same processes for death. Not only in the sense of avoiding death and prolonging life, but dealing with the inevitable fact of death. For all the culturally enwrapped denial, we all still die, and this fact has to be managed. The extent of the denial makes this a difficult task. Lafontaine (2008) sees death as having been removed from a world where it is special and set apart and is now considered within the realms of mundane bio-chemical processes. As a consequence those experts who have the knowledge of ritual and symbol around death are sidelined and the expertise of medicine and biological science takes first place in our cultural understanding of old age and death (Vincent 2003a; Turner 2007; Vincent 2011).

3.4.2 The Sociology of the Body and its Implications for the Older Body

The development of the sociology of the body reflects the way in which the body has been re-evaluated in contemporary culture. The body has always been a site of cultural inscription but bodily aesthetics have changed and become more self-aware. The marks of a culturally acceptable body are scrutinised and contested. People's bodies have become self-help projects to be actively moulded and sculpted (Giddens 1991; Gimlin 2002). The body has also increasingly become the site of personal identity. People understand who they are through their bodies, how they experience them and other people's reaction to them (Oberg 2003). Some writers have looked at the body as cultural capital and the process of aging, particularly for women and for athletes, as loss of capital (Dumas et al. 2005; Tulle 2008). Laura Hurd Clarke (2011) has explored the powerful impact of aging on older women as they experience and interpret their aging bodies. The kind of self-loathing and coping strategies in terms of bodily practices she describes are testimony to how deep the antagonism to the aging body is. The political dimensions of this particular cultural orientation towards the body includes issues of consumerism and the commodification of appearance, and in policy terms the individuals personal responsibility for the condition of one's body. These cultural developments create significant problems for the end of life (Twigg 2000, 2004; Moreira and Palladino 2008). What choices, what responsibilities, what methods of personal fulfilment in bodily terms are possible for an individual with an aging or dying body? (Hallam et al. 1999; Lloyd 2011)

3.4.3 Sociology of Identity and the Extreme Individualism of Contemporary Culture

Historically, the discipline of Sociology starts from concerns about the nature of individual and collective identity in changing society. Many sociologists, anthropologists and historians have identified individualism as a unique characteristic of western culture. Compared to the full global range of contemporary and historical cultures, that of the contemporary West lies at the far end of a continuum furthest away from group ascription and collective identity. Historians and historical sociologists have sought to identify the impact of individualism through time (MacFarlane 1978; Lukes 1973; Badie 1990; Gagnier 2010). Anthropologists such as Louis Dumont have compared cultures (Dumont 1985) including comparisons of old age (Amoss and Harrell 1981; Keith 1994; Holmes and Holmes 1995). A further set have looked at changing ascribed collective identities, including that of old age, when faced with the dynamics of modernity (and post-modernity) (Cumming and Henry 1961; Elias 1985; Gilleard and Higgs 2000; Daatland and Biggs 2005).

La Fontaine's (2008) account of the 'post-mortal society' emphasises the role of individualism and the removal of collective rituals of life and death. She examines how the history of ideas in the west has accentuated the individual as the location of moral worth. The image of the rational individual capable of exercising choice is a dominant post-enlightenment image of the moral person. The dominant values of the West are those of individual self-fulfilment and policy priorities are framed as promoting individual choice (Phillipson 1998; Mishra 1999; Estes et al. 2003). Habermas argues that science and technology and liberal values emphasising individual choice have historically developed together.

> And since enlarging the scope of individual choice fosters individual autonomy, science and technology have, to date, formed an evident alliance with the fundamental credo of liberalism, holding that all citizens are entitled to equal opportunities for an autonomous direction of their own lives. (Habermas 2003, pp. 24–5)

Individualism becomes problematic faced with death. La Fontaine argues that from the point of view of the individualist there is no valid reason to die and leave a place for a new generation (Lafontaine 2008, p. 184). Immortalist urges are fuelled by a lack of value ascribed to anything other than the individual. There is the Warren Buffett solution, the mega rich foundation which will keep your name alive through charity or the arts. Or, some wealthy people use their money in an attempt to preserve themselves through scientific endeavours by endeavouring to create corporeal immortality (Boia 2004).

Thus from the perspective of each of these three extensive bodies of literature we can identify a 'high modern' [historical west in 20th and 21st c.] cultural trend which feeds into the burgeoning of anti-aging movement. These developing features of contemporary western culture which provide the context for 'anti-aging' sentiment are: An understanding of the essence of life as being located at the level of complex organic molecules; a focus on the body as the locus of identity; and an extreme form of cultural individualism. They all feed into the view that aging is no more than an undesirable aspect of the external appearance or internal constitution of a person's body which is capable of manipulation and remedy. They translate into policy dilemmas as (a) finding alternative sources of knowledge, power and authority to the growing command of old age by medicine and biology; (b) finding positive sources of identity for older people other than their bodies; and (c) finding alternatives to the current dominant model of social policy which de-emphasises social solidarity and valorises choice and individual responsibility.

3.5 Celebrating Old Age or Eliminating it—Critiques of Anti-Aging

We can now return to the initial question posed at the beginning of the chapter. The development of anti-aging science and practice challenges us to consider how people should approach old age. While the Third Age movement has to some extent been successful in establishing the idea of a new positive stage in life post retirement, they have failed to overthrow the dominant image of old age as one of illness and decline. The concept of the third age can be seen in some circumstances as an attempt to prolong youth and not necessarily to create a new attitude to old age as a life stage valuable in its own right. But it is clear that despite the success of some of these attempts at re-evaluation, increasingly the dominant contemporary cultural attitude to later life is that of 'anti-aging'; trends within western culture seek not to celebrate aging but to avoid it.

The perspective of critical gerontology has been mobilised to question the nature of the anti-aging enterprise (Vincent et al. 2008). Some are critical of the corporate exploitation of ageism and the use of fear of aging to promote consumer markets in skin products and other commodities which fetishize youth. Others fear that the new biology will add a further twist to the bio-medicalization of old age (Estes and Binney 1989). The problem of aging has become: 'What is a satisfactory/healthy body?' Does it include the aging body (Dumas and Turner 2007; Powell 2010)? Contrasting arguments have been made, on the one hand about the value of non-standard bodies (for example people whose bodies exhibit dwarfism, deafness, disability, or old age), and on the other a desire to eliminate pain, suffering or social stigma. The concern is that anti-aging science will add new scientific tools to discipline old age into something tractable in the interests of the state, commerce and other powerful elites and inhibit the incipient politics of identity amongst older people who seek to reclaim the later part of life as valuable in its own right (Vincent 2009). In the contemporary world, the great investment of time, resources, and cultural ingenuity to find ways to live longer and if possible for ever, have consequences for old age. The fantasies of avoiding old age created through commercialism and the dominance of the medical model are bad for older people. These attitudes identify old age as a problem waiting for a solution, postpone action on current problems of old age, and inhibit research into death as a natural event and the final stage of the life course as a positive meaningful process (Vincent 2003b). Anti-aging can do as much for age discrimination as skin whiteners for race relations, and gender re-assignment surgery for sex equality.

3.6 Celebrating Old Age or Eliminating it—Critiques of Anti- Anti-Aging

The anti-aging protagonists have sought to counter these arguments in a number of ways (Overall 2003). I will deal with three criticisms of the anti-anti-aging position outlined above. Firstly there is the argument presented by a number of philosophers, ethicists and others that 'life' is a self-evident good and therefore all attempts to prolong it are 'good' (Ackerman, 2007). I suggest that the social science evidence of people's attitudes to death show a rather more nuanced and pragmatic picture (Palgi 1984; Gott and Ingleton 2011). In response to the question: "How old would you like to be when you die?" posed in a 2011 UK national survey by ComRes[1] only 15 % replied that they would like to live for ever. Men in the survey were slightly more likely to tick the 'live for ever box' (17 % as opposed to 13 % for women). Or to put it another way, offered the chance to indicate a preference for of living for ever, 85 % of respondents chose not to. There was a systematic relationship to class with those with high status occupations less likely to want to live for ever. A further finding was related to age, where there was a progressive relationship, the older the respondent the less likely they were to indicate they wished to live for ever. I suggest these findings are grounds for scepticism that highly educated, middle aged, male professionals are the right people to introspect on the value of immortality for the rest of us.

Secondly there is the accusation that the anti-anti-aging position is an untenable rationalisation for prolonging pain and suffering (Glaser 2009). A critique that comes from a modernist approach to science, which sees the growth of scientific knowledge as inexorable, inevitably beneficial and therefore aging as merely one more technical problem to be solved. I have presented the case against that critique in two papers (Vincent 2009, 2011). A case which, whilst acknowledging the power of science as a knowledge creating system, does not see it as necessarily progressive or standing outside culture and operating independently from cultural and ethical considerations. Defining old age as pain and suffering is part of the problem, not a way forward.

There is a third and in some ways more powerful critique, which is that the anti-anti-aging position relies on an appeal to naturalism and an untenable reliance on a traditional notion of human nature (Jones and Higgs 2010). However, although many do make such an appeal, it is not necessary to base a critique of anti-aging on assumptions about the essence of human biology. I hold a humanist position that seeks to value the diversity of age-based identities, differences that are based in the cultural (not biological) nature of gender and generation. 'Human nature' is itself a cultural concept. Although the relationship between culture and biology in human behaviour is a large, complex and poorly understood, the appropriate starting point is the diversity of human behaviour and experience and not assumptions about universality. The evolutionary history of our distinctive human characteristics comes from the increasing dominance of culturally acquired behaviour over biologically inherited ones. The fascinating thing about the emergence of human society is precisely the way gender and generation moved from the realm of the natural to become features of culture. Humans have built their diverse kinship societies by culturally moulding the basic building bricks of gender and generation. A society without age would be without generations and therefore less human. The concern is not merely that the powerful forces behind framing old age around failing bodies are detrimental to older people now, it is that a successful anti-aging, science based,

---

[1] "The research, which was carried out for the Dying Matters Coalition by Comres, reveals that although most people are scared of dying, quality of life is viewed as more important than how long we live for… Comres interviewed 2,028 British adults online between 28 April and 1 May 2011. Data were weighted to be representative demographically of all British adults." Survey published 16 May 2011 on http://www.comres.co.uk/dyingmatterssurveymay11.aspx.

immortalist technology would undermine a human cultural process—the succession of generations—and hence compassion and social solidarity. Societies that are without compassion and atomised to a high degree have existed—such societies are not 'unnatural' in any biological sense. It is an ethical and a cultural position that such societies are undesirable.

3.7 Conclusion

The policy conclusion of the humanist position on old age advocated in the above discussion is that increasing power to a diversity of older people's voices is essential for progress. The extent to which ageism is embedded in the cultural and scientific, knowledge-creation institutions has to be acknowledged and countered. Commercial, medical and biological institutions have re-inforced the anti-aging mind set. If the 'Third Age' agenda of a renaissance for later life is to succeed, older people must themselves be in the lead in positively valuing themselves, which means that together they/we have to find ways to live well and die well.

References

Ackerman, F.N. 2007. Death is a punch in the jaw: Life extension and its discontents. In *The Oxford Handbook of Bioethics*, ed. B. Steinbock 324–348. Oxford: Oxford University Press.

Amoss, P.T., and S. Harrell. 1981. *Other ways of growing old: anthropological perspectives.* Stanford, CA: Stanford University Press.

Badie, B. 1990. Community, individualism and culture. In *Individualism: theories and methods*, eds. P. Birnbaum and J. Leca. 95–115. Oxford: Clarendon.

Barnes, B. 1985. *About science*. Oxford: Blackwell.

Barnes, B., D. Bloor, and Henry, J. 1996. *Scientific Knowledge: A Sociological Analysis.* London: Athlone.

Benecke, M. 2002. *The Dream of eternal life: Biomedicine, aging, and immortality* (trans. R. Rubenstein). New York: Columbia University Press.

Boia, L. 2004. *Forever young: A cultural history of longevity.* London: Reaktion Books.

Bostrom, N. 2008. Why I want to be a posthuman when I grow up. In *Medical enhancement and posthumanity*, eds. B. Gordijn and R. Chadwick 107–137. http://www.nickbostrom.com/posthuman.pdf.

Bradley, H. 1996. Fractured identities: Changing patterns of inequality. Cambridge: Polity Press.

Bramstedt, K.A. 2001. Scientific breakthroughs: Cause or cure of the aging 'Problem'. *Gerontology* 47(1):52–54.

Carey, J. 2003. *Longevity: The biology and demography of lifespan.* Princeton: Princeton University Press

Cole, T.R. 1992. *The journey of life: A cultural history of aging in America.* Cambridge: Cambridge University Press.

Cooke, G. 2008. Effacing the face: Botox and the anarchivic archive. *Body and Society* 14(2):23–38.

Cumming, E., and W.E. Henry. 1961. *Growing old: The process of disengagement.* New York: Basic Books.

Daatland, S.O., and S. Biggs. eds. 2005. *Aging and diversity: Multiple pathways and cultural migrations.* Bristol: The Policy Press.

Dumas, A., and B.S. Turner. 2007. The life-extension project: A sociological critique. *Health Sociology Review* 16(1):5–17.
Dumas, A., S. Laberge, and S.M. Straka. 2005. Older women's relations to bodily appearance: The embodiment of social and biological conditions of existence. *Aging & Society* 25(6):883–902.
Dumont, L. 1985. *Essays on individualism: Modern ideology in anthropological perspective*. London: University of Chicago Press.
Dupré, J. 1993. *The disorder of things: Metaphysical foundations of the disunity of science*. London: Harvard University Press.
Elias, N. 1985. *The loneliness of the dying*. Oxford: Blackwell.
Estes, C.L., and E. Binney. 1989. The biomedicalisation of aging: Dangers and dilemmas. *Gerontologist* 29(5):587–596
Estes, C. L., S. Biggs, and C. Phillipson. 2003. *Social theory, social policy and old age*. Buckingham: Open University Press.
Franklin, S., and M. Lock. 2003. *Remaking life and death*. Oxford: James Currey.
Fukuyama, F. 2002. *Our posthuman future: Consequences of the biotechnology revolution*. New York: Farrar, Straus and Giroux.
Gagnier, R. 2010. Individualism, decadence and globalization: On the relationship of part to whole, 1859–1920. Basingstoke: Palgrave Macmillan.
Giddens, A. 1991. *Modernity and self-identity*. Cambridge: Polity Press.
Gieryn, T.F. 1999. *Cultural boundaries of science: Credibility on the line*. Chicago: University of Chicago Press.
Gilleard, C., and P. Higgs. 2000. *Cultures of aging*. Harlow: Prentice Hall.
Gimlin, D. 2002. *Body work: Beauty and self image in American culture*. Berkeley: University of California Press.
Glaser, V. 2009. Interview with John Vincent. *Rejuvenation Research* 12(1):59–63.
Gott, M., and C. Ingleton. 2011. *Living with aging and dying: Palliative and end of life care for older people*. Oxford: Oxford University Press.
Habermas, J. 1987. *The theory of communicative action. Vol.2, lifeworld and system: A critique of functionalist reason*. Cambridge: Polity.
Habermas, J. 2003. *The future of human nature*. Oxford: Polity.
Hall, S.S. 2003. *Merchants of immortality: Chasing the dream of human life extension*. Boston: Houghton Mifflin.
Hallam, E., J. Hockey, and G. Howarth. 1999. *Beyond the body: Death and social identity*. London: Routledge.
Holmes, E.R., and L.D. Holmes. 1995. *Other cultures, elder years*. 2nd ed. Thousand Oaks, CA: Sage Publications.
Hurd Clarke, L. 2011. *Facing age: Women growing older in anti-aging culture*. Plymouth: Rowman & Littlefield.
Isin, E.F., and P.K. Wood. 1999. *Citizenship and identity*. London: Sage.
Johnson, P., and P. Thane. eds. 1998. *Old age from antiquity to post-modernity*. Routledge: London.
Jones, I.R., and P.F. Higgs. 2010. The natural, the normal and the normative: Contested terrains in aging and old age. *Social Science & Medicine* 71(8):1513–1519.
Katz, S. 1996. *Disciplining old age: The formation of gerontological knowledge*. London: University Press of Virginia.
Keith, J. 1994. *The aging experience: Diversity and commonality across cultures*. Sage Publications: London.
Kenyon, G.M., J.E. Birren, and J.J.F. Schroots. eds. 1991. *Metaphors of aging in science and the humanities*. Springer: New York.
Kirkwood, T. 1999. *Time of our lives: The science of human aging*. London: Weidenfeld & Nicolson.
Lafontaine, C. 2008. La Société Post-Mortelle. Paris: Seuil.
Laslett, P. 1989. A fresh map of life: The emergence of the third age London: Weidenfeld and Nicolson.

Lloyd, E. 2011. What do we know about the congruence between what older people prioritize at the end of life and policy and practices? In *Living with aging and dying*, eds. M. Gott, and C. Ingleton. 63–72. Oxford: Oxford University Press.
Lock, M., and J. Farquhar. eds. 2007. *Beyond the body proper: Reading the anthropology of material life*. Durham, N.C.: Duke University Press.
Lukes, S. 1973. *Individualism.* Oxford: Basil Blackwell.
Macfarlane, A. 1978. *The origins of English individualism: The family, property and social transition*. Oxford: Blackwell.
Macnicol, J. 2006. *Age discrimination: An historical and contemporary analysis*. Cambridge: Cambridge University Press.
Mannheim, K. 1952. *Essays on the sociology of knowledge.* London: Routledge & Kegan Paul.
Minois, G. 1989. *History of old age*. Polity Press: Cambridge.
Mishra, R. 1999. *Globalization and the welfare state*. Cheltenham: Edward Elgar.
Moreira, T., and Palladino, P. 2008. Squaring the curve: The anatomo-politics of aging, life and death. *Body and society* 14(3):21–48.
Mykytyn, C.E. 2006a. Anti-aging medicine: A patient/practioner movement to redefine aging. *Social Science and Medicine* 62(3):643–653.
Mykytyn, C.E. 2006b. Anti-aging medicine: Predictions, moral obligations, and biomedical intervention. *Anthropological Quarterly* 79(1):5–31.
Nishimura E.K., S.R. Granter, and D.E. Fisher. 2005. Mechanisms of hair graying: Incomplete melanocyte stem cell maintenance in the niche. *Science* 307(5710):720–4.
Oberg, P. 2003. Images versus experience of the aging body. In *Aging bodies: Images and everyday experience*, ed. C. A. Fairclough 78–102. Oxford: AltaMira.
Overall, C. 2003. *Aging, death, and human longevity*. Berkeley: University of California Press.
Palgi, P. 1984. Death: A cross-cultural perspective. *Annual Review of Anthropology* 13(1):385–417.
Phillipson, C. 1998. *Reconstructing old age: New agendas in social theory and practice*. London: Sage.
Post, S.G., and R.H. Binstock. 2004. *The fountain of youth: Cultural, scientific, and ethical perspectives on a biomedical goal*. Oxford: Oxford University Press.
Powell, J. 2010. The aging body: From bio-medical fatalism to understanding gender and biographical sensitivity. In *Culture, bodies and the sociology of health*, ed. E. Ettorre. Chap. 6. Farnham: Ashgate.
Rose, N. 2001. The politics of life itself. *Theory, culture & society* 18(6):1–30.
Rose, N. 2007. *Politics of life itself: Biomedicine, power and subjectivity in the twenty-first century*. Oxford: Princeton University Press.
Rose, S. 1997. *Lifeline: Biology, freedom, determinism.* London: Penguin.
Spindler, M. 2008. Surrogate religion, spiritual materialism or protestant ethic? Three accounts of the function of religiosity in anti-aging. *Journal of Aging Studies* 22(4):322–330.
Thane, P. 2000. Old age in english history: Past experiences, present issues. Oxford: Oxford University Press.
Ting, P. T., and A. Freiman. 2004. The story of Clostridium botulinum: From food poisoning to Botox. *Clinical Medicine* 4(3):258–261.
Touraine, A. 1971. *Post-industrial society*. New York: Random House.
Triolo, F., and B. Gridelli. 2006. End-stage organ failure: Will regenerative medicine keep its promise? *Cell Transplantation* 15(Supplement 1):S3–S10. http://cfweb.ismett.edu/documenti/pubblicazioni/CellTx_Vol15-S1_2006/CellTx_Vol15-S1_s03–10.pdf.
Tulle, E. 2008. *Aging, the body and social change*. Basingstoke: Palgrave Macmillan.
Turner, B.S. 2007. Culture, technologies and bodies: The technological Utopia of living forever. In *Embodying sociology: Retrospect, progress and prospects*, ed. C. Shilling 19–36. Oxford: Blackwell.
Twigg, J. 2000. Care work as a form of bodywork. *Aging and Society* 20(4):389–412.
Twigg, J. 2004. The body, gender, and age: Feminist insights in social gerontology. *Journal of Aging Studies* 18(1):59–73.

Vincent, J.A. 2003a. *Old age*. London: Routledge.
Vincent, J.A. 2003b. What is at stake in the 'war on anti-aging medicine'? *Aging & Society* 23(4):675–684.
Vincent, J.A. 2006a. Anti-aging science and the future of old age. In *Futures of old age*, eds. J. Vincent, C. Phillipson, and M. Downs. eds. London: Sage.
Vincent, J.A. 2006b. Aging contested: Anti-aging science and the cultural construction of old age. *Sociology* 40(4):681–698.
Vincent, J.A. 2007. Science and imagery in the 'war on old age'. *Aging and Society* 27(6):1–21.
Vincent, J.A. 2008. The cultural construction old age as a biological phenomenon: Science and anti-aging technologies. *Journal of Aging Studies* 22(4):331–339.
Vincent, J.A. 2009. Aging, anti-aging and anti-anti-aging; who are the progressives in the debate on human biological aging. *Medicine Studies* 1(3):197–208.
Vincent, J.A. 2011. Anti-aging and scientific avoidance of death. In *Living with aging and dying: Palliative and end of life care for older people*, eds. M. Gott, and C. Ingleton. 29–41. Oxford: Oxford University Press.
Vincent, J.A., E. Tulle, and J. Bond. 2008. The anti-aging enterprise: Science, knowledge, expertise, rhetoric and values. *Special edition of The Journal of Aging Studies* 22(4):291–294.
Weeks, J. 1994. *The lesser evil and the greater good: The theory and politics of social diversity*. London: Rivers Oram Press.

Chapter 4
Do We Need an Anthropology of the Aging Person and What Should it Look Like?

Wim Dekkers

4.1 Introduction

The classical Greek philosophers called human beings 'mortals' as opposed to 'immortals' who were gods and angels. All individual life—including human life—has limited duration and passes from growth through maturity into a phase, especially in higher forms of life, in which biological functions gradually deteriorate. This is the process that we call aging. Aging is generally considered to be a biological process that is beyond our control. Aging just happens.

Because aging is a feature of all individual life, it is not specific to human beings in the way that self-consciousness, personhood, freedom, and, for example, narrativity are characteristic of the human condition. Plants age as plants, animals age as animals, human beings age as human beings. This means that aging—while not a specific feature of human beings—is still linked to typical human characteristics such as self-consciousness, personhood, freedom and so forth. Aging and mortality have been a subject of reflection throughout the history of philosophy and mankind. Old age and senescence have always been fascinating subjects (McKee 1982).

It is not an easy task to produce a clear and indisputable definition of aging when there are so many perspectives from which the phenomenon of aging can be studied. 'Aging' is a multidimensional concept. A quite simple, but attractive, definition of the aging of human beings is: "a process that converts healthy young adults into less healthy older ones with progressively increasing risks of illness and death" (Miller 2004, p. 228).

Nowadays, aging is a hot issue. Demographic change has increased the proportion of elderly people in the population, and consequently the proportion of the population suffering from the diseases of old age such as dementia. Contemporary biogerontology tries to extend the maximum human lifespan. Policy guidelines aim at stimulating active, healthy and successful aging. Contemporary bioethics pays a lot of attention to the ethical aspects of care for the elderly, and the accompanying

W. Dekkers (✉)
IQ healthcare, Radboud University Nijmegen Medical Centre, The Netherlands
e-mail: W.Dekkers@iq.umcn.nl

M. Schermer, W. Pinxten (eds.), *Ethics, Health Policy and (Anti-) Aging: Mixed Blessings*, Ethics and Health Policy 1, DOI 10.1007/978-94-007-3870-6_4,

biomedical research and policy guidelines. The thesis of this chapter is that we also need a philosophical anthropology of aging. The scientific work that is carried out in the field of aging, the practice of care for the elderly and ethical considerations about aging are based upon—most often implicit—presuppositions about a number of fundamental characteristics of human beings.

The discipline of philosophical anthropology can draw on all kinds of scientific anthropologies such as biological, cultural and social anthropologies, but it aims to go beyond a purely empirical approach. The central question is: What is a human being? The argument of this chapter now is that there is a connection between our understanding of the 'nature' of human beings on the one hand, and of the principles of morality on the other. Philosophical anthropology is not a neutral assessment of a state of affairs, for example, of a human being's fundamental characteristics, but necessarily implies a view of what a human being ought to do in an ethical sense. When one tries to formulate an answer to the question 'what is a human being?', one cannot take an objective stance. Every participant in philosophical anthropology is a stakeholder, a subject and an object of research. Ethics, conceived of as a philosophical inquiry into the principles of morality, presupposes a particular view of human beings. This relationship between philosophical anthropology and ethics has been nicely formulated as follows: "From an anthropological perspective, man is a being that must *become* what he is, while from an ethical perspective, man is a being that *must* become what he is" (Sporken 1983, p. 41, translation by WD).

Any anthropology has ethical implications while ethical models, principles and arguments are necessarily founded on a set of some fundamental presuppositions about human beings. To give a few examples: Taking dignity as a fundamental anthropological category implies an invitation to respect all human beings; seeing human beings from a relational perspective can lead to the ethical principle of solidarity; a particular view of the 'person' has implications for the alleged role of advance directives. In the same vein, any anthropology of aging implies a view of how the elderly should behave and how they should be treated, and any discussion of ethics in the field of aging presupposes a particular view of human beings.

In this chapter I want to answer two questions. First, why do we need a philosophical anthropology of aging? And second, what should such an anthropology look like? In sect. 2, I will present three arguments for an anthropology of aging. Briefly, these arguments consider (1) the ideal of making implicit anthropologies explicit, (2) the consequences of new developments in the field of biogerontology, and (3) the defects of the posthumanist idea of prolongevity and expanding the human lifespan. In Sect. 3, I will deal with some crucial elements of a philosophy of aging, concentrating on the relationship between the bodily and mental aspects of aging, thereby emphasizing the role of the body in our understanding of aging. In sect. 4, I will focus on vulnerability, and the capacity to look after oneself, as both anthropological categories and moral values. In sect. 5, I will recapitulate some central features of the aged body with special attention to the similarities and dissimilarities between illness and disabilities on the one hand and aging on the other. In this section, I will also emphasize the need of an empirically informed and culturally based anthropology of the aging body. In sect. 6, I will summarize the conclusions and discuss some policy implications.

4.2 Understanding, Explaining and Controlling Old Age

The prospect of aging may be greeted with mixed feelings. If one surveys reflections about aging throughout the ages, they can be roughly divided into positive and negative views. A popular wisdom reads 'the longer we live the more we learn' and another says 'age is a heavy burden'.

On the one hand, there is a long tradition of commentary that accepts old age, gives it a place in life and finds meaning in it. This tradition starts (in the West) with Greek philosophers such as Socrates and Plato. Cicero's treatise *On Old Age* (Cicero 1923) is a well known plea for a positive stance toward old age and all the ailments it brings with it, and so are Montaigne's essays about old age and mortality (Montaigne 1993, p. I:19, I:20, I:36) and the writings of many others. Common in these works is the idea that old age, the decline of physical and mental strength, fragility and vulnerability are not necessarily to be seen as negative aspects of human life.[1]

On the other hand, there is an equally ancient tradition in which aging and mortality are not accepted and in which one can dream of eternal life. An obsession with immortality is a central theme in the Babylonian epic about king Gilgamesh who ruled southern Mesopotamia in about 3000 B.C. Perhaps the oldest written record of attempts to reverse aging is in an Egyptian papyrus circa 1600 B.C. which provides instructions for preparing an ointment that transforms an old person into a youth of twenty (Post and Binstock 2004). An example of this utopian tradition is the allegoric painting of Lucas Cranach 'The Fountain of Youth'. At the one side of the painting, elderly, disabled and frail people jump in a swimming pool and climb out at the other side as young, healthy and perfect beings. Francis Bacon also played with the idea of eternal youth. At the end of his utopian essay *The New Atlantis*, he lists among the foreseen goals of medicine: "the prolongation of life, the restitution of youth to some degree, the retardation of age" (Bacon 1996, p. 481). Throughout the centuries, a variety of anti-aging approaches have been tried. Amongst them have been alchemy, the use of precious metals, elixirs, drugs, hormones, dietary supplements, specific foods, grafts or injected extracts from testicles, ovaries, or glands of various animal species. Even sleeping with young virgins has been claimed to have positive effects on the length and vitality of life.

Life expectancy has increased steadily across most of Europe since around 1800. This is not something that anybody set out to achieve, but rather represents the incidental result of many changes introduced for other reasons. These changes included improved socio-economic conditions, public health interventions, changes

[1] A fine example of an older person who accepts his bodily decline is Emperor Hadrian as described by Marguerite Yourcenar in her book *Memoirs of Hadrian*, written in the form of letters to Hadrian's adopted grandson and later emperor Marcus Aurelius. In a pictorial way Hadrian depicts the deterioration of his body—he suffers from weakness of the heart—and his longing for death. After considering suicide and assisted suicide he finally decides to accept his bodily decay and all other "real ailments" of life: "death, old age, incurable diseases, unrequited love, rejected friendship, the poverty of a life that is less grand than our plans and more faded than our dreams: All the misery caused by the divine nature of things" (Yourcenar 1995, p. 107; translation WD).

in lifestyle, new developments in medical technology and progress in medical care. Disease prevention and treatment and slower aging have been two sides of the same coin.

Since the late 70s progress in the molecular biological sciences has begun to tantalize us with the prospect of direct intervention in the mechanisms of aging. In the last three decades the previously utopian ideal of an eternal youth is beginning to loose its utopian character. Prolongevity, defined as a "significant extension of the human lifespan and/or average life expectancy without suffering and infirmity" (Gruman 2003), is nowadays a much-debated topic. Inspired by posthumanist considerations, the American Academy of Anti-Aging Medicine has formulated its goal as: "to extend the time its patients can live without the morbidities of the aging process" (Post and Binstock 2004, p. 321).

Contemporary biogerontology has matured well beyond the traditional attitude that aging was considered as something that just happened. Many scenarios for the extension of life exist. Most of these extensions are attainable in laboratory animals by the combination of genetic, nutritional (caloric restriction), and other interventions, but are entirely impractical in humans (Rose 2004). All that has been managed so far has been to extend the lifespans of some simple animals (nematodes, yeast) by a limited percentage. No currently marketed intervention has yet been proved to slow down, stop, or reverse human aging (Binstock 2004).

In spite of the absence of any practical way to extend the maximum lifespan (MLS) three contemporary models of prolongevity exist (Post and Binstock 2004): (1) compressed morbidity as proposed by Fries (1987): A relative long phase of health, increased average life expectancy (ALE), MLS not increased; (2) decelerated aging: ALE and MLS are increased, medical problems of old age occur at a later stage; (3) arrested aging. The idea of arrested aging which would bring about 'virtual immortality' is simply inconceivable to me—an immortal individual would have to be perfectly self-repairing and thus could not change, learn or adapt, which would make the extended life pointless. Compressed morbidity and decelerated aging sound more realistic, although I do not believe that the MLS of human beings will become substantially longer in the next century. The oldest documented person who ever lived on earth is the French woman Jean-Louise Calment, who was 122 when she died. There is not the slightest evidence that this limitation of our MLS will change in the near future.

Granting the absence of any practical way of extending the MLS, it is striking that recent decades have seen an intense debate about the ethical acceptability of life extension. An explanation for this is that life extension touches upon a number of fundamental philosophical problems such as how to cope with aging, mortality and death. On the one hand, statements such as "few could seriously maintain that an average lifespan of 35 years would be preferable to the 75 enjoyed today even if many do spend their final years weak, demented, or debilitated" (Caplan 2004, p. 272) do have a point and are difficult to criticize. On the other hand, I am inclined to agree with anti-posthumanist thinkers like Jonas, Callahan and Kass, that a longer life is not necessarily a better life. I sympathize with Callahan's notions of 'natural lifespan' and 'natural death' and with his argument that "medicine should be used not for the further extension of the life of the aged, but only for the full achievement of a natural and fitting lifespan and thereafter for the relief of suffering" (Callahan 1987, p. 53). Jonas (1992) speaks not only about the burden, but also about the "blessing of mortality" and Kass even writes: "The finitude of human life is a blessing for every human individual whether he knows it or not" (quotation in: Binstock 2004, p. 30). This statement sounds paternalistic and needs a great deal of argument. However, I do not want to discuss this problem in detail here and confine myself to just one comment: The rather common argument that it is 'unnatural' to live much longer than we do now, is not convincing. To put it briefly: Natural processes are not intrinsically good things and unnatural and artificial procedures are not necessarily bad things (Caplan 2004).

The arguments for and against life extension are difficult to assess because they are mainly based upon a variety of alleged but unproved positive and negative effects of a longer life for individuals and society as a whole. If, for example, the MLS were extended to—let us say—150 years, we simply do not know what the added 30 years would look like and what the effect would be on earlier phases in life. In *Beyond Therapy*, the President's Council on Bioethics (2003) acknowledges this problem, but nevertheless provides a list of six categories of consequences of life extension for individuals: (1) greater freedom from constraints of time, (2) less commitment and engagement, (3) less aspiration and urgency, (4) less renewal and children, (5) changed attitude toward death and mortality, and (6) problems with the meaning of the life cycle. Generally, The President's Council was quite negative about the acceptability of life extension. That most of the consequences mentioned (2–6) are used as arguments against life extension, does not mean that the methodological approach of the President's Council on Bioethics is a purely consequentialist one. In particular, the arguments number 5 and 6 refer to a fundamental discussion from a teleological and deontological perspective regardless of the consequences. Underlying presuppositions about—say—justice and solidarity, the relational dimension of being human and the meaning of life do need further attention (Pijnenburg and Leget 2007).

In conclusion, I would like to answer the question 'why do we need a philosophical anthropology of aging?' as follows. First, in line with the old philosophical tradition of attempts to find meaning in aging, mortality and death, we do have to contemplate aging and mortality. To quote Plato: "examining myself and others is the greatest good to man, and [. . .] the unexamined life is not worth living" (Plato 1966: Apology 38a). Everyone, specifically every participant in the bioethical debate has their own anthropology which—from a philosophical perspective—should be made explicit (cf. Chap. 5). Second, in recent decades science and technology have begun to suggest that what once was just utopian thinking might become a practical possibility. The possible consequences of contemporary developments in biogerontology influence and nuance the sometimes rather abstract debate on aging, mortality and life extension. Third, a (re)turn to the old philosophical tradition of attempts to find meaning in aging and reasons to accept old age and death might function as a counterbalance to the utopian idea of an eternal youth, and the posthumanist idea of prolongevity and expanding the human lifespan. I agree with anti-posthumanist thinkers that we can better focus on the acceptance of aging rather than on its scientific modification. Although it might be possible that in the future we could live a few years longer, we cannot escape the aging process. Aging just happens. Old age will catch up with us and finally we will die. This fact will continue to stimulate the search for a meaning-giving process and, ideally, for an acceptance of the fact that we are all going to die.

4.3 Epistemological and Metaphysical Aspects of Aging

Before answering the question of what a philosophical anthropology of aging should look like, I will first briefly deal with some epistemological and metaphysical aspects of aging.

4.3.1 From a Biological Toward an Existential Approach

There has been a long-standing debate about the relationship between age-related diseases and 'normal' aging, i.e. aging that is seen when individuals with specific diseases are excluded from consideration (Murphy 1986). Nowadays, there is a broad consensus that aging is not a disease, but a natural (normal) process by which human beings become more likely to develop diseases such as cardiovascular degeneration, cancer, chronic neurological disorders and dementia. Yet it is difficult to avoid terms such as 'pathological' as the 'normal' process of replacing worn-out cells generates progressively more imperfect copies derived from ever-older stem cells.

Let me give a few examples taken from a report on the current state of affairs regarding the biology of aging (Kirkwood 2007). Nowadays, this report reads, it has become clear that the most likely avenue of progress toward understanding and intervening in age-related diseases will be "to identify precisely why the aged cell or tissue is intrinsically more vulnerable to *pathology*" (p. 29, italics added). The current consensus is "that aging is driven by the lifelong, gradual accumulation of a broad variety of molecular *faults* in the cells and the tissues that make up our body" (p. 27, italics added), and by a "progressive accumulation of molecular and cellular *defect*" (p. 29, italics added). Aging is prone to genetic predispositions, but many non-genetic factors also play a role. It is considered now that all these determinants lead to "molecular and cellular *damage*" (p. 28, italics added). The damage that will determine our health, vitality and level of independence in later life has been accumulating since we were in the womb. This means that research on the biology of aging concerns not only those who are old but is equally relevant to children of all ages.

Terms such as 'pathology', 'faults', 'defect' and 'damage' reflect that aging can be considered a pathological process which must be approached from a disease model. The question, however, is why aging is 'abnormal' and what '(ab)normal' then means.

From a statistical point of view it is difficult to claim that aging is abnormal because every living creature ages. Aging is entirely natural. Without wanting to discuss in full detail the normal/abnormal debate (Murphy 1986), I would like to stipulate here that aging is a consequence of a 'normal' decline of biological processes. Moreover, I want to hold that a purely biological approach of aging is a reductionist one. It is true that the biological dimension of aging is relevant for any anthropology of aging, but must be part of a comprehensive model in which all the psychosocial and existential dimensions of aging are also taken into account. Let me turn now to an existential perspective.

In *The Coming of Age* De Beauvoir expresses a typical existential approach, when she argues: "like all human situations it [old age] has an existential dimension—it changes individuals' relationships with time and therefore their relationships with the world and with their own history" (De Beauvoir 1996, p. 9). De Beauvoir adds that human beings never live in a state of nature: "In their old age, as at every period of their lives, their status is imposed upon them by the society to which they belong" (ibidem, p. 9). Not surprisingly, De Beauvoir emphasizes that the understanding of old age is time and culture dependent. The image of the aged differs from time to time and place to place. Reminding us that old age is our universal destiny, De Beauvoir argues that its lived meaning is specific to our historical and cultural situation. If we speak of old age as a universal category we will miss the crucial differences among the aged that are hidden behind the myths and images around aging. That any view of old age is time and culture dependent can be illustrated by the fact that *The Coming of Age* can be considered a critique of society's indifference to the elderly at that time. In the fifties and sixties of the twentieth century, old age was not an issue. This neglect of old age is entirely different from the situation today. Nowadays, in many countries, old age is a hot issue and elderly people have become an important policy target group.

The lack of engagement of the aged, De Beauvoir notes, is in part imposed 'from without'—influenced by cultural norms and values— and in part comes 'from within'. For, as we age, the body is transformed from an instrument that engages the world into a hindrance that makes our access to the world difficult. Before turning to the body as an instrument to access the world, I will discuss two questions, that is: How do we know that we are aging and what is the subject of aging?

4.3.2 How do We Know That We are Aging?

The epistemological question 'how do we know that we are aging' is analogous to the problem of personal identity which can be summarized in the question 'how do I know that I am the same person as I was yesterday, one week ago, one month ago and many years ago?'. Human beings not only age biologically, but also conceive of themselves as aging. "I can easily conceive of myself in a variety of physical forms, but I cannot conceive of myself without my thoughts", Esposito (1987, p. 57) writes. How, then, do we conceive ourselves as aging?

First, as De Beauvoir pointed out, we experience 'from within' impaired vision, hearing, strength, mobility, coordination, balance etc. and other bodily complaints such as fatigue, stiffness, and pain. Qualitative research based on interviews with elderly people into factors that, either on their own or together, formed so-called 'boundary conditions' for the experience of aging produced the following list: Deteriorating health, deteriorating sensory perceptions (particularly eyesight and hearing), frailty, pain, impaired memory, mobility problems and loss of human relations, particularly the loss of a spouse through death (Heikkinen 2000). One of the respondents in this study said: "Yes, well, you do feel that you're getting older don't you, when you start to lose your hearing and then you're afraid what will happen if you lose your eyesight as well, because after that there wouldn't be much point in living, would there?" (Heikkinen 2000, p. 469).

Second, we know on a cognitive level that we are growing old, not necessarily in the biological, but primarily in the chronological sense as having been around for a longer time. This happens when we realize that we are already 60 or 70 years of age, that we have a past and rich experience and that we can remember things that happened quite a while ago. For example, when we—as baby boomers—remember our youth without telephone, TV, internet, and remember the day that president Kennedy was shot, we realize that we do have a past and that we are part of history. The other side of the coin is a 'lack of future' and the awareness that our earthly life is running out. Circumstances such as retirement and becoming a grandparent make us realize that we are getting old, because we know that retirement and becoming a grandparent are things that usually happen to elderly people. Also other people's behaviour tends to draw our attention to the fact that we are considered to be an aged person: For example, when someone offers us a seat in a crowded train or bus, or when someone is not on familiar terms with us anymore. The fact that students entering the university seem to be getting younger every year also reminds us that we are getting old.

Third, when confronted with our own appearance we realize—'from without'—that our face and body—posture, decreased muscle mass, alterations of the skin, wrinkles—are not as young and vital as they used to be. An example of the utopian ideal of an eternal youth is Oscar Wildes *The Picture of Dorian Gray* (1890). Wilde even suggests a relationship between being old and infirm on the one hand and morally bad on the other. When the young and attractive Dorian Gray once again stands for his picture, he sighs: "How sad it is! I shall grow old, and horrible and dreadful. But this picture will remain always young [...] If it were only the other way [...] If it were I who was to be always young, and the picture that was to grow old! For that [...] I would give everything [...] I would give my soul for that" (Wilde 1985, p. 49). A 73-year-old retired laboratory technician described her initial disbelief at her 'wrinkly' reflection in the mirror as follows: "I looked in the bathroom mirror and thought 'Gosh you look old.' And I never really noticed it [...] I think it's because usually when I make-up or comb my hair, I'm always in a place where the light isn't very good ..." (Paulson and Willig 2008, p. 111).

4.3.3 What is the Subject of Aging?

These three ways of conceiving ourselves as aging are linked and cannot be separated. Nevertheless, one can ask what it is that grows old, and whether there is a hierarchy within these three forms of perception. I am inclined to think that the first and the second ways of knowing are more crucial than the third one, but I am aware that this depends on what we understand by 'aging'. Aging in the sense of just living longer in time (chronological aging) primarily relates to the second way of knowing, while aging in the sense of bodily and mental decline primarily relates to the first. My presupposition is that human beings as persons are the subject of aging and that aging is primarily something that occurs in our body. People age because their bodies are aging. One can also speak about an 'aging mind' (Prado 1986), but this expression refers to psychological aging in the sense of memory loss, the loss of cognitive flexibility and the like. Our cognitive capabilities can demonstrate signs of deterioration, but only as a consequence of a deterioration of neurobiological processes. The disease of dementia is a good example here. It might make sense to say that someone's mind is deteriorating, as 'dementia' literally means, but the mind as metaphysical principle cannot deteriorate. I am inclined to answer the question 'can the mind age?' in the negative. In the next two sections of this chapter I will therefore concentrate on what I will call the 'centrality of the body in aging'.

4.4 Vulnerability and the Ability to Look After Oneself

If De Beauvoir's suggestion—that we are to develop a time and culture specific anthropology of aging—is to be taken seriously, two contrasting anthropological categories, i.e., vulnerability and the ability to look after oneself, seem to be good candidates for further analysis. As I will illustrate below, these two categories are quite central in recent literature and policy guidelines about aging and the ideal of healthy aging. Moreover, these two categories express a contrast that can be found in many anthropologies. On the one hand, human beings are considered insignificant creatures in the universe because they are prone to disease, fragility, vulnerability, and mortality. On the other hand, a human being is considered to be a thinking, autonomous, and mighty being endowed with self-awareness, the creator of science and art, the 'ruler of the world'. One of Blaise Pascal's *Pensées* perfectly illustrates what I mean:

> Man is but a reed, the most feeble thing in nature, but he is a thinking reed. The entire universe need not arm itself to crush him. A vapour, a drop of water suffices to kill him. But, if the universe were to crush him, man would still be more noble than that which killed him, because he knows that he dies and the advantage which the universe has over him, while the universe knows nothing of this. (Pascal 2001, Sect. 6, p. 347).[2]

[2] Pascal seems to specifically refer here to the Cartesian dichotomy of *res extensa* and res *cogitans*.

4.4.1 Vulnerability and Autonomy

Vulnerability is a fundamental concept with a long history, but a relatively new one in bioethics (Haugen 2010).[3] Schematically three uses of the concept can be distinguished. First, it may refer to a specific category of human beings: The unborn, children, handicapped people, elderly, patients in general, patients with specific diseases (psychiatric and neurological disorders, dementia), patients who cannot give an informed consent, poor people, and people with a low social-economic status. Second, it may refer to all human beings as an anthropological description of (part of) the human condition. Vulnerability, then, is considered to be inherent to human existence. Bodily beings—as human beings are—are mortal by nature. According to Seneca (1979), becoming human means learning to live consciously as finite and mortal beings. Dependence and independence become characteristics of an attitude towards mortal life. They concern the way in which human beings give meaning to their vulnerable bodily existence. Similarly, Levinas (1971) describes human beings as bodily beings who therefore share an inescapable condition of vulnerability. This leads to the recognition that one's receptivity for others can be increased by acknowledging one's own vulnerability. Third, vulnerability may refer to the good life. According to Martha Nussbaum (1986)—following Aristotle—the best human life or human good living (*eudaimonia*) is vulnerable (fragile) in the sense that it depends upon so-called external goods which are factors outside a person's control. According to Nussbaum, vulnerability and dependency are inherent to a good life. If human beings could overcome dependency and vulnerability, they would live as immortal gods, but then nothing would have the value that we attach to it by now. It would not be a human life anymore.

Contemporary Western culture, healthcare and healthcare ethics put much emphasis on the ideal of autonomy. Autonomy, then, means self-sufficiency, self-governance, and independence from controlling influences. The autonomous individual acts freely in accordance with a self-chosen plan. Beauchamp and Childress describe what they take to be essential to personal autonomy as "personal rule of the self that is free from both controlling interferences by others and from personal limitations that prevent meaningful choice, such as inadequate understanding" (Beauchamp and Childress 1994, p. 121). As a consequence of this interpretation of autonomy, it is a quite common understanding that vulnerability and autonomy are mutually exclusive phenomena: The less autonomous we are, the more vulnerable, and the other way around. However, following authors like Seneca, Levinas and Nussbaum, I would prefer to argue that autonomy is not the opposite of vulnerability and dependence (Dekkers 2001). Callahan writes: "The greatness of human life, its most majestic stories and epics, has not always centered on dramas of triumphant independence, standing alone and isolated in the midst of the crowd, though surely

[3] Sometimes 'vulnerability' and 'frailty' are used interchangeably. Tulle, for example, writes: "Human beings are inherently at risk of frailty: A universal condition of the human species [. . .] Any manifestation of frailty, from injury, illness or biological aging, threatens to disrupt our sense of self" (Tulle 2008, p. 5). Most often, however, these two terms refer to different things: frailty then is considered the physical aspect of a more encompassing vulnerability.

there are many such stories. It has no less often centered on the way people worked to share their suffering, to create bonds of interdependence" (Callahan 1993, p. 142). Independent is the one who can give meaning to his or her own mortality and knows to incorporate the finiteness of his or her existence in daily life. Dependent is the one who does not recognize or denies the limitations of his or her bodiliness and therefore lets slip the capacity to give meaning to his or her vulnerable life. People who seem to be dependent because they need care from others can be pre-eminently independent and autonomous—in a broader sense than just self-determination—because they are thoroughly confronted with the vulnerability of human existence (Manschot 1992). The point that I want to make is not that (respect for) autonomy is not important, but that we cannot close out eyes for the flipside of the coin, that is, human vulnerability and dependence.

I consider vulnerability (and frailty, dependency etc.) and the ability to live independently (agency, autonomy, empowerment, self management etc.) as part of the human condition and as normative ideals as well. Vulnerability and autonomy are normative concepts (Kottow 2004). Two recent policy reports about the care for elderly people in the Netherlands illustrate this point in more detail. The one focuses on autonomy and the other on vulnerability.

4.4.2 Healthy Aging

In the report *Prevention in the elderly*, The Health Council of the Netherlands (2009) adopts a broad framework to accommodate the concept that successful or healthy aging can be compatible with disease. "Healthy aging [. . .] is not limited to maintaining good physical and mental health, but importantly also promotes a process that enables elderly people to live, and to continue to live, lives of good quality, as independently as possible, and to continue to participate in society" (p. 17). What the Health Council calls "functioning independently in daily life", "functioning in daily life" or just "functioning" are translations of the Dutch term 'zelfredzaamheid'—literally "self-reliance" or "self-sufficiency". The Health Council continues: "As the health of the elderly—medically speaking—eventually proves lacking, values such as functioning in daily life and well-being become increasingly important. Health, functioning and well-being are strongly interdependent" (p. 17–18). In order to fully utilise the potential for healthy aging, there is need of 'function-oriented prevention' in addition to prevention of disease. According to the Health Council, elderly people can and must play an active part in defining the goals and form of preventive activities. The emphasis is on self-management and empowerment.

The Netherlands Institute for Social Research has published a report on frail elderly focusing on vulnerability (Van Campen 2011).[4] This report is of interest here for two reasons. As is argued by the SCP, there are many definitions of frailty. In a narrow approach the focus is purely on physical aspects of frailty. A broad

[4] Also the Royal Dutch Medical Association recently published a report about medical care for the elderly with a focus on vulnerability (RDMA 2010).

approach also pays attention to psychological and social aspects. The SCP opts for a broad approach: "Frailty in older persons is a process involving the accumulation of physical, psychological and/or social *deficits* in functioning which increase the risk of adverse health outcomes (functional impairment, admission to an institution, death)" (p. 199, italics added). I would like to make two observations here. First, it seems that for the SCP 'frailty' is identical with 'vulnerability' and that, in line with a common understanding of aging, frailty is described in terms of deficits. It is therefore understandable that the SCP-report is interested in identifying which older persons are frail, apparently based on the presupposition that not all older persons are frail or vulnerable. Second, according to the SCP report, older persons do not think about themselves in terms of frailty. They rarely use the word 'frail' to describe their situation. They are mainly concerned with their quality of life. The older persons who were interviewed for this study cited health, life partner, children and grandchildren and other close relatives as being important in their life. Loss of health and relationships, and anxiety about that loss, would severely undermine their quality of life. They also expressed a desire to continue living independently for as long as possible. The broader interpretation and measurement of frailty therefore overlaps with the notions of quality of life and successful aging.

4.5 Phenomenology of the Aging Body

Human beings are a unity of body and mind, they are embodied persons, and persons and bodies are inextricably linked. However, for the reasons given above I will now focus on the human body, drawing attention to the "inescapable and prominent presence of the body in aging experiences" and the "centrality of the body" in understandings and experiences of aging (Tulle 2008, p. 17). Two questions will be addressed. First, what can a phenomenological approach of the aging body—compared to an objective and reductionist approach—tell us? Conceptually, a distinction can be made between old age on the one hand and the diseases of old age on the other. Can we track down this distinction as a phenomenological difference? In other words, what is the difference between a subjectively lived aged—but healthy or normal—body on the one hand and a subjectively lived disabled, handicapped or diseased body on the other? This question arises because many qualitative studies apparently do not make a distinction between these two categories. Second, in line with De Beauvoir's suggestion that any existential analysis of old age must take into consideration time and culture dependent aspects, I will emphasize the importance of an empirically based phenomenology of the aging body.

4.5.1 Disability and Old Age

Biological theories that attempt to unravel the causes of aging focus on the objective body. From a phenomenological point of view the emphasis is on the subjectively lived body which is distinct from other physical or animate bodies in a variety of ways (Leder 1990). I am able to do things with my body, and I am related to all other realities because I have a body. I have access to my body 'from within', kinesthetic sensation gives me information about the posture of my body and a sense of where my body is in space as well. I have also some awareness of the interior of the body through visceral sensations such as stomach-aches. My body has the capacity to be reflexively related to itself, for example, by perceiving one's hand with one's other hand. Bodily capacities reveal the experiencing subject's involvement in the world (Toombs 1999).

Von Gebsattel (1954) argues that we do not have just one subjectively lived body, but many of them, dependent on what we are doing and involved in: Loving bodies, aesthetic bodies, painful bodies, and also aging bodies. In this context he speaks about the 'multi-aspectivity' of the body. What now is the difference between these sorts of subjectively lived bodies, especially between disabled bodies and old bodies? To explore this question I'll take as an example an article written by Campbell (1995) about the relationship between embodiment and diminishment in case of illness, pain, disability, and particularly aging. Taking Tolstoy's *The Death of Ivan Ilyich* as point of departure, Campbell focuses on three 'existential marks of embodied diminishment': (1) magnification and alienation, (2) loss of voice, and (3) contraction.

A first emergent consequence of diminishment is both an increasing awareness of the body and an increasing alienation from the body. When I am healthy, I am often unaware of my body and I am sufficiently at one with my body not to explicitly reject it. But the unity of the embodied self is ruptured by the presence of pain, illness, or attention to our aging. Then the body with which I was identified is increasingly dissected into different body parts that presents a new reality of otherness: That ache in my leg, the burning in my throat, those wrinkles on my skin. In all these cases, there is a dialectical relation between a heightened awareness of embodiment and a heightened alienation from the body. It is striking that Campbell refers to studies of pain and disability, and not of aging per se. From Scarry's *The Body in Pain* (Scarry 1885) he derives the idea of magnification. In pain we experience a 'magnified body'. The body in pain increasingly engulfs other aspects of consciousness. From Toombs' phenomenological analysis of disability, especially in case of multiple sclerosis (Toombs 1992), Campbell derives the idea of an 'oppositional body', that is a body that frustrates our life possibilities and projects and is increasingly experienced as 'other' because it is beyond our control.

A second mark of diminishment in the body, according to Campbell, is the loss of voice. Speech is a primary means of self-extension through which the spatial and temporal limits of embodiment can be transcended. Voice is a mode of enlargement of self in the world. Campbell seems to hint here at three possible situations. People may loose their voice in a literal sense as a consequence of a neurological disease. They may also loose their voice in a metaphorical way, that is, being dependent on caregivers' decisions with insufficient participation in medical decision making. Finally—Campbell again refers here to Scarry's *The Body in Pain*—patients suffering from severe pain may find themselves speechless because the subjective experience of bodily pain makes it virtually inexpressible to others. The person in pain is reduced to moans, groans, and cries to convey discomfort and suffering in the body.

A third mark of diminished embodiment is that of contraction. Even as the body assumes a magnified presence in consciousness, one's world contracts, from experience in a natural environment to being 'home bound', limited in mobility to a room, a bed, and to "the very core of the self" (Campbell 1995, p. 177). Here again, Campbell quotes Scarry: "In very old and sick people, the world may exist only in a circle two feet out from themselves" (Scarry 1985, p. 33).

Campbell argues that the phenomena of (1) magnification and alienation, (2) silence, and (3) contraction can be particularly pronounced in the process of aging. That might be the case, but does this mean that there is a difference between the disabled and diseased (but not aged) body on the one hand and the aged (but not disabled and diseased) body on the other? Is there a difference, from a phenomenological perspective, between illness, disease, and disability on the one hand and aging on the other? The only way to solve this problem is by doing phenomenological oriented empirical research into the lived experience of 'healthy' older people. With 'healthy' I mean possibly experiencing some minor ailments belonging to daily life, but not suffering from a specific disease. In a very general way, one can say that phenomenology takes as its point of departure our intuitive, lived and direct experience. But we should not forget that all our experiences are coloured by all kinds of presuppositions and an already existing complex of socially and culturally dependent meanings. This means that physical experiences such as painful and stiff legs or joints may get a specific meaning dependent on whether we are young or old, healthy or ill and on how people around us treat us. The fact that we—at a cognitive level—know that we are older influences the way we experience bodily ailments.

4.5.2 The Lived Aging Body

There are many differences in the way older people experience their body that depend on their life style. For example, a dependent person has a different experience from someone who has an active or reactive coping style (Hennesy 1989). But the physical alteration that accompanies and determines growing old is commonly regarded as one of its most problematic aspects. The loss that threatens with aging is not the sheer fact of physical decline, but the alienation from our own body (Gadow 1991). Phenomenological analyses of the experience of elderly people suggest that aging itself does not necessarily produce distortion, disruption or devaluation of one's body image, but senescence on the whole is regarded as an accelerating loss of control of the personal body. With increasing age we become more and more aware of things that we used to be able to take for granted. Age brings along a whole host of "annoying companions that are 'part' of us" (Heikkinen 2000).

The results of a study based on qualitative research consisting of open ended, semi-structured interviews with thirteen elderly people showed one general theme running through all the interviews concerning the experience of aging and the aged body (Bullington 2006). This general theme had to do with the experience of a changed life world, reactions to these changes in terms of body and self, and finding ways to feel at home in this changed life situation. Life was no longer as it had been. The experience of the aging body was not always the most salient aspect of feeling like an old person, although all respondents felt that their bodies were a reminder that they were no longer young. This research also showed that there are many different ways of experiencing the aged body and of constituting the meaning of an aged body. For example, although for some, the decline of memory and physical abilities was experienced as terrible, the aging body was experienced just as a surface seen by others, not as something to do with the self. Because the body as a surface is no longer a source of a positive self-image, there is a tendency not to pay so much attention to the body. There is a movement "away from self toward others" (Bullington 2006, p. 27). For others, however, the decline of the body was a much more central experience in the sense that it influenced the self. These respondents dealt extensively with how the aging body gave rise to new experiences, both bad and good. A negative experience was the fear of travelling. An example of a positive experience is that one respondent realized that she no longer felt that she had to worry about prestige and that being an older person gave her permission to be a bit childish. This way of giving meaning seems to correspond with a positive attitude towards vulnerability as described by Levinas and Nussbaum.

Thus, the mere fact of having aches and pains and age-related limitations does not automatically result in negative experiences of self. It is also older people's pessimistic belief about their health and ability to control the decline of their aging body that contributes to the actual loss of function later on in life. The role of activity and agency are important factors that can counter a negative downward spiral. This empirical finding supports the ideal of 'the ability to do things independently' as I discussed above. A general conclusion of Bullington's study was that the importance of the lived body, either as the presented surface of self or the facilitator of activity, suggests that we should focus attention upon the aging body in order to deepen our understanding of the experience of growing old (Bullington 2006).

4.6 Anthropology, Ethics and Policy

In this chapter I have argued that we need a philosophical anthropology of aging for three reasons. First, every person involved in the care for the elderly and every researcher in the field of aging holds an implicit anthropology which should be made explicit in line with the philosophical tradition of trying to find meaning in aging, mortality and death. Second, new developments in the field of biogerontology mark a transition in the sense that what once was just utopian thinking might become reality in the future. These new technological developments demand reflection and deliberation about their impact on the human condition. Third, attempts to find meaning in aging and to accept old age and death might function as a counterbalance against the utopian idea of eternal youth and the posthumanist idea of prolongevity and expanding the human lifespan.

I have also emphasized that such an anthropology of aging should take into account socio-cultural factors. I am thinking here of tendencies such as the medicalization, marginalization and stigmatization of elderly people. We should also be aware of the so-called 'deficit model' of old age (and dementia) in which the focus is on mental and bodily decline rather than on remaining capabilities. Moreover, an anthropology of aging should be informed by the results of qualitative research about how we experience aging, with particular attention to a phenomenology of the aging body. The idea of an all-encompassing, culture- and time-independent anthropology with the focus on an alleged 'essence' or 'nature' of human beings is not attractive. There is no standard anthropology of aging.

The relationship between anthropology and ethics is analogous to the relationship between anthropology and policy. Policy documents in the field of aging should be as clear as possible about the underlying, but often implicit, anthropological presuppositions. Being aware of the normative anthropology that underlies policy documents is paramount. In this chapter I have provided two examples of contemporary Dutch policy guidelines in the field of aging. It was interesting to see that these documents together illustrate a traditional tension in any anthropology, that is, the tension between vulnerability and dependency on the one hand and the ideal of autonomy and independence on the other. Like 'dignity', 'vulnerability' can be understood as an anthropological and normative description of (part of) the human condition and as an invitation to consider care as a fundamental human attitude. Vulnerability suggests the need to develop deontological arguments in support of the protection of vulnerable beings (Kottow 2004). However, vulnerability is just one side of the anthropological coin. Policy documents in the field of aging should therefore attempt to find a balance between the focus on vulnerability and on the ability to look after oneself. That is what policy makers can learn from a philosophical anthropology of aging.

A second balance that should be kept in mind is the one between optimism and utopian thinking on the one hand and fatalism in the sense of 'aging is inevitable, it just happens, we cannot do anything about it' on the other. In this regard, it should be wise for policy makers not to explicitly aim at prolongevity and extension of the human lifespan. They rather should focus on the improvement of the quality of life and the quality of care of elderly people. To put it simply: It is rather the quality than the quantity of life that matters. Any increase in the average life expectancy (ALE) and the maximal lifespan (MLS) should be seen only as welcome side-effects of a better quality of life for the elderly, or of improved care of them.

References

Bacon, F. 1996. The new atlantis. In *Francis Bacon: A critical edition of the major works*, ed. B. Vickers 457–489. Oxford: Oxford University Press.

Beauchamp, T.L., and Childress, J.F. 1994. *Principles of biomedical ethics.* New York/Oxford: Oxford University Press.

Beauvoir, S. de. 1996. *The coming of age.* New York/London: Norton & company.

Binstock, R.H. 2004. The search for prolongevity: A contentious pursuit. In *The fountain of youth. Cultural, scientific, and ethical perspectives on a biomedical goal*, eds. S.G. Post and R.H. Binstock. 11–37. Oxford: Oxford University Press.

Bullington, J. 2006. Body and self: A phenomenological study on the aging body and identity. *Journal of Medical Ethics; Medical Humanities* 32(1):25–31.

Callahan, D. 1987. *Setting limits. Medical goals in an aging society.* New York etc.: Simon & Schuster.

Callahan, D. 1993. *The troubled dream of life. Living with mortality.* New York etc.: Simon & Schuster.

Campbell, C.S. 1995. Marks of the body: Embodiment and diminishment. In *Embodiment, morality, and medicine*, eds. L.S. Cahill and M. Farley. 169–183. Dordrecht: Kluwer Academic Publishers.

Campen, C. van ed. 2011. *The frail elderly [Kwetsbare ouderen].* Den Haag: Sociaal Cultureel Planbureau.

Caplan, A.L. 2004. An unnatural process: Why is it not inherently wrong to seek a cure for aging. In *The fountain of youth. cultural, scientific, and ethical perspectives on a biomedical goal*, eds. S.G. Post and R.H. Binstock. 271–285. Oxford: Oxford University Press.

Cicero. 1923. *On old age.* Cambridge, Mass.: Harvard University Press.

Dekkers, W. 2001. Autonomy and dependence: Chronic physical illness and decision-making capacity. *Medicine, healthcare and philosophy* 4(2):185–192.

Esposito, J.L. 1987. *The obsolete self. Philosophical dimensions of aging.* Berkeley: University of California Press.

Fries, J.F. 1987. An introduction to the compression of morbidity. *Gerontologica Perspecta* 1(1):5–8.

Gadow, S. 1991. Recovering the body in aging. In *Aging and ethics. Philosophical problems in gerontology*, ed. N.S. Jecker 113–120. Clifton, New Jersey: Humana Press.

Gebsattel, V.E. von 1954. *Prolegomena einer medizinischen Anthropologie.* Berlin: Springer Verlag.

Gruman, G.J. 2003. *A history of ideas about the prolongation of life.* New York: Springer.

Haugen, H.M. 2010. Inclusive and relative language: The use of the concepts of autonomy, dignity and vulnerability in different contexts. *Medicine Healthcare and Philosophy* 13(3):203–213.

Health Council of the Netherlands. 2009. Prevention in the elderly: Focus on functioning in daily life [Preventie bij ouderen: Focus op zelfredzaamheid]. The Hague, Health Council of the Netherlands; publication no. 2009/07.

Heikkinen, R-L. 2000. Aging in an autobiographical context. *Aging and Society* 20(1):467–483.

Hennessy, C.H. 1989. Culture in the use, care, and control of the aging body. *Journal of Aging Studies* 3(1):39–54.

Jonas, H. 1992. The burden and blessing of mortality. *The Hastings Center Report* 22(1):34–40.

Kirkwood, T. 2007. Changing expectations of life. In: *AGEACTION. Conference report*, 26–33. http://ageaction.ncl.ac.uk/AgeAction_book.pdf. Accessed 17 Oct 2011.

Kottow, M.H. 2004. Vulnerability: What kind of principle is it? *Medicine Healthcare and Philosophy* 7(3):281–287.

Leder, D. 1990. *The absent body.* Chicago and London: The University of Chicago Press.

Levinas E. 1971. *Totality and infinity. An essay on exteriority.* The Hague: Martinus Nijhoff Publishers.

Manschot, H. 1992. *Levenskunst of lijfsbehoud? Een humanistische kritiek op het beginsel van autonomie in de gezondheidszorg*. Utrecht: Universiteit voor Humanistiek.

McKee, P.L., ed. 1982. *Philosophical foundations of gerontology.* New York: Human Sciences Press.

Miller, R.A. 2004. Extending life: Scientific prospects and political obstacles. In *The fountain of youth. Cultural, scientific, and ethical perspectives on a biomedical goal*, eds. S.G. Post and R.H. Binstock. 228–248. Oxford: Oxford University Press.

Montaigne, M. de. 1993. *The complete essays.* London: Penguin Classics.

Murphy, T.F. 1986. A cure for aging? *Journal of Medicine and Philosophy* 11(3):237-255.

Nussbaum, M.C. 1986. *The fragility of goodness. Luck and ethics in Greek tragedy and philosophy.* Cambridge: Cambridge University Press.

Pascal, B. 2001. *Thoughts*. Harvard Classics, vol. 48, Part 1. On-line edition. Bartleby.com.
Paulson, S., and Willig, C. 2008. Older women and everyday talk about the aging body. *Journal of Health Psychology* 13(1):106-120.
Pijnenburg, M.A.M., and Leget C. (2007). Who wants to live forever? Three arguments against extending the human lifespan. *Journal of Medical Ethics* 33(10):585-587.
Plato (1966). *Plato in twelve volumes,* vol. 1. Cambridge, MA: Harvard University Press/London: William Heinemann Ltd.
Post, S.G., and R.H. Binstock. 2004. Introduction. In *The fountain of youth. Cultural, scientific, and ethical perspectives on a biomedical goal*, eds. S.G. Post and R.H. Binstock. 108. Oxford: Oxford University Press.
Prado C.G. 1986. *Rethinking how we age. A new view of the aging mind.* Westport (CT): Greenwood Press.
President's Council on Bioethics. 2003. *Beyond therapy: Biotechnology and the pursuit of happiness*. Washington, DC.
Rose, M.R. 2004. The metabiology of life extension. In *The fountain of youth. Cultural, scientific, and ethical perspectives on a biomedical goal*, eds. S.G. Post and R.H. Binstock. 160–176. Oxford: Oxford University Press.
Royal Dutch Medical Association [KNMG]. Strong medical care for vulnerable people [Sterke medische zorg voor kwetsbare ouderen]. March 2010.
Scarry, E. 1885. *The body in pain.* New York: Oxford University Press.
Seneca. 1979. *Het Gelukkige Leven.* Baarn: Ambo.
Sporken, P. 1983. *Begeleiding en Ethiek.* Baarn: Ambo.
Toombs, S.K. 1992. The body in multiple sclerosis. A patient's perspective. In *The body in medical thought and practice*, ed. D. Leder. 127–137. Dordrecht etc.: Kluwer Academic Publishers.
Toombs, S.K. 1999. What does it mean to be somebody? Phenomenological reflections and ethical quandaries. In *Persons and their bodies*, ed. M.J. Cherry. 73–94. Kluwer: Academic Publishers.
Tulle, E. (2008). *Aging, the body and social change: Running in later life. School of law and social sciences. Paper 112.* London: Palgrave Macmillan: http://researchonline.gcu.ac.uk/lss/112/. Accessed 11 Apr 2012.
Yourcenar, M. 1995. *Hadrianus' Gedenkschriften.* Amsterdam: Atheneum-Polak & Van Gennep.
Wilde, O. 1985. *The picture of Dorian Gray.* London: Penguin Classics.

Chapter 5
The Implicit Anthropology of Bioethics and the Problem of the Aging Person

Søren Holm

5.1 Introduction

For those of us who are lucky enough to live in affluent societies it is quite likely that we will survive into old age and will eventually die as a result of age related conditions and diseases. The fact that more of us get old is not a problem, it is a sign of economic and medical success!

But the increase in the number of old people does create challenges for society and for bioethics. There is a voluminous bioethical literature on reproduction and the beginning of life, and an almost equally large literature on the very end of life, but the perhaps 20 or 30 years that many of us are hoping to spend between retirement and death are very incompletely theorised in bioethics. Work has been done on issues raised by specific conditions, e.g. Alzheimer dementia; and on the use of age as criterion for resource allocation in healthcare but the situation of the old person as such has received little attention.

In this Chapter I want to suggest that this is because the aging and old person falls outside of the standard, implicit anthropology of bioethics. In order to show this I will first define 'an anthropology' and discuss the role that anthropologies play in bioethical thinking and argument. I will then briefly mention some explicit anthropologies found in bioethics and proceed to excavate the implicit anthropology underlying much of English language bioethics.[1] I will then show how this 'standard

[1] Because of my own linguistic limitations the focus of this paper is on English language bioethics. I can vouch for the fact that a similar implicit anthropology can be found in bioethics writing in the Scandinavian languages, but am unable to say whether it is also typical elsewhere. There may well be an analytic/continental divide in relation to philosophical anthropology as with some many other philosophical topic areas, but investigating whether that is the case is outside the scope of this paper.

S. Holm (✉)
University of Manchester,Manchester School of Law, United Kingdom
e-mail: soren.holm@manchester.ac.uk

M. Schermer, W. Pinxten (eds.), *Ethics, Health Policy and (Anti-) Aging: Mixed Blessings*, Ethics and Health Policy 1, DOI 10.1007/978-94-007-3870-6_5,

anthropology' can help to explain (1) why old age is under-theorised in bioethics, and (2) why old age is primarily seen through a deficit lens when it is discussed.

But before beginning the analysis I need to enter a few caveats. First, the 'standard anthropology' that I will attempt to excavate is an ideal type. I do not claim that any specific bioethicist or school of bioethics hold it in the exact form in which I am going to describe it. What I do claim is that major elements are held implicitly or explicitly by many bioethicists.

Second, I do not claim that those who hold the standard anthropology implicitly cannot on reflection modify their position and adopt a more nuanced anthropology. This is perfectly possible and I make some specific suggestions for how it should be done in the last section of the paper. But as long as explicit reflection on philosophical anthropology is a rarity in bioethics it is unlikely that most will do so.

5.2 The Role of Anthropologies in Bioethics

In the following I will understand the concept of 'an anthropology' in the following way:

1. An anthropology is a set of assumptions about typical characteristics of human individuals, for instance about typical vulnerabilities, interests, motivations, ways of flourishing
2. An anthropology does not necessarily claim that any of the typical characteristics are part of some deeper conception of 'human nature', it just claims that they are typical

This conception of an anthropology means that an anthropology can have both empirical and normative components. It might for instance claim something empirical about human cognitive abilities and something normative about human interests. The conception further entails that, at the limit everything that is claimed to be typical can be the result of social forces and socialisation, nothing need to be seen to be natural. And, finally whereas most anthropologies imply that there is some fixity (either social or biological) to the typical characteristics of human individuals, a transhumanist anthropology claiming total malleability of all human characteristics will still count as an anthropology under this conception. It is, however, important to note in passing that even if every typical human characteristic is malleable, many are also relatively and temporally fixed in the sense that it will either (1) take time to change them, and/or (2) that they cannot be readily changed in adult, fully socialised individuals.

We can define a broader conception of an 'agentology' not restricted to human individuals but encompassing all actual (or perhaps even possible) moral agents, but this is unnecessary for present purposes.

Does bioethics have an anthropology in the sense defined above? Some consequentialists, libertarians and transhumanists might deny it, claiming that they assume nothing about human beings in their arguments, but is that really true?

'Proof texting' is usually not a good basis for a sound argument, but in the present case it may just do the job, since it is possible to show that very prominent bioethicists very explicitly make anthropological claims.

Julian Savulescu for instance writes in a paper on human enhancement:

> This paper argues that we have a moral obligation to enhance human beings. It is argued that if one is committed to the moral obligation to treat and prevent disease, one is also committed to genetic and other enhancement in so far as this promotes human well-being. It is argued that this is not eugenic but *expresses our fundamental human nature*: To make rational decisions and to try to improve ourselves. *To be human is to strive to be better*. (Savulescu 2005, p. 36, my emphases)

And in one of his foundational papers on the concept of 'procreative liberty' John Robertson writes:

> Procreation is a complex activity that develops over time and involves many disparate behaviours. The importance of procreation as a whole derives from the genetic, biological, and social experiences that comprise it. *Reproduction is a basic instinct* that supplies societies with the members who maintain and perpetuate the social order and who provide services for others. *Reproduction also satisfies an individual's natural drive for sex and his or her continuity with nature and future generations*. It fulfils cultural norms and individual goals about a good or fulfilled life, and many consider it the most important thing a person does with his or her life. (Robertson 1983, p. 408, my emphasis)

In these two quotes we see a strongly essentialist anthropological statement in Savulescu, and biological, metaphysical and social anthropological claims in Robertson. In both quotes the anthropological claims provide one of the premises for an argument with the structure

P1. Activities that are important for human beings should be promoted
P2. Activity X is important for human beings because of what they (anthropologically) are
Therefore: X should be promoted

By making P2 rely on the anthropological claim, instead of on the claim that X is important merely because people want to pursue X, P2 is furnished with an air of self-evidence and deep foundation that it would not otherwise possess. Instead of being a normative premise it is converted into a seemingly empirical premise. But as long as the anthropology is implicit and not argued for or further justified, we do in reality not know what weight we should put on P2 (see more below).

By providing an explicit, worked out anthropology we can say more precisely which element of this anthropology it is that makes X important as an activity. But despite this argumentative advantage there are few explicit anthropologies that have gained currency in bioethics.

5.3 Some Explicit Anthropologies

Let us briefly survey 4 examples of explicit anthropologies, 2 inspired by Christian theology and 2 completely secular.

John Finnis has worked out an anthropology as part of his revitalisation of Natural Law theory in ethics and jurisprudence. According to Finnis human beings are essentially characterised by 7 goods (Finnis 2011):

Three substantive goods: (1) human life (health and procreation); (2) knowledge and aesthetic appreciation; (3) skilled performance.

Four reflexive goods:(1) self-integration; (2) authenticity/practical reasonableness; (3) justice and friendship; and (4) religion/holiness.

Everything that is good for a person relates to one of these 7 goods, and nothing can be good if it does not relate to these goods.

Finnis is inspired by Thomistic theology and another explicit anthropology which has gained some currency in European bioethics also has roots in Catholic moral theology. Leuven personalism as developed by Louis Janssens posits 8 basic human characteristics from which an ethic can be developed:

> (1) The human person is a subject, not an object like the things of the world. Since the person is called to self-determination, he or she is a moral subject, deciding on all his or her doings in conscience and consequently in a responsible way. (2) The human person is a subject in corporeality. Our body forms part of the totality that we are: What concerns our human body affects our person. (3) Because of the materiality of our body, our being is a being-in-the-world. (4) Human persons are essentially directed toward each other. (5) Not only because of our openness to one another are we social beings, but also because we need to live in social groups with appropriate structures and institutions. (6) Human persons are fundamentally open to God, and it is the task of moral theology to explain how, according to our Christian revelation, our relationship to God affects us in all the dimensions of our person. (7) Human persons are historical beings since they are characterised by historicity. (8) All human persons are fundamentally equal, but at the same time each is an originality, a unique subject. (Janssens 1990, p. 94)

At this point many will feel an almost irresistible temptation to say something like: 'But, this is not right. Janssens' 6th characteristic is not a characteristic of human beings, and although there is some truth in his 2nd it is not expressed quite right, and...' However, any intervention along these lines merely shows that the person making the intervention has her own implicit anthropology, against which she compares Janssens'. She will, of course often be able to give reasons to prefer her anthropology to Janssens', but she can no longer deny that she actually holds an (implicit) anthropology.

There are also secular anthropologies in bioethics, although they are rarely called anthropologies or identified as anthropologies. In the Stanford Encyclopedia of Philosophy entry on 'Feminist ethics' we can, for instance find the following paragraph:

> *Feminist Ethics* is an attempt to revise, reformulate, or rethink traditional ethics to the extent it depreciates or devalues women's moral experience. Among others, feminist philosopher Alison Jaggar faults traditional ethics for letting women down in five related ways. First, it shows less concern for women's as opposed to men's issues and interests. Second, traditional ethics views as trivial the moral issues that arise in the so-called private world, the realm in which women do housework and take care of children, the infirm, and the elderly. Third, it implies that, in general, women are not as morally mature or deep as men.

> *Fourth, traditional ethics overrates culturally masculine traits like "independence, autonomy, intellect, will, wariness, hierarchy, domination, culture, transcendence, product, asceticism, war, and death," while it underrates culturally feminine traits like "interdependence, community, connection, sharing, emotion, body, trust,* absence of hierarchy, nature, immanence, process, joy, peace, and life." Fifth, and finally, it favors "male" ways of moral reasoning that emphasize rules, rights, universality, and impartiality over "female" ways of moral reasoning that emphasize relationships, responsibilities, particularity, and partiality (Jaggar, "Feminist Ethics," 1992). (Tong and Williams 2009, my emphasis)

Here a number of very significant anthropological claims are made on several levels. And these claims are really anthropological, even though Alison Jaggar who is quoted in this section is careful not to make any of these claims rely on biology. Giving importance to a 'culturally masculine' trait is to invoke a culturally based anthropology, and the claim that those traits are overrated by traditional ethics implies an anthropology that can help us to rate them correctly.

Another example of a secular explicit anthropology masquerading as something else is Powers and Faden's list of 'Essential Dimensions of Well-Being' presented in their book on social justice (Powers and Faden 2006):

- Health
- Personal Security
- Reasoning
- Respect
- Attachment
- Self-Determination

This list is, apart from the absence of a relation to God, quite similar to the lists provided by Finnis and Janssens. This is perhaps not very surprising, except to any philosopher strongly wedded to non-essentialism or post-modernism. Human beings are, after all biological organisms of a specific kind, living in groups with fairly similar characteristics over time, and it would arguably be more surprising if there was no commonality between the things that contributed to their wellbeing. Why do Powers and Faden develop and argue for this list? Their book and its argument is rich and complex, but at least one of the functions of the list is to try to establish that some things are more important to human beings than other things. And, that they are important as basic interests, not only as something we merely happen to want.

5.4 Why so few Anthropologies?

Why are explicit anthropologies rarely seen or used in bioethics? One reason may simply be the publication culture within the field with a predilection for, and preponderance of relatively short papers. This may make it difficult to find space for a fully worked out anthropology, when the paper is primarily about a specific question of moral controversy. A second possible reason is that bioethicists have become used to not finding it strange that ethical arguments begin *in media res* and take a very large number of premises as given, or more often not even as given but just left as enthymematic. A third reason is that by making your anthropology explicit in your arguments, you open up a further potential area of disagreement and contention. People who agree with your conclusions and who saw no problem with your argument, may suddenly realise that they do not agree with some of your important premises. They may, to use Sunstein's useful concept suddenly realise that what you and they have achieved is an 'incompletely theorised agreement' and that further theorising is likely to make the agreement disappear (Sunstein 1994). As long as the anthropological premises were enthymematic the disagreement may not have been evident. It may be the case that discovering a deeper disagreement at the anthropological level does not affect the agreement on the practical level, because arguments from both set of anthropological premises still lead to the same conclusions. But there will also be cases where this is not the case. And before we have uncovered the anthropological disagreement we are not in a position to say whether or not the practical agreement will hold. A final an fourth reason is that philosophical anthropology in itself has a very limited place in Anglo-American analytic philosophy and that many bioethicists trained in this tradition may never have come across any sustained, non-theological reflection on anthropology.[2] In a later section I will provide a more in depth analysis of some the problems that not having an explicit anthropology may create, but before that it is time to excavate the implicit anthropology.

5.5 The Implicit Anthropology of Bioethics

What is the implicit anthropology of bioethics? Given that it is implicit we cannot simply read it off the writings of bioethicists, but we have to infer it from clues in the text. One rich source of material for this inferential work are texts where bioethicists compare groups of subjects to what we can call 'the normal bioethics subject', for instance in writings about vulnerability in research ethics. These texts provide us with a *via negative* to the definition of the normal subject. Ken Kipnis for instance contrasts the vulnerable consenting subject (C-S) with an implicit norm in his definition of different kinds of research vulnerability:

> Cognitive: Does the C-S have the capacity to deliberate about and decide whether or not to participate in the study?
> Juridic: Is the C-S liable to the authority of others who may have an independent interest in that participation?
> Deferential: Is the C-S given to patterns of deferential behavior that may mask an underlying unwillingness to participate?
> Medical: Has the C-S been selected, in part, because he or she has a serious health-related condition for which there are no satisfactory remedies?
> Allocational: Is the C-S seriously lacking in important social goods that will be provided as a consequence of his or her participation in research? [. . .]. (Kipnis 2001, p. G-6)

[2] This is partly substantiated by my experience of giving talks with 'anthropology' in the title at bioethics conferences, where people often ask me before the talk why I am going to talk about social science and are surprised when I say that I will talk about philosophical anthropology.

Here the standard consenting subject must, for instance be characterised by not being 'given to patterns of deferential behaviour' and by not being 'seriously lacking in important social goods'.

Another clue to the contours of the implicit anthropology is to ask what the minimal set of characteristics is that are presented as sufficient for ethical analysis. In early and influential work on the now almost ubiquitous bioethical concept of 'the person' Stanley Benn for instance provides the following description of the kind of entity we need to consider in our ethical thinking:

> By a *person I understand a subject with a consciousness of himself as* agent, one who is capable of having projects, and assessing his achievements in relation to them. To *conceive someone as a person is to see him* as actually or potentially a chooser, as attempting to steer his own course through the world adjusting his behaviour as his appreciation of the world changes, and correcting his course as he perceives his errors. It is understood that his life is for him a kind of enterprise like one's own, not merely a succession of more or less fortunate happenings, but a record of achievements and failures; and just as one cannot describe one's own life in those terms without claiming that what happens is important, so to see another's in the same light is to see that for him at least this must be important. (Benn 1971, pp. 8–9, my emphases)

Like many accounts of what it is to be 'a person' this is a rather rarefied account, focusing almost completely on cognition and a specific account of self-determination. Benn's person does not need a body, and he does not seem to have any relations to other persons, except appraising them and in return being appraised by them. Most bioethicists do realise that human beings have a body, but many are still resistant to the idea from Merleau-Ponty and others that the human person in a significant sense is his body (Merleau-Ponty 1962).

The non-relational understanding of the standard person can probably be traced back to bioethics' initial engagement with and critique of the traditional doctor-patient relationship. A critique and reconceptualization that involved a very strong emphasis on autonomy and self-determination. This emphasis on 'the individual' has embedded itself deeply in bioethics as can be seen, for instance in discussions about much more personal relationships like those within a family.

For many people the fact that they are members of a family or some other closely knit group of people is extremely important, and even in Western societies where the stable nuclear family may no longer be the norm,[3] people tend to live in a succession of family like arrangements. The importance of the family to people shows itself in many ways from an interest in genealogy,[4] to a willingness to help family members to a greater extent than other people.

Family decision making, i.e. decision making where the views of family members are heard and a common, consensual (or at least agreed) decision is reached is also very commonly practised (concerning where to go on holiday, which car to buy, whether to have another child etc. etc.). In many instances of family decision making

[3] It is questionable whether the stable nuclear family has ever been the statistical norm in Western societies, but it has definitely been the type of family held up as the ideal.

[4] An age old preoccupation of human beings already pursued in great detail in the Hebrew Bible, the Norse Sagas and many other ancient writings.

some, or all family members will agree to a decision they would not have made if they were the sole (dictatorial) decision maker.

However, in modern bioethics the family is not mentioned that often and when it is mentioned it is almost invariably as a problem. In the bioethical literature families create conflicts (see for instance the extensive commentary on the Terry Schiavo case) or they prevent family members from pursuing their legitimate and autonomous choices.

The only three places in the bioethics literature family is taken seriously are in discussions about proxy decision making for children, about access to genetic information and about living organ donation (Hallowell et al. 2005), but even in that literature there is a tendency to see joint, family decision-making as an exception and individual decision-making as the norm.[5] The extensive literature on proxy decision making for children does, for instance often proceed as if there was only one parent to take account of.

This is a strange state of affairs, not only because families are important to many people and because family decision-making is so common, but also because many healthcare decisions people make impact directly on the members of their family. A prime example is the lifestyle advice often given to patients with ischaemic heart disease. They are told to alter their diet and take up exercise, but altering your diet when you are living in a family will almost inevitably impact other family members, and suddenly spending more time exercising may also lead to changes in the family dynamic. Family involvement in care of the elderly or persons with disability is also still expected, even in societies where many elderly people are not cared for in the family but in nursing homes. Research, for instance shows that when people are in nursing homes family involvement in their care is expected, even when staff won't let the family take part in decisions about the care (Ryan and Scullion 2000).

A third pointer to the implicit anthropology of bioethics of specific relevance to the topic of this chapter can be found where the majority view rejects any importance of aging, or of the fact that people are at different life stages. Daniel Callahan has for many years argued that aging, and 'the natural lifespan' that follows from aging has implications for what claims people ought to make of the healthcare system (Callahan 1987). Whether or not we agree with the specific implications Callahan draws from his 'natural lifespan' idea, it is undoubtedly worth noticing that many of his opponents do not criticise the specific implications but criticise the basic idea that aging or a natural lifespan could have any ethical implications (see for instance Cutas and Harris 2007).

The excavation of the full implicit anthropology of bioethics is a rather tedious project that could take up many pages of this volume, but on the basis of the analysis so far it is possible to identify sufficient of its main components to enable us to analyse the implications of this anthropology both in general and for the issue of aging. These components are:

[5] With some honourable exceptions like Ross 1998.

Positive components: The human being has primary interests in rational decision-making, reproduction and fulfilment of individual preferences

Negative components: The human being is non-gendered, a-temporal, non-relational, and fully functional

In short the standard human being of bioethics is something like a non-gendered version of Leonardo Da Vinci's geometrically perfect man. It is not quite the 'rugged individualist' claimed by some to characterise much of (American) bioethics, but it is not far removed from that caricature. And although the standard person sees his life as an enterprise and has a life-plan, that life plan often does not seem to involve growing old, whether gracefully or not.

5.6 What are the General Implications of Having an Implicit Anthropology?

Above we have briefly discussed the main function of anthropological claims in bioethical arguments and discourse, i.e. that they lend foundational 'weight' to claims that certain human activities are important. They do this by stating that a person A has an interest in the activity X which is not based on A's preferences, wants or desires, but is based on something that A is *qua* being a human being. And if this statement is taken to be true it converts A's claim to be able to do X from being a claim based on (mere) preferences, wants or desires, to being a claim based in need. And for many this will make the claim look stronger and more worthy of being recognised. This may be the case whether A just claims the liberty to do X or also claims some kind of right to assistance in the pursuit of X. The implicit nature of the standard anthropology thus allows for an elision between different ways of justifying the importance of a particular activity. An elision that is difficult to detect because it is hidden behind a seemingly empirical claim that X in some way links directly to what humans are.

A further effect of this is that claims that are based in the standard anthropology are more likely to become accepted as possible justifications for rights claims and given priority over claims that are not based on the standard anthropology. Because the 'standard' claims are more likely to be seen as self-evident, uncontroversial, justified, or true.

I do not want to deny that any of the positive elements in the standard anthropology are important and are really elements of any well worked out philosophical anthropology. But I do want to deny that they are the only elements. It is only when all the elements of an anthropology are properly identified and foregrounded that we can, for instance say something about how important reproductive freedom is in relation to other freedoms and obligations based on a full anthropology.

That the standard anthropology is implicit also means that it will sometimes hide *a petitio principii*. The importance of X to A is, illicitly implicit in the premises, when it should have been argued for.

5.7 Why is the Implicit Anthropology a Problem in Relation to Aging?

The standard anthropology is a specific problem in relation to theorising about aging for three reasons.

First it invites us to conceptualise everyone who departs from the standard in terms of deficit and it therefore tempts us to understand aging according to a deficit model. The aging person becomes someone who has not quite got what it takes to be a standard bioethical agent.

This does not automatically lead to a devaluation of the older person. There is no direct entailment between a negative evaluation of 'the aged state' and a negative evaluation of 'the aged person'.[6] And there is, for instance a large literature in bioethics decrying and arguing against ageism in resource allocation in healthcare.

But an absence of some of what is standardly accepted to matter may never the less have consequences. If the old are seen through a deficit lens they are more likely to be perceived as subjects of our bioethical solicitude, than as agents themselves.

Second, the fact that the old are typically not completely encompassed within the standard anthropology invites us to see them as a distinct and identifiable group that are afflicted by a distinct and identifiable condition, i.e. 'the old' suffering from 'aging'. But this is very problematic. Aging is not one, single identifiable condition, and this follows whether we are realist or nominalist in our approach to disease classification. This tendency to reify both the group and the condition should be resisted. We do not need this double reification in order to be able to argue that the manifold conditions afflicting some old people must be researched and combated.

Third the non-relational aspect of the standard anthropology may have especially pernicious effects in relation to old age. For many of us it does not matter in our daily lives that the standard anthropology of bioethics does not see relations as a basic feature of the human being. It doesn't matter either because we do not (yet?) feel a need for relations, or because we have valuable relations. But this does matter for anyone who is more than usually reliant on others to facilitate the establishment and maintenance of relations.

5.8 How can We do Better?

We have above identified 4 reasons that may, in combination explain why explicit anthropological theorising is rare in bioethics; and we have identified a range of problems that follow from the fact that the implicit anthropology that is then used in bioethical discourse is rather impoverished. How can we improve this situation, especially in relation to the urgent need to develop a 'bioethics of aging'?

---

[6] Just as there is no direct logical entailment between a negative evaluation of 'the disabled state' and a negative evaluation of 'the disabled person'. But the absence of a logical entailment is not the absence of a possible inference.

The solution is not that every paper written in bioethics should include a section entitled 'My anthropology'. The publication pattern in bioethics involving relatively short articles focusing on very specific topics is not in itself, necessarily dysfunctional.

But that does not mean that there is no need for deeper anthropological reflection on the part of most, if not all bioethicists.[7] The very reason that we can, and do disagree about the explicit anthropologies outlined above is precisely because they have been made explicit and are thus open to reflection and criticism. If I hold my anthropology implicitly, in the sense that I have never made it explicit, even to myself it may, paradoxically, be the case that I do not even agree with my own anthropology. My true 'horizon of understanding' to use a Gadamer term is hidden to me. There is thus a very good pragmatic reason to engage in anthropological reflection. And there may also be reasons of intellectual integrity. Many philosophers feel an obligation to 'follow the argument all the way', but that works both ways. We need to follow the argument not only forward from the premises to the conclusion, but also backward to investigate the status of our premises.

If we engage in that kind of reflection we are likely to notice that the simple, standard anthropology that works well for certain questions is not giving an accurate picture of what human beings are biologically, cognitively, socially etc. There is a reason why many of the explicit philosophical anthropologies outlined above are multi-faceted and complicated, and that reason is that 'the human being' we are considering is multi-faceted and complicated.

In a more in-depth reflection on anthropology and its implications we need to take proper account of the fact that human beings are biological and social beings who normally pass through a complicated set of life stages. Unlike machines we don't roll of the assembly line fully made, and we are not sent for recycling at the first sign of any problem. We therefore need to consider all the life stages in our thinking about anthropology and not valorise one as the most important or most normal life stage. We might wonder whether it is a coincidence that the life stage we valorise in this way is the one that most academics writing on these matters happen to be in. Performing this more 'stage neutral' consideration will not be easy, since it is not only among philosophers that what is perhaps best described as late-early adulthood is seen as the normal state of the human being.[8] In this context it is interesting to note that bioethics is much more interested in the 'bad fortune' that may befall adults (including moral philosophers), than in the 'bad fortune' they have already been through. The teenage years are characterised by various departures from the standard anthropology, not all of them positive. But whereas there is a large literature on 'anti-aging' the literature on 'anti-teenage' is minute.

[7] I have argued in previous papers that even transhumanists rely on an implicit anthropology, although many of the future scenarios they claim to evaluate morally contain no human agents. But, even for such scenarios we need to be able to say something about what is good for the agents in the scenario, and we most often do that by, in this case illicitly, importing assumptions taken from an implicit anthropology. See Holm 2006, 2007.

[8] Also evidenced in the medieval and later depictions of the 'wheel of life'.

But, if we, for instance find that it is an integral part of an adequate anthropology that human beings are (typically) relational and that having relations is thus an integral component of (typical) human wellbeing and flourishing, we need to develop an account of what implications this has across the different life stages. In the infant it is plausible that it may ground an almost absolute right to (at least) one, strong primary caring relationship. And for the old institutionalised person it may ground a right to institutional arrangements that are conducive to the maintenance of old personal relationships and the creation of new ones.

Not valorising one particular life stage above all others also means doing our best to abandon a deficit model of aging. It is true that old age, in our society often involves a decline in a range of physical, social and cognitive attributes. But it is equally true that it often involves an increase in other attributes. The old have more memories, they have a longer perspective and they may even have more time on their hands than middle-aged academics busily writing and reading papers. These are all valuable attributes and we might see them better as valuable, if we can move away from the standard anthropology of bioethics to an anthropology that more adequately takes account of all of the things that go into being a complete human being.

References

Benn S.I. 1971. Privacy, freedom, and respect for persons. In *Nomos XIII*, eds. J.R. Pennock and L.W. Chapman 1–26. New York: Privacy Atherton Press.

Callahan, D. 1987. *Setting limits: Medical goals in an aging society*. New York: Simon and Schuster.

Cutas, D., J. Harris. 2007.The ethics of aging, immortality and genetics. In *Principle of healthcare ethics*, eds. R.E. Ashcroft, A. Dawson, H. Draper, J.R. McMillan. 2nd ed. 797–801. Chichester: Wiley.

Finnis, J. 2011. *Natural law and natural rights*. Oxford: Oxford University Press.

Hallowell, N., A. Ardern-Jones, R. Eeles, C. Foster, A. Lucassen, C. Moynihan, and M. Watson. 2005. 'Men's decision-making about predictive BRCA1/2 testing: The role of family'. *Journal of Genetic Counselling* 14(3): 207–217.

Holm, S. 2006. The nature of human welfare. In *Future perfect?* ed. C. Deane-Drummond and P. Manley Scott 45–55. London: T & T Clark.

Holm, S. 2007, Naturalness and anthropology in modern bioethics, with a special view to trans- and post-Humanism. In *Theology and science—Issues for future dialogue*, ed. H. Kragh 17–29. Aarhus: University of Aarhus.

Janssens, L. 1990. Personalism in moral theology. In *Moral theology. Challenges for the future.* ed. E. Charles Curran. 94–107. New York: Paulist Press.

Kipnis, K. (2001). Vulnerability in research subjects: A bioethics taxonomy. In National Bioethics Advisory Commission. *Ethical and Policy Issues in Research Involving Human Participants*, vol. II. G1–G13. Bethesda, MD.

Merleau-Ponty, M. 1962. *Phenomenology of Perception.* London: Routledge and Kegan Paul (original in French 1945).

Powers, M., and R. Faden. 2006. *Social justice—The moral foundations of public health and health policy*. Oxford: Oxford University Press.

Robertson, J.A. 1983. Procreative liberty and the control of conception. *Virginia Law Review* 69:405–464.

Ross, L.F. 1998. Children, families and healthcare decision making. Oxford: Oxford University Press.

Ryan, A.A., and Scullion, H.F. 2000. 'Family and staff perceptions of the role of families in nursing homes". *Journal of Advanced Nursing* 32(3):626–634.

Savulescu, J. 2005. New breeds of humans: The moral obligation to enhance. *Reproductive Biomedicine Online* 10(1):36–39.

Sunstein, C.R. 1994–1995. Incompletely theorized agreements. *Harvard Law Review* 108:1733–1772.

Tong, R., and N. Williams. 2009. "Feminist Ethics", The Stanford Encyclopedia of Philosophy, ed. Edward N. Zalta http://plato.stanford.edu/archives/fall2009/entries/feminism-ethics/. Accessed Fall 2009.

Chapter 6
On Old Age—Impressions of a Geriatrician

Bert Keizer

Before saying something about ethics and aging I would like to point at the source of my knowledge. I am 63 years old and have been working as a geriatrician in nursing homes and chronic care facilities in Amsterdam during 28 years. This means that I have only seen the less attractive sides of old age, because obviously most elderly people do not end up in a nursing home or chronic care facility. But in my work I did not only have to deal with my mostly elderly patients, but also with their husbands, wives, brothers and sisters. Incidentally, when I speak of old, elderly or aged I am referring to people who are 85 or older.

A second preliminary is the fact that I am writing this and not a person of 80 or 90 years old. This is not because elderly people do not have an opinion about being old, but strangely enough, they are conspicuously absent in the public discussion of their predicament (and mainly absent in this volume?). An absence which I regard as a painful illustration of an aspect of old age that I want to bring to your attention. I mean the horrible experience of being irrelevant socially.

6.1 Why We Don't Like Old People

As an introduction to the subject I would like to call your attention to the fact that we do not like old people. What I mean is the biological basis of what I can only describe as the disgust or at best indifference with which we look at elderly people. I believe this dislike of the old is deeply rooted in our DNA. The reason for this dislike or the slight shudder with which we approach the very old is not only that they are a reminder of approaching death. There is also a biological reason. We are primates, higher apes, if you'll pardon the adjective, who used to live in groups of 20–40 individuals. In such a setting old animals are useless—biologically speaking. They do not procreate, they do not fight when it comes to warding off enemies, but

B. Keizer (✉)
Geriatrician, Amsterdam
e-mail: keizer47@gmail.com

M. Schermer, W. Pinxten (eds.), *Ethics, Health Policy and (Anti-) Aging: Mixed Blessings*, Ethics and Health Policy 1, DOI 10.1007/978-94-007-3870-6_6,

they do eat and they do take advantage of the security a group offers. Caring for the elderly is biologically speaking a wrongheaded initiative.

This dislike of old persons is something they themselves feel as well. In the nursing home where I work we have a rehabilitation ward where people recover after orthopaedic surgery—hip replacement usually. Time and again I am called to the bedside of a fresh arrival, herself 85 years old if a day, who points out to me that she refuses to mix with the permanent residents, who in her eyes, are disgustingly old and as such a great hindrance on her road to recovery, because the sight of them severely depresses her.

I am not saying these things in order to announce my assent, but I believe we shall never arrive at a realistic view of old age if we do not take this biologically based disgust into our stride. I do not know what our chances are to liberate ourselves from these rather base emotions. I fear the chance is nil if we do not acknowledge these feelings in the first place.

Passing on from here I would like to say something about old age and wisdom; fashion; bodily functions; dementia or Alzheimer; and finally of course death and its availability.

6.2 A Few Remarks on Old Age and Wisdom

Possibly one of the few attractive things about getting old is that one experiences a certain relief at having been released from the struggle that life often is. La Rochefoucauld said that old people like giving advice to the young in order to compensate for the fact that they are no longer in a position to make certain mistakes. This remark covers their motivation, but a direr question is if they have any advice on offer that makes sense to the younger generation. My mother was very skilled in feeding, clothing and grooming her husband and six children with rather limited means in such a way that we could appear decently clad and fed in school, in church or at our jobs. These skills of hers are useless for my generation.

I was born in 1947 in a lower middle class, maybe artisan class, family and the problem for my generation was how to escape from the life your parents led and which they thought was good enough for you as well. How to make the most of your chances was our problem, and we were quite nifty when it came to finding our way out of the milieu we grew up in.

This eagerness to seize an opportunity to get away from your parents is something that mystifies my children, who don't want to get away from us at all. Their problem is that they are absolutely drowning in chances and opportunities. They can study what they like, travel to wherever they want to go and they have no need of my eagerness to seize upon any chance coming along.

Our children are also faced with a veritable onslaught of opportunities in the realm of money, alcohol, tobacco, and drugs, my experience of which is tiny and piecemeal. When I was 20 we could hardly afford much alcohol and apart from that we sometimes succeeded in laying our hands on hash, LSD, speed and very occasionally some opium to smoke. It all resembled the small neighbourhood store. Our children live in the present day supermarket when it comes to drugs. Alcohol is cheap, cocaine and XTC are rife and the marihuana they smoke these days is so strong that you easily become psychotic. You can imagine what my advice to them is: Don't touch the stuff. And you can also imagine how effective this suggestion is.

Parents are always 40 years behind the times with their advice, and this is even worse in the case of grandparents.

6.3 A Few Remarks on Old Age and Fashion

To my parents the absolute height of impressive celebrations was a Christmas Night Mass with a mixed choir (that is boys and men) and on the altar three priests and six acolytes. To me the hair of The Beatles, Mick Jagger's accent, the desperate recklessness of Jimi Hendrix, Dylan's incredible style of singing, represented what I thought of as stylish, elegant, beautiful and true.

But the world moves on. More than forty years have passed since then and all those things that you regard as lovable, maybe even sacred, have lost their relevance in today's world. It is all still revered, but it has been shoved aside and is now to be seen behind glass, in a museum. Look at the Resistance Movement in World War II. Who does not feel respect for some of the things they achieved? But at the same time what is it to us now, when we are no longer threatened by obviously murderous armies marching into our country, but by banking people who do ungraspable things with the most awful consequences?

That is what is so unpleasant about getting old: The things you liked, knew, wished or feared, are slowly becoming more and more irrelevant. Very gradually but unmistakeably you are being shoved aside. It is a relentless process, it is wholly unintentional, and the result is an ever-deepening sense of social irrelevance.

6.4 Bodily Functions

Another well-known and rightly feared aspect of getting old is the loss of bodily functions. I don't think there's any need to go into a long disquisition on the fact that as we grow older, lots of things that we do in daily life become more and more tiresome, difficult or downright impossible. I don't think anyone is going to say that as they get older, and I am talking of the years beyond eighty, they get better and better all the time in the performance of: Walking, stair climbing, skating, lovemaking, cycling, seeing, hearing, digging in the garden or driving their car.

And now that we are on to the body anyway, there is a gradual retreat from the erotic scene which Sophocles famously experienced as a case of good riddance, but which most of us, being a little less philosophical, do experience as an irretrievable loss. I am not saying that one cannot be an attractive man or woman in old age, but that particularly fateful variety of attraction has faded irredeemably. Which inevitably leads us to the ultimate consequence of having a body: We have to die, and we know it.

When I was eleven, I knew only four dead people: Two grandparents, one neighbour and my mother. Now that I am 63 I would need a couple of hours at least to arrive at an estimate of the number of dead people I know. It must be more than a hundred and the number grows every year.

The older you get, the more dead people you know. This is a strangely hurtful kind of knowledge. Knowing that Shakespeare is dead doesn't affect us in the least, but knowing that your dear mum or unforgettable uncle Arthur has passed away and will forever remain in that state, is, when you come to think of it, a crushing realization, to which you'll have to add the inescapable conclusion that you yourself are heading that way.

I need to correct myself here. I sound too desperate. In the many years during which I have dealt with elderly people I have discovered that one of the few things that seem to be reasonably well arranged on this planet is the zest for life in the elderly. This is often more or less in accordance with their physical powers. I'm not saying that elderly people have no zest for life, but their anxiety about the approach of death has not the piercing sharpness that is felt by people in their forties.

6.5 Alzheimer and the End of Life

We have succeeded in prolonging what is called old age to such an extent that two dangers are lurking there. The first is Alzheimer. Once you are 85 or older, the risk of losing parts of your mind, or even most of it, is as high as 40 %. Quite a few people would like to forego this descent into a premature oblivion by ending their lives. In the Netherlands this is possible but not often enacted because not many doctors are convinced of the intensity of the anguish which is caused by mentally falling apart in the course of Alzheimer.

The treatment of Alzheimer patients offers a convincing illustration of that dislike of the elderly with which I started this chapter. For imagine that Alzheimer was not a problem of old age, but that we had 250,000 adolescents in our midst who were affected by a similar brain disease. Do you think it likely that we would dump these youngsters in care homes on the outskirts of our community so as to be able to get on with our lives? Yet that is precisely what we do with Alzheimer patients. Of course losing so many 18 year olds to a lethal brain disease is quite a different thing than losing the aged in that manner. The difference being that the aged are, biologically speaking, not a loss, when they die. I do not mean to applaud this, I merely point out that this is what we feel and it shows in the way we care for them.

The second danger of a prolonged old age is that it may simply last too long. It is difficult to say how many old people would like to die. I met quite a few who didn't feel like waiting around until the aging process would destroy their power to act independently, some of them taking their own lives before it was too late. I am not talking here of desperate suicides committed in horrible secrecy and utter loneliness.

I am speaking of what a Dutch psychiatrist, Boudewijn Chabot, has proposed to describe as auto-euthanasia—do it yourself euthanasia. In such cases the last act is perpetrated in the company of sons, daughters or friends.

Usually such initiatives are condemned outright, and this condemnation comes from those in their forties, for it is they who determine under what conditions people in their eighties or even older will be allowed to leave the planet. When a person of 85 years old is dizzy, loses urine, is short of breath, is no longer capable of climbing a stairway or walking in a street, cannot see the television, hear music, write or type a letter, or even read a book or put on his own clothes, it is the 40 year olds who say that asking for death on account of all this is an irresponsible request, forgetting that this list of incapacities would be seen as unbearable in someone of their own age. But their answer is: well that's all part of getting old.

There's nothing we can do about old age and death as such, but I do think that the last part of life's journey is unduly influenced by people who are themselves still at a much earlier stage. The entire healthcare in our countries is pervaded by the wishes and anxieties of those in their forties. It is their zest for life and it is their fear of death which are stealthily forced onto the very old when these fall into their hands. I think this is wrong. I don't propose to let the very old run our hospitals and care homes, but I do plead for them to be listened to attentively so that they can utter what it is they really want when it comes to questions of prolonging their lives or accepting and even organizing their deaths.

Chapter 7
You Don't Grow Old on Your Own

Frans Meulenberg and Wim Pinxten

> Although there is plenty of space
> on a gravestone to contain, bound in moss,
> the abridged version of a man's life,
> detail is always welcome.
> (Vladimir Nabokov)

Age is a metamorphosis (particularly physical). A competition (in infirmities, illness and length of life). The bill you pay for your life. A liberation. A job you have to do. A confused script. Involuntary house arrest. A challenge. A punishment (at least), a disaster (at worst).

People often use metaphors about age to make pithy comments about the process of aging. It is important to listen to these metaphors, and to what older people themselves say, when thinking about age, ethics, and policy. It's one thing to look at what happens when we get older (or at how we cope with the process); the actual experience is something else again:

> Young people have no idea what old age is. I didn't when I was younger. You have to see for yourself what growing old means. There is no training to prepare you. And there's no avoiding it either.

That is why this chapter looks at how older people themselves see their aging process, and why there is plenty of room for their comments.

How do older people experience the reality of aging and being old, something that happens to them regardless of their own wishes, and often catches them unprepared? How do they see individuality, autonomy, relationships and the passing of the years? What do they think gives meaning to the specifics of their existence? What doesn't? This chapter provides impressions of aging on the basis of interviews with fifteen

F. Meulenberg (✉) · W. Pinxten
Erasmus MC/University Medical Center Rotterdam, Department of Medical Ethics and philosophy of medicine, The Netherlands
e-mail: frans.meulenberg@woordenwinkel.nl

W. Pinxten
e-mail: w.pinxten@erasmusmc.nl

M. Schermer, W. Pinxten (eds.), *Ethics, Health Policy and (Anti-) Aging: Mixed Blessings,*
Ethics and Health Policy 1, DOI 10.1007/978-94-007-3870-6_7,

older people: Eleven women and four men, all born between 1925 and 1940. The group included two couples. During the rounds of interviews, a distinction was made between two groups: Older people living on their own (including those living in sheltered housing) and residents of nursing or residential homes. Educational backgrounds varied: From a manual worker to university education; and employment histories also varied: From charlady or housewife, through to journalist and teacher. The considerable variety was not restricted to background and employment history. The way individual people thought about old age proved to be very heterogeneous, with contrasts between statements like *'I think it's wonderful to not do anything'* and *'Everything is always "finished". It drives me mad!'*. The numerous metaphors about aging from the first-hand accounts also revealed the internal conflicts between the positive and negative aspects of aging. Below the surface, there is a tension between liberation (the relief from burdens and obligations), humility (putting your own position into perspective, modest gratitude) and humiliation (loss of freedom and identity).

7.1 Dynamic Life Stories

'Life stories' take on many shapes: From relatively limited forms (contact ads, curriculum vitae and tombstones) to specific versions (medical histories, for example) and genuine, fully fledged autobiographies. Life stories are individual repositories of identity and life meaning. The way the story is told changes continuously depending on the time, the aim, the audience and the perspective. Life stories are dynamic entities, full of colour and then draining of colour, like gobstoppers.

The multitude of metaphors and first-hand accounts show that how people perceive aging is an ambiguous process, and that it does not squeeze us into a single, predetermined path. Every older person still has to settle on the conclusion of their own life story. What do older people tell us about their experiences in this respect? Which *faits accomplis*, challenges and decisions do they face?

Throughout the interviews, a clear undertone is that aging is something that happens to you, but that it is certainly not a passive process. On the contrary, aging induces conflicting behaviours: Preservation and adaptation, rebellion and resignation, persistence and farewells. In addition, the people interviewed also described different strategies for making sure that their lives were not hollowed out by the deterioration that comes with age: *Confirmation, integration* and *continuity*.

7.2 Confirmation

Aging is a job that has to be done. More and more, older people are confronted by a loss of things in life that previously seemed to be so obvious and natural. And as freedom, autonomy and independence become slowly but surely more restricted, physical strength is also affected. There is a slow, reluctant decline. So old age is a suspect venture:

> Even so, I'm not sure whether there's anything good about getting old. Of course, I try to make the most of it. Who doesn't? But however you look at it, my time is running out. The emptiness gets emptier, and it gets more and more difficult to fill it up'. [After a hesitant pause:] Aging is a sort of job you have to do, that you can't get out of.

What can older people commit to? How can they avoid the complete erosion of their life? What do they want and/or what can they accept? How do they organise their resistance and/or orient themselves? How can they maintain their presence in a world where their being and functioning is slowly but surely declining? During the interviews, three areas of action emerged: (1) discipline and routine, (2) coping with role reversal and (3) monitoring living space.

7.2.1 Discipline and Routine

How do people actually get older? 'It happens all on its own', said most of the people we interviewed. This does not imply that no effort or adaptation is required. The less you can depend on your own body, the more discipline self-preservation requires:

> You do nothing at all, it happens to you whether you want it to or not, you can't avoid it, it's inevitable. You have to accept that you can count on your body less and less; the only thing you can do is to make the most of it.
> You need discipline at all times. Especially in the morning. Before, you could ask your body to do anything, now you have to do everything for your body so you can keep using it as long as possible. And it takes more and more time to become yourself every morning. The time when you could just rub a wet flannel over your face is long gone.

Aging means we have to invest to remain ourselves. Rather than achieving what is possible, older people have to preserve what they have, using numerous routines:

> Alongside the enjoyment of things around you, you also have to make what there is. For me, routine is important. I live to what may be ingrained habits: From eating fruit at a particular time to clearing up. And the idea of having "all the time in the world" is relative. There's more than enough time to get everything done. Maybe that's the problem... that everything is "finished", even the things that don't need to be'. Does that mean unruffled days? 'That's one way of putting it, but I think "monotonous" is a better word. Even though I'm not complaining. At my age, I shouldn't complain.

Routines would seem to be a particularly unwelcome restriction on freedom in nursing and care homes in particular:

> Everything here is done to a timetable. In one way, that is pleasant, but I also feel sometimes that I'm being treated like a child.

7.2.2 Role Reversal

Older people also have to confirm and redefine their own rules in social relationships. There are often shifts in social relationships, and certainly between generations. For example, the relationship between children and adults is reversed:

> I used to warn the children about crossing the street. Now they do that with me and I think they do it a lot more than I used to: 'Mum, are you being careful on the stairs?', 'Watch out, this floor is slippery', 'No, you can't do that any more, forget about it. . . ' or 'You're starting to go deaf'. These are little things, but even though I'm an old lady I carry on doing the things I used to. I've always liked window shopping. I still do, preferably on my own. The children don't see it like that. They seem to think that somebody in her eighties doesn't need to be active any longer. At least, not like I am. I don't find that pleasant. And it's very restrictive; I sometimes get the feeling that I'm not free any more.

Just like the process of aging itself, changes in the relationships accompanying old age are a source of conflicting emotions. As a rule, concern from others is not felt to be out of place but the dividing line between concern and meddling can sometimes be very thin indeed. When older people stand up to meddling from their children, annoyance is the result:

> I even refused to pick up the phone a few times when I saw it was my youngest daughter calling. That just made it worse. She turned up at the front door ten minutes later. . .

> A while ago, I arranged to go and see an old friend in Groningen by train. A lovely idea: Two hours sitting quietly in the train. My eldest daughter was meant to be going with me. But she called off at the last minute because she went down with the flu. And what do you think she said? "You shouldn't go on such a long journey alone. . . " I mean: Perhaps she's right, but it's not good. I can still decide for myself what I want to do or not. I don't always need someone to hold my hand?' [A brief pause, and then with a smile on her thin lips:] 'Did I go? You bet I did. . .

Nor do residents of nursing and residential homes escape from meddling or patronising, even though they are unhappy about it:

> If I don't feel like play therapy, surely I'm allowed to say: "Not today, thank you!" But I don't even dare to say that. Because three carers descend on me in no time, trying to convince me how "good for me" it is, and that it gets me "seeing other people". That's as may be, but for heaven's sake, let me decide who I want to sit next to at mealtimes. I do have that privilege, don't I?

7.2.3 Life Space

Getting older for many people means that they need to make a greater effort to maintain links with the world around them. Even so, people generally think this freedom of movement is very important:

> My daughter picks me up on Sundays and we go to the heath for a walk. It's the high point of the week. I almost feel like a little girl again. I wouldn't want to miss those trips out for anything.

Even so, and this is something that is highly frustrating for them, a long life means a loss of freedom of movement for many older people:

> If you hardly get out any more, just because it is too much effort to get dressed properly, to get the Zimmer, to walk to the lift and walk through the hall to the outside. . . then I feel locked up. It's compulsory house arrest.

> I really felt old a few years ago, when I didn't dare to get on my bike after I'd broken my arm. It's a terrible loss: You suddenly feel cut off from everything outside. You're like a goat tied up with a rope: You can't move outside the circle dictated by the length of the rope. All the cycling I do now is in my dreams. It's lovely, but a big disappointment when I wake up.

In addition to the loss of physical freedom of movement, some older people find it more difficult to be a part of the world around them:

> I would like to be a genuine part of this world, and I read the papers every day, so don't get me wrong. And I never miss the news on television. I really do know what's going on in the world. But I sometimes get the feeling that a pair of scissors has come along and cut my links with the world.

7.3 Integration

Old age demands effort. At least, it does if older people want to use confirmation strategies to stem the tide that can wash in with old age, submerging their freedom, autonomy and independence. As an ongoing effort to counteract the crumbling of one's own powers, old age would seem to be an unavoidable Sisyphean challenge. However, the first-hand accounts from older people show that this is not the whole reality. Older people can, for example, feel stronger when they have secure links with the social context, and more particularly in the form of relationships with people and animals.

7.3.1 Relationships

As the world gets smaller with increasing age, the social network generally shrinks as well. Children, grandchildren and great-grandchildren can constitute a basis which can be firm in some cases, and shaky in others:

> I've been parked on the margins of society. That's something I see almost every day. Fortunately, the children come round, as do the grandchildren. It's so nice to see the curiosity and eagerness to learn of the youngest children. Looking for answers to the hundred "why's" that my great-grandchildren ask. And if I get the names mixed up, they correct me gently and cheerfully. Silly granddad.

Of course, childless people can't fall back on their children. But they can have social relationships as well:

> I struck up a friendship with a girl from two doors down. She is 16 and she comes round almost every day. She certainly rings every day. I sometimes feel like a sort of very-late-adopted-child. But I can always count on her. That's a very basic feeling for me.

The loss of a partner, however long ago it may be, is an enduring feeling.

> Of course, our marriage wasn't always rose petals and sunshine. But you're always stronger together than alone. That's just the way it is. When he was very ill, I could do a lot for him.

> That was a sort of consolation, but not the sort of consolation I asked for. When he died, the shield protecting me from the outside world went with him. You only really understand how valuable things are when they have been lost. I miss him, above all as a companion. And that feeling of missing him stays with me, even though he used to drive me round the bend.

Some of the people we interviewed take a more active approach to their links with society and actively take on responsibilities:

> It's not a question of charity but of solidarity, interest in your fellow human beings. That's a belief I share with all my brothers and sisters, past and present. That's the upbringing our father gave us. And it's become my motto, the thing I have passed on to my children. I think feeling responsible for something keeps you on your toes. And that's not something that stops when you get older.

During the course of the interviews, it emerged clearly that these links with society imply a lot of personal significance.

7.3.2 Pets

As well as people, animals also play an important role in terms of care, company and a sense of security. They are loyal companions who are always present.

> I got my first cat when I was seven or eight. Since then, I've never been without a cat for more than a couple of days. They are beautiful, sociable animals. They're no trouble at all. And fortunately you don't have to take them out for walks like dogs. She has her own regular spots around the house and she's very loyal. They say that cats don't get attached to people. But nothing could be further from the truth! She's much more than just company for me. It's a close friendship. We are inseparable. That's right! I never feel alone.

Older people sometimes establish contact with other people through their pets:

> I take my dog for a walk three times a day. It's wonderful. It's great to take off the leash and to watch the dog running around and playing on the field. It doesn't matter what the weather is like, I have to go. It's a pleasant sort of duty. A dog's love is unconditional, just like small children. I can't imagine my life without my dog. And it's because of the dog that I have so many social contacts. Because there are always other people letting out their dogs. And we always end up chatting. Maybe only about the weather, but it's still very pleasant.

People in nursing or care homes have enduring memories of their pets:

> It really is too sad for words that people aren't allowed to have pets here. Not even a budgie. I think that is so miserable and unhappy. For me too, that's right. Surely everybody wants to look after somebody else or a nice animal? That's it: Here, they take away your chance to give love. I think that is degrading. There is no other word for it. Eyes blazing with fury as these words were spoken.

7.4 Continuity

A third strategy for a strong old age is to enrich it with memories of the past and prospects for the future.

7.4.1 *Continuity*

Age is a metamorphosis that can result in a radical transformation of a person's abilities. However, in the interviews, older people said there were other ways of maintaining valuable ways of experiencing life:

> I love gardening. I could spend every day trimming, hoeing, digging, whatever. That outdoor person has been shut up inside. But I still love caring for my plants. And there are the freshly cut flowers every week, of course. They give my life colour, literally.

> I've had plenty of boyfriends and lovers throughout my life. And I still don't feel like I'm out of the picture. Because there are still men who have that glint in their eyes when they look at me. Yes, I like that. But I still haven't met that one man who could be the definitive solution for my existential loneliness'. With a laugh: 'And time is starting to run short...

This continuity strategy can embrace different generations :

> I think people are just the way they are. I mean: My father taught at a junior school and later he was a headmaster. My brother, my sister and I all went into education. All of us in our own ways, but all in the same branch. Apparently, things like that get passed on from generation to generation. I hope I've kept some of that inquisitiveness and thirst for knowledge. I keep an eye on what's going on in the world and I talk about it a lot with other people. I make up my own mind first and then I try to convince other people. It may be a bit missionary, but it's part of the essence of my character.'

Finally, the continuity strategy certainly need not imply a grim attachment to the past:

> I've read a lot all my life. As a teacher at a secondary school, of course, I had to. I'm still very fond of reading. Decades ago, I promised myself that, after retiring, I would finally get round to reading all those literary masterpieces that I never read, like Musil's The Man Without Qualities. Well, I just haven't got round to it...

7.4.2 *Looking Back*

Looking back to the past is a cherished activity.

> Almost everything takes me back to the past. When the children come and visit, I always end up thinking about when they were little or about their years at school. It's like a silent movie.

> Remembering keeps me fit! All those good things from the past... They're something to hold on to, certainly.

7.4.3 *Survival*

The here and now is something else that takes on another dimension in old age. From being something self-evident, survival is transformed more and more into a competition. In those terms, life continues to be a *struggle of the fittest*:

> There is a certain relief in seeing other people being affected by all sorts of major and minor infirmities that come with age. Every funeral or cremation turns into a sort of reunion. At the reception, we older people get together and draw up the balance. Paul's already gone, and I hear that Marianne and William are not doing so well. It's more common among men than women, I suspect. Life's competition is a source of satisfaction down to the grave.

7.4.4 Taking Leave and Best Wishes Looking Ahead

Memories are indispensable for personal and human identity, and a common activity. Looking ahead is much less widespread, even though it is definitely a feature. In virtually all cases, it is linked to people's own children and friends in the form of wishes for the future:

> I hope my children and grandchildren are happy and stay healthy'. Concerns about 'the world at large' are felt particularly by people who keep a close eye on the news through newspapers or television: 'I just don't know where the world's headed. I worry a lot about that.

And the couple in a residential home:

> Of course, we realise that this is our last stop in the journey through life. We are happy that we are still together. Not many people are this lucky. We are grateful to our creator. We have had a good life and we are ready for death. We have both made our own bereavement cards: We wrote the words and chose the pictures. And we have also worked out the details of the requiem. The music, the readings, everything. It was quite a labour of love.

But is this really looking forward? Is it not more like making preparations for saying goodbye? Only a few people start genuinely new initiatives: 'Well, why go to all the fuss?' Even though one woman was learning Russian: 'I love languages, and it's a challenge for me. You're never too old to learn, as they say. That's right!'

7.5 Conclusion

That aging just happens is not an issue. But the issue of whether to let it just happen at the cost of giving up autonomy, freedom and independence is all the more so. Because even though acceptance and resignation are not altogether strangers to older people, many of them also find antidotes to the forces that sometimes inevitably affect them. Confirmation, integration and continuity strategies allow them to re-invent themselves again and again, despite the changing terms and conditions of their lives.

Together, older people create a place for old age. Here, it is striking how often the word 'time' came up in the interviews. Often in the shape of familiar expressions like 'having all the time in the world', 'time flies', 'time heals all wounds' or 'everything in its own time'. Life—read: 'time'—needs ordering and discipline. One interviewee talked about the phenomenon of time in more depth:

> When you're young, time doesn't exist; it's a tyrant during your working life and it loosens its grip as you get older. Time turns out to be very flexible and elastic, like an elastic band.

> In other words: Time is what you make of it, is something I came to understand. That's a consoling thought.

In conclusion. Another remarkable fact is that the word 'destiny' wasn't used during the interviews. By contrast, the word 'consolation' came up with amazing frequency, often phrased as: 'Oh well, I may be old, but I console myself with the idea that. . . ' These consoling thoughts include the openings people still have: Children and grandchildren, the presence of friends and inclusion in social networks, excursions, activities and skills. And, last but not least, memories that are often rich. One thing is certain: 'You don't grow old on your own'. For some people, this will be a consolation; for others, it won't.

Part II
Taking Care: Caring for Elderly in an Aging Society

Chapter 8
Aging Beyond Frailty: The Future of Old Age

Cees Hertogh

What I say is no subtle doctrine, but a thing that all of us, Greeks and foreigners alike, in some way perceive—that from the beginning existence is difficult for every living creature: First, partaking of the state of things conceived, then again being born, and further, being reared and educated—all these processes involve a vast amount of toil, we all agree. And our time must be a short one, I do not say in the reckoning of the wretched, but on any supposition of what is tolerable. It does seem to give just a breathing-space about the middle of human life: Yet swiftly old age is upon us, and must make any of us loth ever to live our life again, when one reckons over the life one has lived. (Plato, Epinomis)

8.1 Introduction

Growing old is a personal challenge for every human being and a positive appreciation of men's final years has never been self evident, on the contrary. There is a long, deeply rooted belief, that old age is an inevitably dark perspective awaiting everyone, given he has time enough to life. Consider, for example, Plato's words, quoted here above and taken from his book *Epinomis*. In this dialogue a nameless Athenian voices a view on the human condition almost entirely focused on our passivity as a fundamental mark of our vulnerability as (inter)dependent human beings. Only in the middle stage of our life, a limited period of activity and self-sufficiency gives us some solace, but only as a prelude to the culmination of our passivity in the frailty and progressive dependency of old age. And yet, we all favour and strive for longevity, while hoping and striving to escape from the weaknesses, both physical and mental, that may accompany high age. That's why Cicero in his famous essay *Cato Major de Senectute* intensively searches for good examples, starting with Cato himself, of people who knew how to grow old and retain their resilience and dignity: Nestor, Solon, Diogenes and many others. Much efforts are made, in past and present times, on scientific and pseudoscientific levels, to extend the human lifespan and to overcome, what Descartes, at the beginning of the scientific era, once termed 'the

Cees Hertogh (✉)
VUmc University Medical Center Amsterdam, Amsterdam, The Netherlands
e-mail: cmpm.hertogh@vumc.nl

M. Schermer, W. Pinxten (eds.), *Ethics, Health Policy and (Anti-) Aging: Mixed Blessings*, Ethics and Health Policy 1, DOI 10.1007/978-94-007-3870-6_8,

enfeeblement of old age'. In fact: One of his principal ambitions, as he emphasized in the first lines of his *Discourse on method*, was to develop a science of human health and longevity. Although he has been credited with being one of the founders of modern science and medicine, in his days this could scarcely be more than a dream.

Much has changed since the times of Plato, Cicero and Descartes, although the most drastic change is of a rather recent date. In fact, it was the past century that witnessed an impressive transformation of old age through the advent of what Laslett has termed 'the third age' (Lasslett 1989). Today, more people than Plato, Cicero and Descartes could ever dream of age successfully and remain active participants of societal life. However, this transformation has not resulted in a falsification of Plato's rather grim point of view, nor has it changed our ambivalence with regard to aging and longevity. Rather, it resulted in a deepening and restructuring of this ambivalence into a duality between two faces of old age: The one of vitality and successful aging—the third age script—and the one of frailty and deep old age, denoted by Laslett as 'the fourth age'.

My intention in this contribution is to sketch both faces of old age and to reflect on the future of old age as it is envisaged by protagonists of successful aging and modern gerontology. In this connection, I will give special attention to the notion of 'frailty' with its multiple meanings and connotations. My thesis will be, that both the focus on successful aging and the present approach to frailty represent a new, scarcely hidden form of ageism related specifically to the other side of old age. As a consequence, care for and research directed towards improving the quality of care for the oldest old is a neglected and underfunded realm.

8.2 Frailty as a Paradigm of Natural Death

First let us have a look at the future of old age from the angle of the optimists. Theoretically, this optimism is best expressed in the so called 'Compression of Morbidity' paradigm, which appeals to our imagination as much as it is attractive for policymakers (Fries 1980, 2005).

According to its founder, James Fries, our concepts of health and illness underwent impressive changes since the beginning of the twentieth century. The first decades of the past century were characterized by a high prevalence of infectious disease and very high perinatal and infant mortality. These plagues were combated successfully, but only to give way to the new morbidities we presently have to face. Today we live in the era of chronic disease. These illnesses badly fit into the classical medical model, in fact they question this model in a fundamental way, for the traditional, so called nosological approach and its focus on causal curative therapy is no longer valid here. Chronic illnesses are multifactorial in causation, they cannot be cured and many of them, such as arteriosclerosis or osteoarthritis, are to a certain degree universal, that is to say: They can affect any individual. So far, Fries summarizes a commonly shared epidemiological and clinical insight (Tinetti and Fried 2004).

However, he prophesizes that we are presently on the verge of entering yet another era with dramatically new characteristics of health and illness. More specifically, the promise of the Compression of Morbidity paradigm is, that the chronic disease era will slowly decline in importance as a result of effective postponement of ill-health through preventive and life style interventions, leaving a third era in which our major health problems will be directly related to aging.

At several occasions after its introduction as a hypothesis of healthy aging in 1980, Fries reviewed his paradigm and presented evolving lines of evidence to establish proof of concept (Fries 2002; Fries et al. 2011).

In essence, the Compression of Morbidity paradigm comes down to the following 'syllogism for aging':

1. Given that the natural lifespan of a species is relatively fixed, it
2. secondly must be possible to increase the date of onset of aging manifestations and signs of chronic disease by more years than life expectancy, as
3. a consequence of which duration of morbidity and disability will be shorter and compressed around the end of life.

Even if the first thesis, that of the fixed lifespan, would proof to be false—and much of the criticism on Fries' hypothesis is directed at this basic assumption—this criticism is misdirected according to Fries, because fundamentally "the Compression of Morbidity paradigm does not depend upon whether the human lifespan is fixed or rising. It depends on relative changes in mortality rates and in morbidity /disability rates. Compression of Morbidity can occur with falling life expectancies, or with rising ones" (Fries et al. 2011). In other words: Compression occurs, as long as the date of onset of disability is postponed with more years than death. It basically works through the twin strategies of early detection and subsequent medicalization of chronic disease (which amounts to a compression of disability through an expansion of morbidity) combined with patient education and life style approaches. In addition, also an investment in the plasticity of aging can help: "We can swim against the current of senescence by training our faculties" (Fries 2005). Additional postponing strategies mentioned by Fries are function-enhancing medical strategies such as hip-replacement surgery and improvements in built environments. As said, at several instances Fries reported on new evidence and data to support his argument, but most impressively this support is expressed graphically in the so-called 'rectangularization' of the survival curve: In successive decades demographic survivorship curves have begun to bend upwards and to the right, thus becoming increasingly rectangular and allowing a visual prediction of future trends. Ideally, this curve is moving towards a situation, where lifetime morbidity is squeezed that much and into such a short period before death, that a situation is realized in which people can die a natural death, not caused by disease, but by the aging process itself. It is then that we will truly enter the prophesied era of the senescence process.

Fries paradigm thus holds a optimistic promise for the future of geriatric medicine and Rowe's and Kahn's (1998) concept of 'successful aging', emphasising independent physical functioning, absence of disease and active engagement with life, is much indebted to this theoretical framework and its underlying ideas. Also—and I will get back to this further on in this chapter—Fries' theory stipulates a concept of frailty that seems to run counter to present day conceptualizations of frailty, at least when it comes to its practical and moral consequences. For example, Grimley Evans (1998), a prominent representative of modern geriatric medicine, identifying compression with healthy or successful aging, embraces the prospect of a good death it proposes. In his view and that of other protagonists of Fries' paradigm, with successful compression, aging will come to the rescue of medicine, because it allows to reconcile the twin moral aims of medicine: That of postponing death and of minimizing suffering from illness: "By delaying the onset of disabling diseases to later ages when intrinsic aging has raised fatality by reducing adaptability, the average duration of disability before death will be shortened... In brief: We shall spend a longer time living and a shorter time dying."

Intrinsic aging is defined here as 'homeostenosis': A progressive narrowing of the capacity to adapt through a decline of reserve power, resulting in a loss of resilience to maintain and restore an equilibrium. According to the protagonists of the Compression of Morbidity paradigm this is in essence what frailty is: A general loss of adaptive capacity (Fries 2005; Tallis 2004; Grimley and Evans 1998). It is associated with an increased probability to death, but also with a death that comes 'without fear or fury' and that is more easily achieved than that of a person with homeostatic mechanisms intact. Death from frailty will thus be close to death without disease, that is to say: A natural death. Such a death will occur when the characteristics of the host resistance become more important than the specific nature of the insult to the homeostatic equilibrium. As a consequence, frailty should not be seen as a state of affairs or a process to be combated: Instead we should respect and even welcome it (Tallis 2004). So conceived, frailty seems to be a more outstanding candidate than Osler's pneumonia to deserve the qualification of 'the old man's best friend'. To make a variation on Dylan Thomas: In the era of the senescence process, we will at last be able to 'go into the good night of death, without having to rave at close of day, without having to rage against the dying of the light'. As a consequence, there is no urgent need to invest in research to improve the quality of care in the last stage of life of older people, no need either to further develop geriatric medicine: In the senescence era old age will be easy to bear and death will be calm and without suffering.

So far for the optimistic point of view. There are numerous criticisms to make here, but it is an irrefutable fact that the Compression of Morbidity paradigm remains a very attractive ideal. Perhaps this is also the most important criticism: That it is indeed an ideal, the prospect of an idealised future which we are believed to approach in an asymptotical way. However, although everyone will have experience in his personal surroundings with old people whose life and death fits into this prophesied senescence era, for a vast majority of older persons this ideal is far from realised. More realistic is to acknowledge, that aging comes down to growing old with extensive co-morbidities and that there is a huge overlap between so called normal aging and the pathogenesis of age-related disease (Kirkwood and Ritter 1997). This point of view results in a rather different, not to say opposite approach to frailty, but it is in this opposite way that frailty is conceived of in today's healthcare research and clinical practice.

8.3 Frailty: Origin, Development and Transition of the Concept

A word is not a concept: This is a basic principle in the study of the history of science. The meaning of a scientific concept depends on the theoretical framework or model within which it functions. Thus, the use of the term in a specific context does not mean that the concept is present and vice versa: Similar concepts can be denoted by different terms (Canguilhem 1968). The problem with frailty is, that it resists clear and unconfounded definition and that the term functions in very different theoretical contexts with the same word being used to denote different concepts. As Hogan et al (2003) lament: "One is struck by a Tower of Babel quality of recent writings on frailty." Already in daily life, the word has an ambiguous meaning. The term not only refers to weakness in terms of a liability to break or a susceptibility to be wounded, but also to moral weakness in terms of an instability of mind or an inclination to immorality (Grenier 2007). This is true in the original English as well as in Dutch translations of the term (Puts 2006). Furthermore, this ambiguity in meaning, charging the factual sense of the word with a covert moral or moralistic undertone, is somehow also implicated in the more scientific and professional use of the word, as we will see further on in this chapter.

A close examination of how scientific thinking about frailty and the use of the term in professional healthcare settings and research have evolved, demonstrates an intriguing transition in meaning and application from a service directed concept in the 1970's to an interventionist concept in the course of the 1990's (Bergman et al. 2007; Hertogh 2010).

8.3.1 Origin: A Service Directed Concept

During the 1970s the heterogeneity of the older population became more widely recognized and the first one who came to use the term 'frail elderly' was a priest. His name was Monsignor Charles Fahey and he was the chairman of the US Federal Council on Aging (FCA) (Hogan et al. 2003). The phrase 'frail elderly' when first used was not felt to have any specific originality, but was selected because the Council felt that there was a need for a dramatic term, in order to focus much needed attention of the general public and policy makers on a very special group of elderly. In 1978 the FCA defined frail older people as "persons, usually but not always over the age of 75, who because of an accumulation of various continuing problems often require one or several supportive services in order to cope with daily life" (Tavani 1978). According to the FCA, core services for this population should consist in an assessment of needs, the development of a care plan and case management. However, except for criticism, this initial use of the term 'frail elderly' hardly met with any response from the site of the geriatric medical community and was disposed of as 'typical jargon of the Council'. In these days, frailty, disability and chronic disease were more or less used as interchangeable terms.

8.3.2 *Towards an Interventionist Concept*

In the course of the 1990's definitions of frailty began to appear, which did not depend on the presence of chronic disease, disability or a need for healthcare services. On the contrary, frailty was redefined as a state of vulnerability and risk, that needed intervention instead of supportive services. Within this more recent field of application, we can further distinguish between biomedical or clinical concepts and more broader epidemiological or public health concepts, such as the one coined by Deeg and Puts (2007) and the concept adopted in the recent report of the Netherlands Institute for Social Research (SCP) (Van Campen 2011; Puts 2006). I call them both interventionist, because they all focus on early detection of frailty and the initiation of interventions to postpone, delay or even reverse its progression (Lang et al. 2009).

Biomedical concepts are often called narrow, because they focus predominantly on physical frailty. Examples are the frailty phenotype, proposed by Linda Fried—who defined frailty as a geriatric syndrome—and the concept of the frailty index, advanced by Rockwood, who posited frailty as a state variable, resulting from the accumulation of deficits (Fried et al. 2001; Rockwood 2005; Jones et al. 2004).

Among the biomedical concepts, the one proposed by Fried and co-workers is best known and most widely used. They define frailty as "a biological syndrome of decreased reserves in multiple systems that result from dysregulation that can occur with aging, disease, and/or lack of activity or adequate nutritional intake" (Fried et al. 2001). Of note is, that Fried distinguishes between primary frailty—frailty as a result of the aging process—and secondary frailty, which can be caused by external factors, such as disease, malnutrition and inactivity. Secondary frailty may be temporary. It was this second category that fuelled the idea that frailty may be—at least in certain forms—distinct from aging and hence potentially reversible. Fried further stresses that frailty is not a state, but a dynamic process that—when a certain threshold is crossed—can result in clinical signs and symptoms to be grouped in a syndrome: The so called 'phenotype of frailty'.

Her concept in part echoes the frailty concept implicit in the Compression of Morbidity paradigm, but its focus is quite different, because whereas the latter conceives of frailty as an end stage of the senescence process to be respected, the former is oriented towards the development of strategies and interventions (such as physical exercise and nutritional interventions) aimed to cure, reverse or slow down the progression of frailty.

As said, epidemiological concepts opt for a multidimensional approach which also take into account psychological and social frailty. Thus, in her recently published report, entitled *The frail elderly*, the Netherlands Institute for Social Research (SCP), conceives of frailty as a heuristic term to identify risk groups. Frailty is broadly defined here as "a process involving the accumulation of physical, psychological and/or social deficits in functioning which increase the risk of adverse health outcomes (functional impairments, admission to an institution, death)" (Van Campen 2011).

Following this definition The Netherlands presently counts approximately 700,000 frail persons (on a total population of plus minus 16,000,000 inhabitants).

The majority of them is living independently in the community. The bad news is that this number will increase with 300,000 in the next decades, but the good news is that this number will be significantly lower than expected earlier. This expected positive effect is almost entirely due to an increase in education level of successive cohorts, as a result of which people progressively live healthier lives. Another piece of support, so it seems, for the Compression of Morbidity paradigm.

The most prominent recommendation of the report is to implement a policy of early detection and intervention to prevent the negative outcomes that are part of the definition of frailty. To this end, monitoring of older persons by means of simple measurement instruments that are also suitable for self-administration is advocated. According to the SCP-report, this activity of monitoring is not only the responsibility of health professionals and municipal officers responsible for the Social Support Act. It is also and primarily a responsibility of the older person herself. A positive score on the frailty index should lead the frail older person to contact her GP and to contemplate, amongst other things, a change in life style, an investment in organizing a social network and a strengthening of ones resilience and capacity for self management. As for the GP: Besides attending to the frail older person's health status, his/her role should be to empower the patient in accomplishing the desired change in life style.

All these recommendations fit in perfectly well with a government policy that focuses strongly on personal responsibility, self management and participation, concepts that are very much 'a la mode' in today's healthcare policy (cf. Chap. 18). They are also in accordance with recent proposals to revise the WHO definition of health in a direction that better fits the challenges of chronic disease. Such proposals move away from the static WHO formulation of health in terms of complete wellbeing towards more dynamic concepts based on the resilience to cope and maintain one's integrity, equilibrium and sense of wellbeing, resulting in a view on health as "the ability to adapt and to self manage" (Huber et al. 2011). It is in this focus on self management and life style approaches that we also recognize the moralistic connotation inherent to the daily life meaning of frailty, where physical vulnerability relates to moral weakness and an instability of mind to live in accordance with generally held norms and standards (such as the norms of healthy aging).

In summary, what we have witnessed in the past decades is the transformation of frailty as a dramatic (service directed) term to focus attention on the special needs of a specific group of care dependent seniors, to frailty as a dramatic (interventionist) term to focus attention on the prevention of care dependency. Frailty is no longer a term for the weak and care dependent elderly; the term has been given a status on its own, separated from impairment, disability and morbidity. It has become a label to identify a particular group of older persons who are at risk for adverse outcomes (Gilleard and Higgs 2010). Indeed: Many interventions developed under the National Care for the Elderly Programme (NCE) funded by the Dutch government, were directed to detecting and influencing this state of frailty. (www.nationaalprogrammaouderenzorg.nl/english/the-national-care-for-the-elderly-programme/)

The question is whether this is a realistic and well balanced approach to aging and its challenges. First of all, there is still no solid evidence for the usefulness of screening for frailty and for the type of interventions advocated by the SCP. Although in modern literature on frailty, frequent reference is made to the theoretical framework of the Compression of Morbidity paradigm to legitimize all kinds of life style interventions and educational programmes, it would be more in accordance with the internal logic of that paradigm to focus more strongly on younger generations and pre-frailty states, where the results will supposedly be far more substantial and the costs more in balance with the gains to expect (Kuh 2007; Vaillant and Mukamol 2001). There is even, as we have seen, some tension between strategies to intervene in the process of (physical) frailty and the Compression of Morbidity paradigm, which foresees an idealised notion of frailty as a not per se unpleasant state, offering the possibility of a good death and hence to be respected rather than opposed. Secondly, the frailty concept itself can be criticized, because it suffers from certain ageist connotations, due to the fact that it is construed as an antonym of successful aging, the latter being conceived of as a way of life that comes close to an imitation and/or continuation of the behavior typical of active and autonomous adults (Kaufman 1994; Chater 2002). The focus on frailty as a state of unspecified risk for negative outcomes tends to reduce the adaptive challenge of frail older persons to a struggle between being independent and self supportive—as characteristics of a good and successful old age—and becoming dependent on others—as a sign of failing self management and an undignified old age. Those deemed frail have to be empowered to take their responsibility and 'swim against the current of time' while they still have a chance to do so, possibly their last one.

In this connection it is important to note, that the whole literature on frailty grossly neglects the perspective of the allegedly 'frail' older persons themselves. 'Frail' is what they become as a consequence of the gaze of others, who view them as balancing on the edge of a downward directed trajectory of ill health and dependency (Markle-Reid and Browne 2003; Grenier 2007). However, research directed at the subjective perspective, although still rather scarce, does seem to indicate that frail older persons generally don't think about themselves in terms of frailty (Becker 1994; Puts 2006). This is not to say that they neglect or deny the functional changes in the body during later life. What it does mean, is that they might struggle more with the emotional and social implications of the change than with the change itself, a struggle that may result in a maladaptive psychological response and thus give rise to what Fillit and Butler (2009) have termed: The frailty identity crisis. In addition, what also complicates their adaptation is the pervasive negative societal value attached to an old age in dependency (Levy et al. 1999).

This brings me to the third and most important criticism. A unilateral focus on risk reduction and postponement of the negative outcomes that are part of the definition of frailty, such as loss of independence and institutionalisation, fails to recognize that for most older people frailty is a transitional stage in life between the third and the fourth age. This transition may be slowed down somewhat, but is better not denied, for a as the proverb says: Forbearance is no acquittance. This calls at least for a double strategy. Instead of unilaterally focussing on preventing and combating frailty, more emphasis should be placed on active anticipation and on assisting people in finding an adaptive response to the implications of frailty. As a transitional stage, frailty implies that the older person must psychologically and emotionally adapt to the (imminent) loss of independence, but not necessarily to the immediacy of death—for surely we are still a long way from the senescence era prophesied by the protagonists of compression! Becoming frail differs however from other developmental life passages, in that this transition is associated with a limited future and a trajectory of progressive dependency (Filler and Butler 2009). This calls for an adjustment of goals in life, but also for appropriate support and medical care tailored to the individual needs and to the consequences of progressive disability. To assist people in finding a successful adaptive response there is a need for a more positive approach towards frailty and old age. For example, realistic but positive role models could operate as a counterpart to the stereotype that the only life worth living is that of continuous vitality and physical robustness (Spielman 1986; Hertogh 2010). The dominance of this model works out as a form of ageism that complicates and intensifies the psychological challenge to find an adequate personal response to a life beyond frailty. In this connection ethics and spirituality might have an important role to play in stipulating ways of a good life in dependency. Especially ethical approaches that focus more strongly on (human) interdependency, relational autonomy and responsivity, such as the ethic of care, offer a promising perspective on what responsible and responsive dependency might mean (Tronto 1993; Walker 1998). Worth mentioning in this respect is also an ethic of the art of living (Foucault 1988; Hadot 1995). Earlier in this chapter I already referred to Cicero who—in his essay on old age—is constantly seeking for good examples and role models that may help us all to anticipate and meet the challenge of high old age: For in order to succeed here, we must learn to emulate the example set by others.

But also the physical aspects of aging in dependency, such as the multiple dimensions of pain, anxiety and other forms of discomfort that may accompany man's final years need to be addressed. It is an impressive and at the same time shameful finding that so many burdensome symptoms remain under recognized and under treated in older persons with complex care problems, while at the same time they are over treated with medicines (notably psychotropic medication, but many other drugs as well) that have barely been tested on safety and effectiveness in their age group, with all related negative consequences such as hospital admission and excess mortality (Bayer and Tadd 2000; Zuidema et al. 2007; Hertogh 2009; Achterberg et al. 2010). To my opinion, adequate symptom relief, combined with restrictive and tailored medical treatment should be an important focus of elderly care medicine and geriatric palliative care. However, just as we all tend to look away from the other site of old age, so do policy makers. Geriatric palliative care is still an under recognized domain of palliative care and a neglected field of medical research. Illustrative in this respect, is that the Dutch NCE, although designed to improve care for elderly people with complex care needs, totally excluded older persons who were, so to say, 'beyond frailty', in other words: Who had crossed the threshold to enter the stage of the fourth age.

8.4 The Fourth Age

The nameless Athenian in Plato's dialogue *Epinomis* warned us not to focus too much on the breathing-space that we are given in the middle of our life, for soon old age will convince us that life isn't worth to be lived again. In today's greying society, that breathing-space has become substantially wider and has come to include a relevant part what we nowadays label as 'later life', a successful life stage relatively free from impairment. As a consequence, all the prejudices with regard to old age now seem to concentrate on the oldest old and the fourth age. The dominant focus on participation, citizenship and self management heightens and accentuates the meaninglessness of a life 'beyond frailty' and also fuels societal debates on the dignity of deep old age and on the right of older people to put an end to their life before they get caught in the trap of dependency. In the Netherlands this has led to a legislative proposal, developed by a social movement called 'the civil initiative on a completed life' and presented to the Dutch parliament, to adapt the laws on voluntary euthanasia accordingly (cf. Chap. 10). Of note is, that the most prominent protagonists of this civil initiative are highly educated representatives of the birth cohort which was once responsible for the youth culture and the student revolts of the 1960s. Holding on to their values of free choice, authenticity and self realization, they vigorously turn their back to the stage of life they are approaching. In addition, also the Royal Dutch Medical Association formulated a policy on physician assisted death in case of suffering related to high age and acknowledged, that the accumulation of multiple deficits and limitations that may occur in old age can be a legitimate ground for physician assisted death (KNMG 2011). Remarkably, these developments are frequently interpreted as expressions of a growing emancipation of senior citizens and hence as an illustration that ageism is loosing ground in our society, while to my opinion there are good reasons to argue that—on the contrary—ageism is more alive than ever. This ageism specifically manifests itself in our attitude towards the representatives of the fourth age and in the care and treatment we offer to them.

It is not easy to delineate and delimitate the fourth age. Roughly sketched, it can be said that the fourth age signals the collapse of 'the third age project'. However, writing from a conceptual and sociological perspective, Gilleard and Higgs (2010) persuasively argue that the fourth age is not simply the end of the third age, nor is it a label to be attached to those who age unsuccessfully. Deep old age, they contend, defies all attempts to define its onset chronologically, just as it is impossible to situate the end of the third age at a more or less precise moment in time. The fourth age more or less functions as the shadow of the third age, just as it was meant by Lasslett (1989). He presented his third age concept as the generalization of an entirely new stage of the life-course against a background of decline and decrepitude that to him and many others is best marginalized to the edges of life. In this sense the fourth age is more or less a product of the third: Both concepts are mutually dependent upon each other when it comes to their meaning and normative implications. Furthermore: Notwithstanding the term, the fourth age concept has less to do with time and chronology, than it has with agency. Again according to Higgs and

Gilleard (2010), it is when people become third persons in the discourse of others, when they are no longer able to self-manage their daily life and become objects of care, that they become candidates for the fourth age. They refer to Charles Taylor's concept of a social imaginary (2004)—the vaguely articulated ways in which people imagine their social existence, the expectations they hold and the deeper notions that underlie these expectations—to conclude that "the fourth age functions (in this way), because it represents not so much a particular cohort or stage of life but as a kind of terminal destination—a location stripped of the social and cultural capital that is most valued and which allows for the articulation of choice, autonomy, self-expression, and pleasure in later life."

If this pessimistic conclusion is correct, it will indeed not be easy, not to say extremely difficult to turn the tide. On the other hand: Imaginaries can be mitigated and transformed once they are brought to consciousness. Thus, they can become a topic for societal debate and procure a source of arguments in favour of positive investments in the future of aging.

In this connection—and before finalising this chapter—it is relevant to also pay attention to more empirical definitions of the fourth age and to findings from empirical research directed towards the outcomes of investments in extending the human life course. The Berlin Aging Study (BASE) is, until now, one of the few multidisciplinary epidemiological studies designed to investigate the questions we have discussed in this chapter: It focused on the pertinence of the distinction between the third and the fourth age script, on the question whether the boundary between them may be dynamic, as is suggested by the Compression of Morbidity paradigm, as well as on identifying the specific health problems of the oldest old (Baltes and Mayer 1999). The study was initiated in 1989 and collected both cross sectional and longitudinal data from a representative (German) sample of both men and women aged 70–100+ years. Besides discipline specific topics, the study was guided by four integrative theoretical orientations: (1) differential aging, (2) continuity versus discontinuity of aging, (3) range and limits of plasticity and reserve capacity, and (4) aging as a systemic phenomenon.

In one of the final papers he co-authored, the director of the study, Paul Baltes, summarized the findings and implications of the BASE study (Baltes and Smith 2003). First of all, this paper acknowledges that the ideas of a third and fourth age, introduced by Neugarten and Laslett, are phenotypic expressions that are themselves subject to evolution and variation. Examination of differences between developing and developed countries for example suggests, that the period of deep old age begins and ends at younger age in developing countries than in developed countries. This suggests the pertinence of a population-based definition of the third and the fourth age. Following this perspective, the transition between the third and the fourth age can be thought of as being the chronological age at which 50 % of a birth cohort is no longer alive. A more differentiated, but still population-based criterion, excludes from the calculation all those persons who died at younger ages and defines the transition as the age at which 50 % of the people who attained age 50–60 have died subsequently. For our highly developed western countries, this differential strategy would put the beginning of the fourth age somewhere close tot 80–85 years and it is the latter definition of the fourth age that was used in the BASE study. In their résumé of the BASE findings and based on these criteria, Baltes and Smith argue, that virtually all the good news from the aging front stems from research with persons and groups which represent the third age. It is here that we find a continual rise in life expectancy of successive cohorts, combined with gains in mental and physical fitness. Research in these age groups also shows evidence for cognitive and emotional reserves, that allow for new learning, successive adaptation and high levels of self-plasticity: with advancing age, the discrepancy between objective medical status and personal evaluations of health increases impressively, but to the effect, that subjective estimates of health hardly differ with age, notwithstanding declines in objective health and increases in morbidity. The question is however, whether this good news—based on studies performed on subjects between 60 and 80 years of age—can be translated to the oldest old and here the answers of the BASE study tend to be negative. When it comes to the oldest old, all the strategies that proved to be successful up until the third age lose their effectiveness and tend to fail. In short: The bad news from the aging front specifically relates to the fourth age, which is characterised by a sizeable loss in cognitive function and learning potential, a loss in self-plasticity and a very high prevalence of dementia, frailty and multimorbidity. With regard to this stage in life, growing old still has its costs: Medically, emotionally, psychologically, as well as economically. According to Baltes and Smith (2003), these data provide serious grounds for reflection. One central issue is that "pushing the limits of aging and its health-related support structures further into advanced old age may actually decrease rather than increase the state of human dignity for many older persons" and hence "the critical question is whether the continuing major investments into extending the lifespan (. . .) actually reduce the opportunity of an increasing number of people to live and die in dignity." The BASE researchers are crystal clear in their conclusion: Instead of pursuing in the direction of further life extension, they make a plea for age fairness in research policy and resource allocation: The aging enterprise is better served with an investment in the young. However, in my opinion, accepting such a policy change would still withhold our oldest old the necessary care and the attention they deserve. Also, such advice runs counter to the present ongoing investments in interventions directed towards reducing and postponing frailty and extending a human life in resilience and vitality. By way of conclusion, I would therefore like to rephrase the question of Baltes and Smith as follows. Indeed the impressive addition of years to people's life creates new challenges for the future of aging and calls for reflection as well as continued investment. But the fundamental question to be answered here is, whether an ongoing investment in strategies to postpone or reduce frailty and to compress morbidity in the next generations is morally acceptable and worth striving for, as long as we don't equally invest in research aimed at an improval of the quality of life for the oldest old, our fellow travellers on the path of life who have passed the threshold of frailty. It all comes down to finally making serious work of the frequently quoted, bur never consequently followed geriatric adage, according to which adding years to life is only acceptable and worthwhile in combination with a policy that equally adds life to years. We cannot dismiss ourselves of our responsibility for the quality of care for the oldest old and for the quality of our own old age, by a continued belief and a narrow minded investment in (the feasibility of future) compression.

References

Achterberg, W.P., G. Gambassi, H. Finne-Soveri, et al. 2010. Pain in European long-term care facilities: cross-national study in Finland, Italy and The Netherlands. *Pain* 148(1): 70–74.

Baltes, P.B., and J. Smith. 2003. New frontiers in the future of aging: from successful aging of the young old to the dilemmas of the fourth age. *Gerontology* 48: 123–35.

Baltes, P.B., and K.U. Mayer., eds. 1999. *The Berlin aging study: Aging from 70–100*. New York: Cambridge University Press.

Bayer, A., and W. Tadd. 2000. Unjustified exclusion of elderly from studies submitted to research ethics committee for approval: descriptive study. *BMJ* 321(7267): 992–993.

Becker, G. 1994. The oldest old: autonomy in the face of frailty. *Journal of Aging Studies* 8(1): 59–76

Bergman, H., L. Ferruci, J. Guralnik, et al. 2007. Frailty: an emerging research and clinical paradigm. Issues and controversies. Journal of Gerontological Medical Science 62(7): 731–737.

Campen, C. Van. 2011. *Kwetsbare ouderen (frail elderly)*. Den Haag: Sociaal Cultureel Planbureau.

Canguilhem, G. 1968. *Études de l'histoire et de philosophie des sciences*. Paris: Vrin.

Chater, K. 2002. Aging: a body of resistance. *Nursing and Health Science* 4(3): 123–129.

Cicero. 2009. *Cato Major, or discourse on old age*. Charleston NC: Bibliolife.

Deeg, D.J.H., and M.T.E. Puts. 2007. Het kwetsbare succes van ouder worden (The vulnerable succes of aging). *Tijdschr Verpleeghuisgeneeskunde* 32: 147–151.

Descartes, R. 1960. *Discourse on method and meditations*. New York: The Liberal Arts Press.

Fillit, H., and R.N. Butler. 2009. The frailty identity crisis. *Journal of the American Geriatrics Society* 57(2): 348–52.

Foucault, M. 1988. *The care of the self.* New York: Vintage books.

Fried, L.P., C.M. Tangen, J. Watson, et al. 2001. Frailty in older adults: evidence for a phenotype. *Journals of Gerontology A Biological Sciences and Medical Sciences* 56A(3): 146–156.

Fries, J.F. 1980. Aging, natural death, and the compression of morbidity. *New England Journal of Medicine* 303(3): 130–135.

Fries, J.F. 2005. The compression of morbidity. *The Milbank Quarterly* 83(4): 801–33.

Fries, J.F. 2002. Reducing disability in older age. *JAMA* 288(24): 3164–3166.

Fries, J.F., B. Bruce, and E. Chakravarty. 2011. Compression of morbidity 1980–2011: a founded review of paradigms and progress. Journal of Aging Research. doi: 10.4061/2011/261702.

Gilleard, C., and P. Higgs. 2010. Aging without agency: theorizing the fourth age. *Aging and Mental Health* 14(2): 121–128.

Gilleard, C., and P. Higgs. 2010. Frailty, disability and old age: A re-appraisal Health. doi: 10.1177/1363459310383595.

Grenier, A. 2007. Constructions of frailty in the English language, care practice and lived experience. *Aging & Society* 27(3): 425–445.

Grimley, Evans J. 1998. Aging. In *Clinical futures*, ed. M. Marinker, and M. Peckham, p. 174. London: BMJ Books.

Hadot, P. 1995. *Philosophy as a way of life*. Oxford: Blackwell.

Hertogh, C.M.P.M. 2009. PREDICT: ontwikkeling van een handvest voor ouderenparticipatie in klinisch geneesmiddelenonderzoek. (PREDICT: Development of a charter for older people in clinical trials.). *Tijdschr Gerontol Geriatr* 40: 180–183.

Hertogh, C.M.P.M. 2010. *De Senectute: Ehtiek en kwetsbaarheid*. Amsterdam: Vrije Universiteit (De Senectute: ethics and vulnaribility. Inaugural Address. Amsterdam: VU University).

Hogan, D.B., C. MacKnight, and H. Bergman. 2003. Models, definitions, and criteria of frailty. Aging Clinical and Experimental Research 15(3): 3–29.

Huber. M, J.A. Knottnerus, L.G. Green, et al. 2011. How should we define Health?. *BMJ*. doi: 10.1136/bmj.d4163.

Jones, D.M., X. Song, and K. Rockwood. 2004. Operationalizing a frailty index from a Standardized Comprehensive Geriatric Assessment. *Journal of the American Geriatrics Society* 52(11): 1929–1933.

Kaufman, S.R. 1994. The social construction of frailty: an anthropological perspective. *Journal of Aging Studies* 8(1): 45–58.

Kirkwood, T.B.L, and M.A. Ritter. 1997. The interface between aging and health in man. *Age and Aging* 26(4): 9-14.

KNMG position paper. 2011. The role of the physician in the voluntary termination of life. Royal Dutch Medical Association (KNMG), Utrecht (www. knmg.nl/voluntary-termination-of-life).

Kuh, D. 2007. A life course approach to healthy aging, frailty and capability. *Journal of Gerontological of Medical Science* 62A(7): 717–721.

Lang, P.O., J.P. Michel, and D. Zekry. 2009. Frailty Syndrome: a transitional state in a dynamic process. *Gerontology* 55(5): 539–549.

Lasslett, P. 1989. *A fresh map of life: the emergence of the third age*. Cambridge: Harvard University Press.

Levy, B., O. Ashman, and I. Dror. 1999. To be or not to be: the effects of aging stereotypes on the will to live. *Omega (Westport)* 40(3): 409–420.

Markle-Reid, M., and G. Browne. 2003. Conceptualizations of frailty in relation of older adults. *Journal of Advanced Nursing* 44(1): 58–68.

Puts, M.T.E. 2006. *Frailty: biological risk factors, negative consequences and quality of life*. VU Amsterdam: Thesis.

Rockwood, K. 2005. What would make a definition of frailty successful?. *Age and Aging* 34(5): 432–434.

Rowe, J.W., and R.L. Kahn. 1998. *Successful aging*. New York: Pantheon Books.

Spielman, B.J. 1986. Rethinking paradigms in geriatric ethics. *Journal of Religion and Health* 25(2): 142–148.

Tallis, R. 2004. *Hippocratic oaths, medicine and its discontents*. Atlantic books: London.

Tavani, C. 1978. *A staff report. Public policy and the frail elderly*. (DHEW publication no. (OHDS) 79–20959). Washington: Department of Health, Education and Welfare.

Taylor, C. 2004. *Modern social imaginaries*. Durham NC: Duke University Press.

Tinetti, M.E., and T. Fried. 2004. The end of the disease era. *American Journal of Medical* 116(3): 179–185.

Tronto, J.C. 1993. *Moral boundaries. A political argument for an ethic of care*. New York: Routledge.

Vaillant, G.E., and K. Mukamol. 2001. Succesful aging. *American Journal of Psychiatry* 158(6): 839–847.

Walker, M.U. 1998. *Moral understandings. A feminist study in ethics*. New York: Roultledge.

Zuidema, S.U., E. Derksen, F.R. Verhey, and R.T. Koopmans. 2007. Prevalence of neuropsychiatric symptoms in a large sample of Dutch nursing home patients with dementia. *International Journal of Geriatric Psychiatry* 22(7): 632–638.

Chapter 9
Former Wishes and Current Desires

Demented Patients and Their Family Members' Effort to Decide What They Would Have Wanted

Dorothea P. Touwen

9.1 Introduction—Representation of Incompetent Patients

By generally getting older, we stand a greater chance of in the last phases of our lives not being able to decide for ourselves anymore, either through dementia or through other diseases that affect cognitive functioning. Not everybody ages successfully, after all. Demographic estimations show that with the increase of life expectancy and of the percentage of elderly people in Western societies, the prevalence of dementia and other age related diseases increases greatly too (Plassman et al. 2007). Since in our Western culture we attach great worth to being able to decide for ourselves, this is a frightening prospect. People fear loosing their identity and tend to want to control what will happen to them and therefore may write down advance wishes. Even when they did not, their loved ones, having to take over the promotion of interest of the person who became decisionally incompetent, will try to do what they think the person himself would have wanted, herewith trying to take over self determination. When having to decide on regular medical treatment for elderly persons, the endeavour to represent the incompetent patient in the decision making process may touch on the complicated matter of weighing the patient's actual interests in his incompetent state versus the ideas and wishes he expressed earlier. This problem concerns society as a whole since we try to design the doctor–patient relationship in a way that does optimum justice to the individual situation and the wishes of the patient. In the Netherlands this is done by giving informed consent a prominent place in the Medical Treatment Contract Act; in other Western countries the informed consent of the patient is regarded as an important necessary condition for medical treatment too. The requirement of informed consent however necessitates a provision for when the patient is incompetent to decide for himself, thus necessitating a procedure of representation or surrogate decision making, asking the surrogate decision maker to try to reach the decision the patient himself would have made, had he been competent.

D. P. Touwen (✉)
Department of Medical Ethics and Health Law, Leiden University Medical Centre,
PO box 9600 postal zone J1-P, 2300 RC Leiden, The Netherlands
e-mail: D.P.Touwen@lumc.nl

M. Schermer, W. Pinxten (eds.), *Ethics, Health Policy and (Anti-) Aging: Mixed Blessings*, Ethics and Health Policy 1, DOI 10.1007/978-94-007-3870-6_9,

Many Western countries promote the drawing up of advance wishes or a living will, hoping to increase the reliability of the surrogate decision making. Advance wishes however turn out to be difficult in their practical application: they lead to the moral problem of the tension between new interests and former wishes, which has important implications for our thinking of patient self determination and the promotion of individual interests.

9.2 Incompetence and Representation

When people suffer from dementia like Alzheimer's disease, generally their capacity to decide is compromised by their diminishing cognitive abilities. In the progressed stages of the disease most patients with dementia are considered incompetent to decide. In most Western countries a proxy or surrogate decision maker will participate on their behalf in the decision making process about treatment and care decisions [from now on I will refer to the surrogate decision maker as 'surrogate']. These persons, often related to and intimately acquainted with the patient, will try to reach the decision that the patient would have made him- or herself [from now on I will refer to the patient as 'she'], would she have been competent. Former decisions, her way of life, expressed wishes, statements concerning others, may all help to make the decision that suits her best. In the thinking about decision making on behalf of someone else, this is called the 'substituted judgment standard' (Buchanan and Brock 1990). However, surrogates are asked not only to try and make the decision the patient would have made, but also to defend the patient's interests. These interests may not always coincide with the patient's formerly declared wishes. A tension may occur between expressions of will before the process of dementia and perceived interests during the disease. This tension is subject of much debate, especially in the light of advance directives, that are meant for just such a situation. How should surrogate decision makers weigh the formerly expressed wishes and the current interests of the demented patients? And what is the role of the healthcare professional in this decision making process? The theme we want to discuss in this chapter is the possible tension between past wishes and actual interests of people suffering from dementia, and the way surrogate decision makers and healthcare professionals should balance these sometimes conflicting values.

Case History 1: Mr Fischer[1]

Mr Fischer is a former scholar in political science who suffers from dementia and is admitted to the psycho-geriatric ward of nursing home Riverside. He is a widower whose two sons visit regularly. Mr Fischer turns out to be a friendly, pleasant man who enjoys the company of his fellows, is cooperative

[1] Both cases derive from empirical qualitative research done by the author in different nursing homes in the Netherlands, in the years 1999–2008.

in the care given to him, likes to attend the classical music meetings in the nursing home and loves his food. He enjoys his meals, particularly the meat products. When alcoholic beverages are served, he likes a glass of wine. His sons are shocked when they hear about their father's preferences. "Our father was always a principled vegetarian!" they proclaim. "He abhorred eating dead animals and strongly disapproved of alcoholic drinks. This behaviour is not appropriate for our father, it is just not like him! We want the nursing home to serve him vegetarian food and fruit juices, no wine." Since this change in policy Mr Fischer takes the meatballs from the plate of his neighbor.

Case History 2: Mrs Gardiner

Mrs Gardiner suffers from severe cardiac dysfunction. After admittance to the long term care ward of the nursing home, she adamantly demands all possible care. It is her conviction God asks her to go through everything possible to prolong her life. Should she defy His will, she will not be admitted to heaven and not see her sister again. Besides, she wants to remain with her beloved grandson as long as possible.

In due time, Mrs Gardiner's condition deteriorates. She doesn't speak anymore, and scarcely reacts when spoken to. When given her medication she turns away and expresses pain. In the past she always extended her arm willingly for the injections, now she retracts her arm and protests the advance of the needle. Her grandson declares to have promised his grandmother never to agree to anything that might shorten her life. But he admits to recognizing her suffering. He confesses not to know what to do.

The ageless ethical question 'what is good for a person?' is, at least in Western society, nowadays mostly answered with 'what the person herself prefers'. People in Western society reserve the right to decide for themselves. Within the confines of societal acceptance and hindrance of others, this is what we agree people are allowed to do. It is even accepted that people decide on important matters for future selves, by formulating advance wishes or writing an advance directive. This appears to take care of possible future dilemmas. Unfortunately, the promotion of interests of incompetent patients is more complicated than the execution of formerly expressed wishes. Our research in Dutch nursing homes (Touwen 2008) shows many examples of changes in preference, shifts in interests and interpretational difficulties, of which Mr Fischer and Mrs Gardiner are two. Let us look at their respective situations.

Mr Fischer's behaviour, now that he is demented, shows a contrast with his principles in his former life. With the diminishing of his cognitive abilities his capacity to adhere to his principled stances seems to have eroded. His actions are guided more by his impulses than by his rational decision how to lead a good life. That does not mean that his former ideas have become irrelevant, nor that he is incompetent to have preferences that are of value to him. However, his preferences now seems to be more of an impulse, whereas his principled vegetarianism was well thought out and strongly argued, as his sons testify. The caregivers are confronted with a dilemma: should they help him adhere to his former values by prohibiting him from eating meat or drinking alcohol? Or should they promote his physical wellbeing by offering him the food he likes best?

Mrs Gardiner was absolutely and eloquently clear about the way she wanted to be treated. In several conversations with the nursing home physician she stated not to be deterred by the possibility of suffering at the end of her life. This she considered her moral and religious duty. Even when de nursing home physician pointed out the limitations of modern medicine, she remained adamant. Now this time has come, and Mrs Gardiner suffers great pain and discomfort. Although not competent to motivate her actions verbally, her behaviour indicates refusal of treatment. The nursing home physician, who has repeatedly told Mrs Gardiner she would not feel obliged to give treatment she in her professional view would consider futile, is nevertheless in doubt. Should she try once more or should she leave Mrs Gardiner to die in peace? Is she allowed to give palliative care in order to relieve Mrs Gardiner's suffering, even if this means she might die a little sooner? Mrs Gardiner's grandson faces a comparable dilemma. His grandmother's wishes are clear, but would he promote her interests when pressuring the nursing home physician into continuing the treatment? What is good for his grandmother: To have her former wishes adhered to, or to be made comfortable for the last few hours of her life?

9.2.1 Perspectives

In both the case of Mr Fischer and the case of Mrs Gardiner we see an incongruence between the wishes the patient expressed when still able to articulate them, and the expression of what we cannot know for certain is a will by the otherwise not verbally articulate patient. The question is which expressions of will should determine what is best to be done: The ones expressed formerly, when anticipating on a new hitherto unknown situation; or the ones now, possibly less competently argued but based on the experience of the new situation. This question is of daily relevance because the perspectives of the different parties involved greatly influence their vision on what is best to be done. Healthcare professionals have to focus their efforts on the actual person in care: The patient who presents herself to them with her particular needs, her particular diagnosis and prognosis, her particular experiences. Since healthcare professionals most often haven't known their patients in their former, competent lives, the professional view is primarily a-historic, concerned with the actual situation. Furthermore the view of a healthcare professional is influenced by his professional experience: When working with sick people all the time, one's norms of what is acceptable tend to adapt. People working in a nursing home understandably view the situation of most residents as fairly acceptable, while an inexperienced bystander may be shocked. This difference in perspective is often a source of misunderstanding in the dealings between healthcare professionals and family members of nursing home patients.

If the professional perspective of healthcare professionals is coloured by their experience, the view of family members is coloured through their distress of seeing their beloved relative suffering from dementia and becoming care dependent. For family members who have a long term intimate relationship with the demented elderly person, the dementia is a process of loss and continuous adaptation to a new state of being. This reflects necessarily on the relationship: roles change, responsibilities shift, and contrasts are felt between how it was and how it is now. Accordingly the view of a relative is intrinsically historic, and thus differing from the professional view. What's more, this historic perspective is enforced by the role assigned to a family member who acts as surrogate decision maker (Livingston et al. 2010). A surrogate decision maker is asked to promote the interests of the incompetent patient and to do this by trying to reach the decision that the patient herself would have made, had she been competent to do so. We call this the substituted judgment: to try to decide according to the patient's former opinions, trying to come as close as possible to the choice she would have made herself. In this way surrogate decision making is a form of substitute self-determination. The surrogate decision maker has to try to think back of what he knows about the incompetent patient's ideas, opinions, and wishes. He will have to think back also of the way the patient lived, the life choices she made, the values and norms that were important to her. This is the assignment one gets when asked to act as surrogate decision maker. A more explicit historical and biographically based perspective is scarcely possible. So there we are: Two parties try to reach a decision on what is in the interest of the patient, and their perspectives diverge fundamentally.

9.3 Interests

In the process of trying to decide what is best for the demented patient, family members and healthcare professionals have to weigh the interests of the patient. These interests lie in the present, demented state, but, as we saw when sketching the historical perspective of the surrogate decision maker, they are also grounded in the way the patient lived her life before her dementia started. Most people have ideas about what they would want for themselves in the future, and it is these ideas that the surrogate decision maker will try to draw on when trying to find the answer that the by now demented patient would have given herself. However, these ideas may not always comply with what seems to promote the wellbeing of the patient in her new state, as the case of Mrs Gardiner shows. She wanted to be treated maximally while this turns out to increase her suffering. So what defines a demented person's interests, her former ideas on how she wants to live and finish her life, or what bystanders observe now? To clarify this fundamental tension Ronald Dworkin's distinction between critical and experiential interests may help (1994).

Critical interests concern the more fundamental choices a person makes concerning the way he wants to live his life and how he wants to be remembered. These choices relate to the things a person considers crucial in his life and his way of living. A person's experiential interests concern those things that make life pleasurable: The things a person values because of his personal preferences, his favourite food, favourite music, favourite holiday destinations (Dworkin 1994, pp. 201–202). In Dworkin's way of thinking the critical interests rank higher on the scale of importance than the experiential interests. It comes as no surprise that when a person turns incompetent, Dworkin considers the critical interests to prevail over the experiential interests. The decisional incompetence implies the absence of capacity to formulate (new) critical interests. Therefore a person's critical interests, developed in his former competent life, continue constituting his most important interests in his new state of being. For Mr Fischer this means, in Dworkin's view, that his experiential interest to enjoy food containing meat should be overridden by his critical interest to avoid eating dead animals. Mr Fischer's obvious enjoyment of meat products may point to an experiential interest (don't we agree that one of the objectives in care for demented patients is having positive or pleasurable experiences?) but does no justice to the person he was. Understandably the sons of Mr Fischer, who wish to represent their father and defend his former self, are upset by this change of heart. They do not trust Mr Fischer's behaviour to be a valuable new choice, but attribute it to his illness.

Mrs Gardiner shows behaviour that does not fit her formerly declared wishes. Her physical expressions of protest to having her medication injected, conflict with her former wish to be treated maximally. In her case it is not a mental illness that causes her change in attitude, but the deterioration of her physical state. Being maximally treated turns out to lead to uncomfortable, even painful treatment. Her grandson struggles with the incompatibility of her former wishes and the now shown protest to being treated. The question he faces is what is more important: to force Mrs Gardiner to receive the treatment she demanded in the past, or to avoid additional suffering in these (probably) last days of her life? Adhering to her professed beliefs would do justice to her critical interests, if we follow Dworkin to the letter. Mrs Gardiner has adamantly chosen a way of dying in which life is prolonged as long as possible, even though this end might be painful or full of suffering. We cannot be sure that her current expressions of resistance are the result of a change of heart. Since Mrs Gardiner's mental state is clouded by her grave illness, some might argue her behaviour should be ignored as being less competently reached and therefore, following Dworkin, of a lesser significance than her well-considered wish to be treated maximally.

9.4 Identity

When family members try to represent their loved one who has become incompetent to make decisions for him, they have to find a way to apply what they know of the patient from before, to the new situation. A patient suffering from dementia undergoes a process of change that sometimes leads to the painful question whether the state of the demented patient is congruent with who the patient was. "This is not my mother anymore" is an often heard outcry. It seems as if the process of dementia by destroying mental functions erases or fundamentally changes identity, at least in the eyes of those who knew the person with dementia before (DeGrazia 2005). In the life of Mr Fischer a great change has taken place. From being a man who had his own firm ideas on how to live his life, he has changed into a man who is dependent on care, who is less explicit in his motivation of his actions and who functions on a profoundly different cognitive level. The contrast between the university professor and the man in his geriatric wheelchair is clear. So what is the link between Mr Fischer then (Mr Fischer 1) and Mr Fischer now (Mr Fischer 2), apart from his name and his biography? Many philosophers have addressed this tricky question, as David DeGrazia shows in his book *Human Identity and Bioethics* (2005). For our discussion of the problems healthcare professionals and family members face when trying to find a balance between former wishes and actual expressions of a demented elderly patient, it suffices to say that the biographical continuity of Mr Fischer and Mrs Gardiner is convincing enough to conclude that we cannot solve the problem of conflicting interests by stating they have changed into a new person, and that therefore their former wishes do not apply anymore. Even though relatives may have the sad experience that their loved one has changed to a seemingly other person, we must note that identity is a fluctuating thing in any person's life (just like at 44 we are not the same person as we were when 16). Sharing name and biography with oneself at different stages of life leads to a relevant continuation of values and connections throughout that life. However, at different stages of that same life we may have changing interests.

9.4.1 Shift

The problem how to weigh wishes expressed in the past in comparison to the new situation, is central to the ethics of surrogate decision making, and even more so when the patient made his wishes explicit in the past. There is a difference between trying to reach the right decision given the situation (best interest), and trying to decide what the patient would have wanted (substituted judgment). But there are even greater problems once the former wishes of the person involved are clear but seem to lead to an undesirable situation, as the case of Mrs Gardiner shows. Advance wishes or living wills are meant to make things simpler (it is clear what the person wanted at the time the living will was written), but prove to complicate matters. For in the new situation the interests of the patient may shift; the situation may turn out to be different from what the patient envisioned; or the consequences of the advance wishes may be harmful to the patient in the new situation. It is this change and subsequent shift in interests that constitutes the greatest hurdle in the debate on advance wishes and living wills.

Rebecca Dresser is one of the most adamant critics of the feasibility of a living will (Dresser 1994, 2003, 2005). She argues that it is impossible to decide for yourself beforehand because circumstances change and, most importantly, interests change. The demented patient is not any longer exactly the same person as she was, and therefore there is no self-evident continuity in interests. It is the signs we gather from the demented patient that should lead our decision making. As long as the incompetent patient has pleasurable experiences, we cannot let her former conviction that to be demented is going to be horrible, decide what is to be done. Mr Fischer should be allowed his meatballs, Dresser would argue. In this matter it is obvious she disagrees with Dworkin (Dresser 2005). As Dresser and others point out, in dementia the experiential and critical interests grow towards each other and become less distinct; it may very well be that in dementia the former critical interests are of less value and significance since the experiential interests (the pleasurable experiences) grow in significance (Fagerlin and Schneider 2004). The strict hierarchy Dworkin designates between critical and experiential interests, diminishes (Hertogh et al. 2007). Jaworska adds to this by stating that demented people retain their capacity to value, and therefore retain their capacity to form new critical interests (Jaworska 1999). Even though one's thought processes may take place at a different level, the capacity to enjoy things or to find experiences or circumstances important, has not necessarily faded. Through the fact that not one demented person is the same we can see that individual preferences still have their influences (Hertogh 2009). Should we follow this train of thought, we have to conclude that Mr Fischer's predilection for meat signifies a new value, namely to eat delicious food, which from now on takes precedence over his former objection to eating dead animals.

It is questionable however whether this reasoning takes away Mr Fischer's sons' concern that, since their father has lost the ability to restrain his more basic urges, it is up to them to defend his principles. They will probably not deny the importance of pleasurable experiences, but might very well reason that by prohibiting him to eat meat they will do justice to what their father always considered important. By doing this they not only fulfil their job as surrogate decision makers (decide what their father would have decided) but also make sure that in his illness he does not transgress norms that were of vital importance to him. The importance of not eating meat may not convince everybody (as is shown by the widespread consumption of meat worldwide), but to Mr Fischer it was very important, argue his sons. Possibly the weight of Mr Fischer's vegetarianism may increase when we compare Mr Fischer's vegetarianism to certain life rules of religious nature. Should nursing home staff respect the norm that a good Muslim must not eat pork? Of course the situation with Mr Fischer would have been easier in such a case, since his desire to eat meat could be satisfied by offering him beef dishes. But the point is obvious: What may seem an unreasonable request by Mr Fischer's sons, gains weight when supported by known religious dogmas. Apparently it is of influence whether or not the request is based on generally accepted norms. In fact, we might be tempted to plead in favour of allowing Mr Fischer his meat simply by the fact that we don't think his vegetarian principles worth it. It might be our own normativity that leads us to think: "Oh well, let the poor man enjoy his meatballs!" The situation would probably be much different when, to use yet another example, the course of Mr Fischer's illness would lead him to uncover his body and make sexual advances. Then the arguments in favour of allowing him his pleasures would be considered far less strong. One could say that sexual lack of restrain is less acceptable because it might cause harm to others, but a principled vegetarian might answer that eating meat harms others too. I am sure it is the conflict with generally accepted norms that makes the case of sexual transgressions so much more clear. So we should be careful not to regard fundamental stances that do not easily comply with our own norms and values as being less important. Mr Fischer's sons have a good point in wanting the nursing home staff to respect his former convictions. Rather than offering Mr Fischer meatballs they should try to offer him appetizing vegetarian food, and in doing so help him live his demented life like he would have done had he been competent.

9.5 New Circumstances

This leads us to another question however. Even though we like to think of ourselves as having firm ideas about what is important and what we would want for ourselves, in reality people often change their minds. People adjust to changing circumstances in ways they did not foresee. Being more flexible than thought beforehand, people often find themselves to be able to bear more suffering than they had expected to (Ubel et al. 2005). They turn out to be able to live in circumstances that they beforehand judged to be unacceptable, and overestimate the negative impact on their quality of life of illness and disability (the so-called disability paradox, Albrecht and DeVlieger 1999; Kutner et al. 2003). So is changing one's mind the prerogative of competent people or should we allow incompetent people to do so too? The trouble is that we cannot ask them whether they have changed their minds. We can try to read their reactions and expressions, but this needs hefty interpretation, which will be influenced by our perspective of either family member or healthcare professional.

In his thinking of experiential and critical interests Dworkin proposes to solve this problem by grounding the critical interests in a lifetime. Since life is really a narrative, a more coherent narrative is a better narrative, and so a more coherent life is a better life (Dworkin 1994, p. 210). Critical interests concern those things that we recognize as being representative of our lives, of the way we want to be remembered (Dworkin 1994, p. 230). These are not things one lightly changes one's opinion about, he states (Dworkin 1994, p. 202). Others fight this conclusion. From empirical evidence it is clear that people do adjust to new circumstances and often accept situations they formerly abhorred. This can concern things they felt really strongly about. Apparently it is difficult to judge beforehand what it will be like to, for example, sit in a wheelchair or to be dependent on others. People 'make do', they cope and go on living, even when their life is changed fundamentally. So the tension between the former wishes and current desires is not easily solved by making a distinction between different kinds of interests, ones that don't change and ones that do. People do change their minds concerning their critical interests, and so we have to consider the possibility that in dementia people may do so too. Former critical interests may stop being applicable, or at least we cannot rule out the possibility. If we cannot be sure people with dementia don't adjust, then we cannot be sure either that their former ideas must be regarded as still firmly applicable.

9.6 Authenticity

The debate on the relation between former wishes and actual experiences is ongoing. The general consensus is that for a demented person actual experiences of pleasure are of importance, but that to continue living according one's most felt values and principles is important too. How to weigh these possibly opposing values is no easy matter. The Nuffield Council (2009) proposes to strive for a sensitive balance in every individual case. In their lengthy report on ethical issues concerning dementia even this body of wise people turns out to be unable to clinch the matter: nor within the council nor in their public deliberative event a firm decision could be reached to favour the one stance over the other. In their concluding remarks the following recommendation shows the ongoing basic ambivalence:

> We recommend that the mental capacity Codes of Practice should be amended to provide additional guidance on how past and present wishes and preferences should be taken into account where these appear to conflict. This guidance should emphasise that neither past nor present can automatically take precedence, but that the relative strength of the person's wishes, the degree of importance of the decision, and the amount of distress being caused should all be important factors to consider. (Nuffield Council 2009, p. 83)

Interestingly the Nuffield Council concludes that the weight attached to former wishes should increase when the patient used to be adamant about his ideas. However intuitively right, this contrasts with Dworkin's claim on critical interests that may not be argued explicitly:

> (. . .) even people whose lives feel unplanned are nevertheless often guided by a sense of the general style of life they think appropriate, of what choices strike them as not only good at the moment but in character of them. (Dworkin 1994, p. 202)

The Nuffield Council's recommendation might be based however on the fact that the more explicit a person's wishes were, the easier it will be to define his past wishes and their relevance to the new situation. Daniel Brudney further explores this theme in his analysis of the surrogate decision making process (2009). A surrogate decision maker tries to decide what the demented person himself would have wanted. This process contains a fundamental ambivalence, Brudney claims. Only when the patient in her former competent life, was explicit and clear in her wishes concerning a specified situation, the role of the surrogate decision maker can be characterized as a form of delegated self determination: only then the surrogate can decide what the demented person herself wants. As soon as interpretation is needed or the former wishes were not clear or to the point, the decision making process changes from a form of (delegated) self determination to an act of trying to decide on behalf of the other person, making use of the value of authenticity. In such cases we try to decide what would suit the patient best, what would be in line of the way she lived. It is,

Brudney states, a misconception to regard this as a question of self determination. The necessity to interpret and to apply what is known of the patient's preferences and values change the character of the decision, moving from an application of autonomy or self determination to the value of authenticity. All in all Brudney's analysis complies with the conclusion of the Nuffield Council, that the degree of explicitness of the expressed former will is of influence on the weighing of the different interests involved, when deciding for a demented patient.

9.7 Back to Mr Fischer and Mrs Gardiner

So how do these conclusions help us in the two cases of Mr Fischer and Mrs Gardiner? We are in no doubt what Mr Fischer would have wanted, had he been asked beforehand what would have been the right reaction to his sudden wish to eat meat. The trouble remains that we are not able to distinguish which is the wish we should abide with: The one from the past or the one expressed now. I have argued elsewhere that the most weighty assignment for care givers is to strive after the wellbeing of the demented person, while at the same time trying to do justice to his former self (Touwen 2008). This means that complying with former wishes should be aimed for as long as no additional harm is caused. This complies with the conclusions of the Nuffield Council, who state that if maintaining a past belief is causing major distress, then it is likely that the person's current wellbeing and not their previous autonomy interests should take precedence (Nuffield Council 2009, p. 83). In the case of Mr Fischer this means that appetizing vegetarian dishes should be offered to him, in order to help him live by the standards he lived by in previous parts of his life. There is no harm done if we limit his meat- or alcohol consumption, as long as it is not necessary that he dines in isolation to prevent him from stealing the meatballs of his neighbour.

In the case of Mrs Gardiner it might mean that her wish to be treated maximally is replaced by her interest in not being caused pain in the administration of her medication. One could argue, as predicted by her nursing home physician, that it leads to questionable medicine to adhere to her former wishes. Even if she consciously chose to take the risk to suffer pain and discomfort when she asked to be treated maximally, the behaviour Mrs Gardiner shows now, gives a clear indication to restrict treatment. It looks as if she has changed her mind now that she experiences what she has chosen. Obviously dying, she shows bodily protest to the administration of her medication. Overriding her protest only aggravates her actual suffering. The argument that treating her would do justice to the person she was, falls short when taking into consideration the ethical obligation to do good and avoid suffering.

Medical intervention always needs explicit justification. Not the sheer possibility to do it, but the careful consideration that for this patient in this situation at this moment, this treatment will probably or possibly benefit the patient, justifies medical intervention. Suffering caused by treatment should always be compensated by (the possibility of) benefits. It is an important belief in medical ethics that for any treatment to be justified, not only the consent of the patient is needed but also the conviction of the physician that he does what is considered to be good medicine. This means that not just Mrs Gardiner's wishes (or her grandson's) but also the assessment by the physician of the merits and consequences of ongoing treatment is decisive concerning the treatment of Mrs Gardiner. Considering the fact that she is dying and further treatment would not substantially prolong her life, and added to this the physical suffering she shows in combination with her bodily resistance, the nursing home physician follows medical professional standard when she does not force her patient to be treated. The nursing home physician should not feel forced to carry out medical treatment she professionally considers maltreatment.[2]

9.8 Concluding Remarks

In accordance with the strong emphasis in Western society on self-determination and autonomy, people often have the conviction that it is necessary to keep control over their future selves, in this way attempting to preserve a sense of identity. The generally accepted desirability of the availability of a written advance directive, does not seem to lead to a *de facto* increase in the number of advance directives (De Boer et al. 2010), not even in countries with legislation to stimulate people to write advance directives, like the United States (the so called Federal Patient Self Determination Act, 1999) (Fagerlin and Schneider 2004). However, this may still be a generational phenomenon that may change when the need of a discussion about one's wishes for the future becomes more widely known. On the whole, people often express the desire to influence their fate. The assignment of surrogate decision makers to reach the decision we would have made ourselves, corroborates this idea. However, it is clear that it is no simple matter to transpose wishes of a competent person to a new situation in which she is not competent to decide anymore. There is no reason to assume that demented people have no capability to have wishes and opinions anymore; nor does dementia mean the absence of interests in as good a life as possible (Schermer 2003). So a competition may ensue between the old and the new interests. The question which of the two should win, is unanswerable in general terms. The shift in interests is too big, and it is impossible to predict how we will react to the new state of being. Considering the importance of identity and authenticity we must respect the old values and wishes, while at the same time striving after wellbeing and pleasurable experiences in the new situation. This means that within boundaries the values and norms according to which a person lived her life, should prevail in het demented sate, but not if it harms her new interests. Mr Fischer should be tempted to eat tasty vegetarian food, unless he consistently expresses dissatisfaction and refuses to eat. In the medical care for Mrs Gardiner the nursing home physician should try to find a balance between palliative care, including pain relief, and the

---

[2] The issue of the professional responsibility and the relation between the physician's judgment and the wishes of the patient, exceeds the scope of this paper, and is extensively discussed elsewhere (e.g. Touwen 2010).

continuation of other life sustaining treatment, as long as she does not unnecessarily add to Mrs Gardiner's suffering. If dementia does not mean the loss of identity but the continuation of a fluent identity, we should try to help the demented person to keep hold of that identity, by defending her values and norms as much as is possible within the confines of the avoidance of increased suffering.

The implications of these conclusions for healthcare policy are not evident. Advance directives do not offer the solution, as we saw, and might even complicate matters. At the same time, an advance directive will give both surrogate decision maker and healthcare professional a firm indication of what the patient considered important in life. So as long as it is clear the advance directive cannot strictly determine what will be done, it has its uses. Another way to feel in safe hands will be to have a timely discussion on wishes and values with loved ones. The more that is known about a person's wishes, the better equipped family members will be to weigh pros and cons of treatment alternatives in dementia care. Policy makers might consider promoting that people appoint their own favoured representative (healthcare power of attorney)—that is: To write an advance directive not concerning what one wants to be decided, but who one wants to be the principal representative and promoter of one's future interests. When one fears the loss of self, the knowledge to be in good and trusted hands may help, especially if this person is intimately acquainted with one's wishes and values and one's way of life (Shalowitz 2006). Trusting that this person will do everything possible to decide what is best to be done considering both former wishes and current interests, may help people to accept the less pleasant consequences of living to an advanced age.

References

Albrecht, G.L., and P.J. Devlieger. 1999. The disability paradox: high quality of life against all odds. *Social Science & Medicine* 48(8): 977–988.

Boer, M.E. de, C.M.P.H. Hertogh, R.M. Dröes, C. Jonker, and J.A. Eefsting. 2010. Advance directives in dementia: issues of validity and effectiveness. *International Psychogeriatrics* 22(2): 201–208.

Brudney, D. 2009. Choosing for another. Beyond autonomy and best interests. *Hastings Center Report* 39(2): 31–37.

Buchanan, A.E., and D.W. Brock. 1990. *Deciding for others. The ethics of surrogate decision making*. Cambridge/New York/Melbourne: Cambridge University Press.

DeGrazia, D. 2005. *Human Identity and Bioethics*. Cambridge/New York: Cambridge University Press.

Dresser, R. 1994. Missing persons: Legal perceptions of incompetent patients. *Rutgers Law Review* 46(2): 609–719.

Dresser, R. 2003. Standards for family decisions: Replacing best interests with harm prevention. *American Journal of Bioethics* 3(2): 54–55.

Dresser, R. 2005. Dworkin on dementia. Elegant theory, questionable policy. *Hastings Center Report* 25(6): 32–38.

Dworkin, R. 1994. *Life's dominion. An argument about abortion, euthanasia, and individual freedom*. New York: Vintage.

Fagerlin, A., and C.E. Schneider. 2004. Enough: The failure of the Living Will. *Hastings Center Report* 34(2): 30–42.

Hertogh, C.M.P.M. 2006. Advance care planning and the relevance of a palliative care approach in dementia. *Age and Aging* 35(6): 553–555.

Hertogh, C.M.P.H. 2009. The role of advance euthanasia directives as an aid to communication and shared decision-making in dementia. *Journal of Medical Ethics* 35: 100–103.

Hertogh, C.M.P.H., M.E. de Boer, R. Dröes, and J.A. Eefsting. 2007. Would we rather lose our life than lose our self? Lessons from the Dutch debate on euthanasia for patients with dementia. *The American Journal of Bioethics* 7(4): 48–56.

Jaworska, A. 1999. Respecting the margins of agency: Alzheimer's patients and the capacity to value. *Philosophy and Public Affairs* 28(2): 105–138.

Kutner, J.S., D.E. Nowels, C.T. Kassner, J. Houser, L.L. Bryant, and D.S. Main. 2003. Confirmation of the "disability paradox" among hospice patients: Preservation of quality of life despite physical ailments and psychosocial concerns. *Palliative and Supportive Care* 1(3): 231–237.

Livingston, G., G. Leavy, M. Manela, D. Linvingston, G. Rait, E. Sampson, S. Bavishi, K. Shahriyarolki, and C. Cooper. (2010). Making decisions for people with dementia who lack capacity: qualitative study of family carers in UK. *British Medical Journal* 341: c4148

Nuffield Council on Bioethics (2009) *Dementia.* Ethical issues.

Plassman, B.L., K.M. Langa, G.G. Fisher, S.G. Heeringa, D.R. Weir, M.B. Ofstedal, J.R. Burke, M.D. Hurd, G.G. Potter, W.L. Rodgers, D.C. Steffens, R.J. Willis, and R.B. Wallace. (2007). Prevalence of dementia in the United States; The aging, demographics, and memory study. *Neuroepidemiology* 29(1–2): 125–132.

Schermer, M. 2003. In search of 'the good life' for demented elderly. *Medicine, Healthcare and Philosophy* 6(1): 35–44.

Shalowitz, D.I., E. Garrett-Mayer, and D. Wendler. 2006. The accuracy of Surrogate Decision Makers. *Archives of Internal Medicine* 112: 493–497.

Touwen, D.P. 2008. *Voor een ander. Beslissingsverantwoordelijkheden in de verpleeghuisgeneeskunde.* Amsterdam: Aksant.

Touwen, D.P. 2010. *Tijdschrift voor Geriatrie en Gerontologie.*

Ubel, P.A., G. Loewenstein, N. Schwarz, and D. Smith. 2005. Mismanaging the unimaginable: The disability paradox and healthcare decision making. *Health Psychology* 24(4): 57–62.

Chapter 10
Death Wishes of the Elderly: Is There a Task for Doctors?

Govert den Hartogh

> Spring, summer and autumn
> the happy poet enjoyed.
> Finally against the winter
> The mound protected him.
> (Goethe, Anakreon's Grab, cf. song by Hugo Wolf)

10.1 Introduction

Mr. Kalisvaart is almost 90 years old. He has been a very independent and active man all his life. Since 15 years he suffers from glaucoma in both his eyes, and by now he is almost completely blind. His hearing has also much declined in recent years. He is therefore unable to listen to classical music which used to be his passion. Moreover, as a result of disc degeneration he has serious back pain. The fact that he is sitting in his chair all day, waiting for nothing, causes him a lot of additional suffering. Only with the accompaniment of staff members of the nursing home where he is living he is able to move a bit around with a walker. In doing so he has fallen down a few times. Being dependent on others, knowing that his situation will not improve anymore, increases his suffering. For these reasons he has asked his physician to provide him with the means for committing suicide, and he repeats this request every time the physician visits him.

This is not an uncommon situation for people in very high age. Nor is it a recent phenomenon, the product of growing prosperity, medical progress and the resulting lengthening of the average lifespan. As we read in Jesus Sirach.

> O death, how welcome is your sentence
> to one who is in need and is failing in strength,
> very old and distracted over everything;
> to one who is contrary, and has lost his patience![1]

[1] Ecclesiasticus 41: 3, King James' translation, cf. one of Brahms' Vier ernste Lieder.

G. den Hartogh (✉)
University of Amsterdam, The Netherlands
e-mail: G.A.denHartogh@uva.nl

M. Schermer, W. Pinxten (eds.), *Ethics, Health Policy and (Anti-) Aging: Mixed Blessings*, 119
Ethics and Health Policy 1, DOI 10.1007/978-94-007-3870-6_10,

It is a fact of all times that for some people at high age death becomes a friend.

We are talking about explicit and permanent death wishes of old people who don't have any life-threatening illness and do not suffer severely from pain, breathlessness or other physical symptoms. In some cases the death wish is the result of a psychiatric disorder which the person already had earlier in life, or of traumatic experiences in that period, sometimes even in childhood (Rurup et al. 2011). That is a different phenomenon which merits a separate discussion. I want to focus on death wishes which are clearly the product of aging. In the Netherlands such death wishes are by now commonly referred to as cases of a completed life. I don't like that euphemism: The notion of completion suggests a nice rounding off of a certain pattern in one's life, but the burdens of old age will more often cause an unwelcome cutting off of that pattern. Sometimes a loss of interest in life gradually develops through the years, but even that condition is better described by detachment than by completion. And neither completion nor detachment requires that one's life takes an end: What one is able to release without regret need not be a burden which one can bear no longer. Because I need a name for the class of cases I will discuss, I will normally refer to them as tired-of-life cases.

How often do old people long for death? Research in several European countries consistently gives us percentages of 15–20 % of the older population. But this percentage concerns people who occasionally have or have had death wishes, they need neither be fully explicit nor permanent ones. Recent research by Mette Rurup provides us with more information (Rurup et al. 2010). 3,4 % of Dutch people beyond the age of 58 report to have had death wishes during the last week before being interviewed and to have no wish or only a weak wish to live. 0.7 % report to have a strong desire to be dead at the moment of the interview. (This group includes people with chronic psychiatric disorders.) We do not really know how many of these people actually plan to end their lives, and start acting on those plans. One Italian study gives a percentage of 0.14 % (Scocco and de Leo 2002), and this corresponds to the number of people who reported to Rurup a strong wish to die and not even a weak wish to live. The number of cases in which Dutch doctors annually receive requests for assistance in ending a person's life in the absence of so-called severe diseases has been estimated as 425 (Rurup 2005, Chap. 2), but this number is based on reports from doctors, and we know from other sources that they often do not recognize or remember a request which the patient clearly intended as such. So the actual number may be much higher (Van de Vathorst et al. 2011).[2]

Should doctors be allowed to grant such requests? And if this is not considered to be a proper medical task, should the law make it possible for old people to receive non-medical assistance?

These questions are at present intensively debated in the Netherlands because of a law proposed by a Citizen's Initiative which aims to create for elderly people a non-medical route to the self-chosen death in addition to the medical route available since

[2] An interesting result of this study is that doctors sometimes do not recognize a request as such because they don't consider it really intelligible in the circumstances. This should make us wonder whether the assessment of the competence of the requester and the evaluation of his request can be as independent of each other as the dominant view of competence in medical ethics holds.

euthanasia and physician-assisted suicide became legally recognized options in 1984. I will describe the context of this debate in Sect. 2. It could be questioned whether the special character of death wishes of the elderly is relevant to this debate. It would not be relevant, if it would be permissible for all competent people requesting assistance in ending their lives to receive it. Therefore I argue in Sect. 3 that such assistance cannot be justified by a mere appeal to the principle of respect for autonomy. The Dutch euthanasia law does not only make requirements on people's request (that it should be voluntary and well-considered) but also on their condition (that they should be in a state of unbearable suffering without any prospect of improvement) and the conclusion of my argument in this section is that it is justifiable for the law to make these additional requirements. The law, perhaps less justifiably, also requires this unbearable suffering to be primarily caused by a medically classifiable disease or ailment. In Sect. 4 I therefore discuss whether a request like Mr Kalisvaart's can satisfy these requirements considering both the nature and cause of the suffering which leads elderly people to such requests. My conclusion will be that it can. However, not all requests will satisfy the requirements, and even if they do many doctors will not be prepared to grant them. Therefore the question is still relevant whether assistance by non-doctors should be legally allowed. In Sect. 5 I will argue that there is no reason to broaden the exception to the legal prohibition of assisted suicide in this way because people, and in particular elderly people, have sufficient opportunities to end their own lives in a carefully prepared and humane way.

The debate about such issues is to a large extent dependent on local circumstances: The state of the law, political feasibilities, the principal fears of the public. For that reason I have chosen to discuss these questions against the background of the present Dutch debate. But I hope that my arguments and conclusions will have a wider relevance.

10.2 The Issue in the Dutch Context

In the Netherlands the discussion about these issues started when the former vice-president of the Supreme Court, Huib Drion, wrote a short piece in a newspaper in 1991, under the title *The self-chosen end of elderly people* (Drion 1992). In this article he advocated to give people beyond the age of 75 the legal right to buy or receive lethal drugs. On his view many old people would wish to have access to such drugs, but not in order to use them immediately. Rather, the mere possession of them would give those people the assurance they wanted that they could use them, if necessary. This proposal is usually referred to by the name of 'Drion's pill'.

When Drion wrote his article, euthanasia and physician-assisted suicide had already been legalized in the Netherlands by judge-made law. Ten years later the legislator took over and codified the law in the shape it already had taken, with only minor changes (2002). A doctor can grant a request for euthanasia or assistance in suicide, if a number of substantial and procedural requirements of due care have been satisfied. The substantial requirements include that the patient's request should be voluntary and well-considered, and that the patient should be in a state of unbearable suffering without any prospect of improvement. The procedural requirements include that another physician should be consulted who visits the patient and gives a written opinion about the extent to which the substantial requirements have been met, and that the doctor should report his action to one of five regional review committees, each consisting of a lawyer, a doctor and a ethicist.[3] The committee assesses in each case whether the doctor has satisfied the requirements of due care.

None of these requirements seems to exclude the request of a person who is tired of life from being granted. In particular the law does not require, as the Oregon Death with Dignity Act does, that the patient should be in the final stage of a terminal illness. But in retrospect we can see that a paradoxical outcome of Drion's intervention has been that in the Dutch debate a separate category of death wishes had been singled out, by now known as completed life cases. This is remarkable, for during the development of the law in the seventies and eighties, such cases were not seen as being a-typical in any way. As a matter of fact the two most path-breaking court cases, the cases of doctor Postma (1973, NJ 1973, nr. 83) and doctor Schoonheim (1984, NJ 1985, nr. 106), concerned such cases.

But when finally the new law was presented to parliament, one of the main issues in the parliamentary debate was whether or not it should cover tired-of-life cases. At that very moment there happened to be a relevant case in court, the Brongersma-case, and in its verdict on it (2002) the Dutch Supreme Court concurred with the government that doctors should not be allowed to assist suicides in tired-of-life cases.[4] Both the government and the Court, however, used a very restrictive definition of the relevant class which would, for example, exclude Mr Kalisvaart. Hence, when Mr Kalisvaart's doctor granted his request, the regional review committee could judge his action to have satisfied the legal requirements of careful action, without recognizing any deviation from the Brongersma-verdict.[5] In § 4 I will consider whether any principled reasons exist to think that requests of assistance from elderly people are not covered by the euthanasia law, given its justifying grounds, which I will discuss in § 3.

As a matter of fact, however, doctors only rarely grant such requests, and part of the explanation is that most of them still believe that the Brongersma-ruling has outlawed physician-assisted suicide in all tired-of-life cases. Until recently this was also the position of the Royal Dutch Medical Association.[6]

Since a recent change of the Dutch constitution it is possible for a group of at least 40,000 Dutch citizens to bring a proposal for new legislation to parliament. In 2010 such a Citizen's Initiative has been started, under the name 'Of their own free will', with the active support of the Dutch right-to-die-society NVVE, for creating a non-medical path to the self-chosen death for people of seventy years and older. The Initiative received 116,000 signatures in a few weeks time, and in March

[3] For an authoritative overview of the present Dutch law and practice see Griffiths et al. 2008.

[4] HR 24/12/2002, NJ 2003/167; *Tijdschrift voor Gezondheidsrecht* 2003/29

[5] *Jaarverslag* (Annual Report) *regionale Toetsingscommissies Euthanasie* 2009, case 9.

[6] The Society has changed its position, however, in a position statement of November 12 2010, in recognition of the case law of the review committees.

2011 the proposed bill has been send to parliament. It follows the general structure ofthe Dutch euthanasia law, but with some important exceptions.[7] The most salient deviation from the euthanasia law is that the new bill proposes to create a new category of non-medical professionals, to be called 'providers of dying help', who are to be trained and certified by a special organisation. People of seventy years and older can ask these persons for assistance in ending their lives. In that case a procedure has to be followed, including consultation and review, which is identical to the procedure of the euthanasia law. But the substantial requirements of careful action are different; that is the second major deviation from the existing law. The provider of dying help only has to make sure that his client's request is voluntary, well-considered and permanent, there is no requirement as regards his condition similar to the unbearable suffering requirement of the euthanasia law, and it is not required either that there should not be an alternative solution to his problems.

The debate about the bill is going on. Many of the issues discussed are of local interest only, in particular the relation between the new proposed law and the old euthanasia law. Will the medical route still be used for older people when the non-medical one makes fewer requirements? Some of the issues, however, may have a wider relevance.

Whether they really do, depends on the basic justification for providing assistance to people in executing their death wish. The most common argument of advocates of legalization everywhere appeals to respect for self-determination,[8] but that doesn't seem to be the moral basis of the Dutch euthanasia law. Rather it is a principle of mercy, allowing doctors to help people avoid unbearable suffering.[9] The new bill on the contrary is clearly and explicitly inspired by respect for self-determination only, although this respect is not to be shown to minors, people younger than seventy years. That is odd because, when you are arguing in terms of respect for self-determination, there seems to be no good reason at all to consider death wishes of elderly people as being in a special class. The reason the proposal suggests is that elderly people are in a better position to determine the balance sheet of their lives. But this ability goes beyond the competence which is the only relevant consideration for applying the principle of respect for self-determination. And the actual distribution of this ability over people of any age is so unequal that it seems preferable to consider it on an individual basis. Hence it is doubtful whether the way in which the elderly are being privileged by the proposal is compatible with equality for the law.

---

[7] Cf. http://vorige.nrc.nl/international/Features/article2478619.ece/Citizens_group_argues_right_to_die. In March 2012 Parliament decided not to take over the Initiative at the moment, but to reconsider the matter in the context of the evaluation of the euthanasia law, which is scheduled for the end of 2012.

[8] A classical source is Dworkin et al. 1997.

[9] I have argued for this interpretation in den Hartogh 2012. The interpretation is confirmed by the travaux préparatoires of the euthanasia law and the parliamentary debate, as analysed by Pans 2006.

10.3 Why Indirect Paternalism is Different[10]

Competent people have the right to end their own lives; this right has been legally recognized, e.g. by the European Court for Human Rights.[11] Doesn't it follow that other people cannot be forbidden to assist them? That would only follow if we should accept the maxim *Volenti non fit iniuria* (you cannot be wronged by actions to which you have consented) without any restrictions. However, the very fact that euthanasia and physician-assisted suicide are forbidden almost everywhere shows that the law normally doesn't recognize the Volenti-maxim without restrictions. The same, as we have seen, is even true of the Dutch euthanasia law, because it only permits doctors to help people to end their lives, and this only in cases of unbearable suffering without any hope of improvement. Similar points can be made as regards the other so-called inalienable rights, the rights to personal freedom and to bodily integrity. Indeed, according to the professional morality of doctors they are never justified to decide upon a certain treatment by the mere request of the patient; in addition they should be convinced that the treatment is required by the best interests of that patient. In this respect both the law and the professional morality of doctors are characterised by indirect paternalism.

The distinction between direct and indirect paternalism we owe to Joel Feinberg (1986, p. 9). Direct paternalism is a two-place relation: A interferes in the execution of B's plans in order to prevent harm to B. Indirect paternalism is a three-place relation: A interferes in the execution of B's plans in order to prevent harm to C, even though C has validly consented to B's actions. According to Feinberg paternalistic considerations can never be a good reason for the state to criminalize any action, and in his view this is as true of indirect as of direct paternalism. This is explained, or at least asserted, by the Volenti-maxim. The question I want to discuss in this section is whether there really is no morally relevant difference between the two kinds of paternalism. For, as I suggested, in that case the Dutch euthanasia law is mistaken in its focus on unbearable suffering, and the question whether it should take the characteristic death wishes of the elderly into account can be dispensed with. The only relevant question about requests by the elderly, as of everyone else, would be the question whether it has been made voluntarily and competently.

What is basically wrong with paternalism? Feinberg is at his rhetorical best in replying: If people are treated paternalistically, "their grievance is not simply that they have been unnecessarily inconvenienced or 'irked', but rather they have been violated, invaded, belittled. They have experienced something analogous to the invasion of their property or the violation of their privacy." (Feinberg 1986, p. 27) That paternalism is insulting is an important point, and certainly the point which people care most about, for nothing is as important to people as their social status. But it must, as the analogy with property and privacy shows, be dependent on other points. For why should people believe that their status vis-à-vis each other depends on the extent to which they are left alone to govern their own actions, unless they would

[10] This section is largely taken from den Hartogh 2013.

[11] *Affaire Haas c. Suisse*, 20 janvier 2011, Requête no 31322/07, cf. den Hartogh 2011.

already attribute great value to this authority? So we have to consider two reasons why paternalism is wrong: The intrinsic value of self-governing authority, and the status conferred to people by having this power.

So my question is: Are these reasons at stake with indirect paternalism as well?

First consider the value of self-governance. It is important to see that the question whether indirect and direct paternalism are different with respect to that value is not identical to the old question about negative and positive liberty. I fully agree with Locke against many so-called Lockeans that to let an innocent person starve when you can feed him is equivalent to murder. So we presuppose a tolerably just system of property rights and of claims on resources. With this background in place my claim is that the value of self-governance is given its full due when competent people are allowed to set their own aims and strive for their realisation with their own resources, within the limits set by the harm and the offence principle and (perhaps) legitimate moralism, if that isn't an oxymoron. It is irrelevant whether or not they succeed in this enterprise. Whether such success is valuable is entirely a matter of the relevant values, including their own real interests: If the enterprise is misguided, we should hope that they do not succeed. So nothing is wrong if other people refuse to help them because they believe such help would only harm them. And nothing is wrong either if people are not allowed to provide such help.

This view would be reinforced if Joseph Raz is right in believing that autonomy is a satiable or threshold concept: If you are allowed to strive for your own aims with your own resources, you are as self-governing as you can be (Raz 1986, Chap. 14). But even if we do not think in that way about the value of self-governance, we should do so about the right to self-governance, or, rather the bundle of rights which protect that value. Perhaps you would be more autonomous if other people were allowed to help you, but it doesn't follow that the prohibition violates any of your rights to make your own choices.

Secondly, there is the question of status. If other people are prevented to provide help to you, is this insulting to you? Of course, it means that some people believe that your beliefs about your own interest are wrong, that you are set on a way on which you will cause a setback to them. That is not an insulting belief; in fact it may be entirely right. Is it insulting that you are prevented to receive aid on those grounds? For example, if you are not allowed to take your own life for any reason you find compelling, you may feel insulted, because you are no longer your own person. But if other people are prohibited to help you, unless they are convinced on reasonable grounds that it is in your best interests to die, is this equally belittling? My suggestion is that it is not. The basic difference here is that direct paternalism by its very nature creates an asymmetry between the two people involved, A and B. A is prevented from acting on his own judgment about his own interest, because B uses his power to act on his contrary judgment of that same interest. But in the case of indirect paternalism A and B are both allowed to use their own resources in symmetrical ways, they are both responsible for the harm they create, both to themselves and to the other party.

Suppose that you are a painter and at a certain moment you have come to the conclusion that the paintings you have made until now are no good at all, they are inauthentic caterings to popular taste. Being for the moment in hospital you ask your best friend to go to your studio and destroy all your work. But your friend disagrees with you, he believes that destroying your work would be a loss for humanity, but also for you. So he says to you: Look, do it yourself when you have recovered, if you still think it is a good idea, but I will not do it. Such a refusal, it seems to me, shows concern for you and in no way any lack of respect. He leaves you full responsibility for your own actions, and only takes an equal responsibility for his own. Now, if a moral code exists which enjoins both of you to do precisely this, to take full responsibility for your own actions, how can such a code be interpreted as showing disrespect?

Certainly, on many occasions indirect paternalism would show lack of proper respect. If we make a perfectly valid contract, and you are prevented from performing your part of the agreement by a third person who thinks you are harming me, that is insulting to me. But that is, I suggest, because the contract actually is to be considered a valid one on independent grounds, so I already have a right to expect your performance. That right is already part of my moral assets; that is why the action of this third person is analogous to "the invasion of my property or the violation of my privacy". That doesn't rule out certain categories of harmful consequences to invalidate contracts to begin with.

What the example suggests is that the Volenti-principle is related to the principle of respect for autonomy in an indirect way only. To begin with, its aim is to protect legitimate expectations, arising from consenting acts. If a person in good faith performs his part of an agreement, he should not normally be held responsible for any harm befalling his partner to the agreement as a result of that. So the Volenti-principle does not primarily aim at protecting the autonomy of the consenting party, but rather at shielding the other party from liability. Only if it is already established on independent grounds that my consent is sufficient for justifying your action, even if the action harms me, is the interference of a third person disrespectful to me. The Volenti-principle doesn't help us to define the boundaries of the domain of the agent's sovereignty, because its validity itself is limited to that domain.

I conclude that the Dutch euthanasia law is right in making requirements, not only as regards the nature of the patient's request for help in dying, but also as regards the reasons for that request. It should be added that, even if you argue in terms of self-determination only, you may see reason to add requirements concerning the requester's situation to requirements concerning the quality of his request, because the former may be judged the most reliable sign of the latter. The law often uses such indirect means for achieving its ends. For such reasons the laws of the American states of Oregon and Washington, which allow physicians to assist people to commit suicide, have restricted their permission to people suffering from a terminal illness and expected to die from it within six months. Even if those laws have only been justified in terms of respect for self-determination, the law considers that it is often difficult to verify whether a person's request is truly voluntary and well-considered, the probability of error being lowest in the case of terminal illness. A similar argument has occasionally been used in the debate about the Belgian, but not the Dutch, euthanasia law, in order to justify the requirement of unbearable suffering (Nys 2007).

For such reasons one might wonder whether the bill proposed by the Citizen's Initiative is satisfactory, even on its own terms. Shouldn't it at least have required the care provider to assess what the reasons are for a person to want to end her life, and to verify that these reasons cannot be accommodated in any alternative way? And this irrespective of her age? Some losses can partially be compensated, some past failures partially be accepted, either without or with the help of others.

10.4 Outside the Doctor's Area of Competence?

I have argued that it is defensible for the law, in the case of a request for euthanasia or assisted suicide, not only to protect the requester from violations of his autonomy but also from harm, and therefore to make it a requirement of due care that the requester should be in a state of unbearable suffering without any prospect of improvement. The Dutch euthanasia law, however, adds two restrictions to this requirement. In the first place it only allows physicians to consider helping people to end their lives. And, secondly, in its Brongersma-decision the Dutch Supreme Court has added the requirement that the patient's suffering should be primarily caused by a medically classified disease or ailment. These two restrictions are interconnected. For, the court argued, if a person's suffering is not caused by a medically classified affliction, it is not a proper task of doctors to help the patient by alleviating or preventing that suffering, and they cannot be expected to have a special expertise for fulfilling such a task. As we have seen, doctors commonly believe that this condition takes tired-of-life cases out of their hands.

If that belief is true, the very principle of mercy which is the rationale of the euthanasia law, would possibly make it imperative to allow non-doctors to provide assistance to the elderly. For, as the same Supreme Court argued in an earlier case, as regards the principle of mercy it is not the cause of the suffering which matters, but its severity.[12] Hence, even if the Citizen's Initiative would be mistaken in rejecting the requirement of unbearable suffering, it might still be right in providing for a non-medical route to the self-chosen death.

But is the common belief of doctors correct? In § 2 we saw already that the regional review committee which considered Mr Kalisvaart's case didn't think that the Brongersma-decision committed them to reproach the doctor who granted Mr Kalisvaart his request for a lack of due care. Did the committee somehow miss the point?

According to recent insight aging is a process of accumulated random damage to a complex system which as a result gradually loses its reserve capacity, hence its ability to compensation and to recovery from stress. As a result there is an increase and accumulation of diseases and ailments, and an acceleration of both the increase and the accumulation. These afflictions include arthritis, Parkinson's disease, glaucoma, macula degeneration, diabetes, stomach problems, heart attack, stroke, hypertension, flu, broken hip, broken bones, infections of the urinary tract, dementia, angina,

[12] Chabot-case, HR 21/6/1994, NJ 1994, nr. 656, translated in: Griffiths et al. 1998, 329–340.

incontinence, impotence, problems with teeth and feet, with vision and hearing, as well as depressive symptoms.

The process results in an increase of functional disabilities which has the same 'cascade' characteristics of accelerated increase and accumulation. These include difficulties with eating, dressing, walking around, climbing stairs, getting in and out of bed, bathing, toileting, using the telephone, going out, shopping, cooking, doing light house work, reading, looking television, taking medicine and managing money, but also sleeplessness, daytime drowsiness, fatigue, loss of energy, loss of appetite, anxiety, anhedony and other negative affective states.[13]

It is sometimes suggested that such disabilities do not belong to the medical domain, because they are the result of a statistically normal process of physiological aging. But the diseases which cause them may occur at any age, although as a result of the 'cascade' process they are most characteristic of old age. It makes no sense to count them as medically classified diseases at other ages, but not at old age, because at that time they occur more often. All these conditions help to explain functional disabilities and give direction to medical efforts to treat them or to compensate for them; therefore they are to be considered "medically classified diseases or ailments". It can hardly be disputed that losses of eyesight, hearing and mobility are conditions which justify or require medical attention at any age, irrespective of their cause.

Moreover, what is—as a result of the process of aging understood in the way I described—most characteristic of old age is the plurality of health deficits, interacting with each other, and creating a decreased resistance to environmental stressors. This condition by now is known as 'frailty' or even the 'frailty syndrome' and it is increasingly measured by means of a 'frailty index' (cf. Chap. 8).[14] The concept of frailty presents itself as an important new medical classification, because it has both predictive value for incomplete recovery from acute ailments, falling down, loss of control, hospitalization, and death, as well as guiding value for medical efforts, e.g. as regards nutrition and physical activity, and in particular for adapted treatment of the component diseases.

This process of aging is the causal background of the emergence of the characteristic death wishes of the elderly. The most prominent factors seem to be declined eyesight, hearing and mobility, as well as depressive symptoms. That is clear in the case of Mr Kalisvaart, and also in all four cases decribed as typical cases of a completed life on the website of the Citizen's Initiative. The effect of these factors is reinforced by other losses, of energy, memory and concentration etc. It is true that often biographical factors are also relevant, the loss of a partner, of a job or of other meaningful occupations, of peers, all of which may result in a sense of emptiness and loneliness, often made worse by disappointment about the perceived lack of attention

[13] Rurup et al. 2010 found a significant number of people with death wishes to have speech impairments.

[14] Some descriptions of the syndrome focus on functional disabilities (Rockwood et al. 2000), others include diseases. Some descriptions are restricted to physiological conditions (Fried et al. 2001), but there is no reason not to include psychological conditions as for example mood disorders or loss of cognitive abilities, Gobbens et al. 2010.

and care from significant others (Rurup et al. 2011). But the effect of such factors is often dependent on functional disabilities. When you loose your comrades when you are 30, you may be able to build a new social network, but probably not when you are 90. It is true that people at all ages need to cope with losses and failures, but what frailty usually (not always) means is a loss of resilience or coping ability. One of the most common elements in the sense of suffering of people with current death wishes is a low perceived level of mastery (Rurup et al. 2010).

There is a more subtle argument for the conclusion that requests for physician-assisted death in tired-of-life cases should not be considered to belong to the medical domain. Such requests, it is sometimes said, are properly addressed to doctors when the suffering which leads to them consists of pain, lack of breath, nausea, exhaustion, and similar physical symptoms, as they are often experienced by patients who are dying from cancer or neurological diseases. Doctors should be professionally equipped to deal with such symptoms. But in tired-of-life cases the suffering is of a different kind: It is existential suffering, and doctors have no special competence to deal with that kind of suffering.

The nature of the suffering in tired-of-life cases has been well expressed in the announcement of the Citizen's Initiative. "The reasons for longing after death are divergent. In some cases we see no possibilities to continue life in a meaningful way, we have the sense of surviving ourselves. Everything of value is behind us and only emptiness is left. Sometimes we are completely dependent on the care of other people, losing all control over our own life. And sometimes we are confronted with the irreversible loss of our personal dignity".

But if we consider the main reasons why patients in a terminal stage of cancer request euthanasia, we get exactly the same list: Loss of meaning, of control and of dignity.[15] This is even true when physical symptoms are highly prominent, which is not always the case. Take for example cachexia. Surely it is a very unpleasant experience by itself to feel completely exhausted, but the consequences of that condition are even more important: That you are unable to do anything, and even hardly to experience what happens in your immediate surroundings. In a report to a regional review committee the counsellor writes: "In his consultation request the doctor informed me that the patient is suffering from cachexia, nausea and decubitus. But the patient himself is only talking about the absence of any perspective. 'The only thing I still can do is to wait for death'."

You can only understand the suffering of terminal cancer patients if you do not only take their physical symptoms into account, but also consider the meaning which those symptoms have for them, and the emotions which they evoke because of this meaning: Sorrow, anguish, despair, sometimes regret or anger. In some cases this meaning is evident. If a person has a strong and continuous sense of suffocation because of lack of breath, you do need not to know anything else about her to understand that she is in a state of extreme suffering. But in the majority of the cases you can only understand the nature and severity of the suffering against the background of the

[15] Van der Wal et al. 2003, p. 51. Jaarverslag regionale Toetsingscommissies Euthanasie 2009, p. 20, cf. case 7.

patient's personality, values and life history. That is the basic task which the Dutch euthanasia law requires to be performed in all cases of euthanasia and physician-assisted suicide, by the doctors involved, the counsellors and the review committees. Doctors should be equipped for performing that task, and many of them are.

In his classical study *The Nature of Suffering* (1991) Eric Cassell argued that pain is not by itself a form of suffering: The suffering really consists in the attack which the pain makes on your person (Cassell 1991). On Cassell's view suffering always has a phenomenological component: It means experiencing something with distress. It also always has a cognitive or interpretative component: The distress results from understanding the experienced condition as threatening the integrity of your person. Hence all suffering is existential suffering. 'Physical suffering' is really an unfortunate term: It can only mean existential suffering of which physical symptoms are an important element of the phenomenological component.

It could be objected to Cassell's view that on that view newborn children, people in an advanced state of dementia and most animals do not suffer, even if they experience severe pain (Edwards 2003). But even if we accept that objection, it is still right that the suffering of persons who understand themselves as living their own lives, is qualitatively different because it has the interpretative component. The condition which they experience with distress is understood by them as a threat, perhaps not always to the intactness of their persons, but at least to some highly valued aspect of the life they live. "It must cast a pall on other activities engaged in, affect one's capacity to enjoy life." (Edwards 2003)

Cassell's view should be amended in one other respect. According to him[16] a person has a privileged access to her own suffering: Because she is the one who suffers, she knows she is, and to what extent. Other people can at best have an empathic understanding of that suffering by imagining what it would be like to be in her condition, an understanding which almost necessarily depends to some extent on their own outlook on life. But except for its phenomenology suffering also has a cognitive dimension, and on this dimension the suffering person is not in an absolutely authoritative position to assess her own suffering. This is confirmed by the fact that people usually don't assess the severity of their suffering by silent introspection, but rather by talking with others, most often their friends and relatives, but sometimes also their doctors. Each partner in this conversation can be mistaken in his interpretation, and be corrected by the other party. One relevant consideration always is whether other ways are still available to protect the valuable aspects of one's life which are being impaired or threatened.[17]

---

[16] And to Edwards 2003.

[17] That suffering is intersubjectively accessible in this way is stressed by Wijsbek 2012. He also refers to research by Patrick Wall a.o. showing that the phenomenological and the interpretative component are mutually dependent on each other: soldiers wounded at the battlefront feel less pain when they know they will now be allowed to return home.

10.5 What Kind of Assistance is Needed?

A euthanasia law like the Dutch or Belgian one[18] would make it possible for at least some old people to receive barbiturates for ending their own lives from their physicians, and be guided by them through the process. But because that law makes requirements on the condition of the patient and not only on his request, not every request will be granted. Moreover, such laws do not require doctors to give assistance, which they are allowed to refuse for their own reasons. In the Netherlands such requests are being refused at present between 75 and 300 times a year[19], and they are being granted, I guess, only about 10 times.[20]

As we saw, it is common for doctors to believe that the law forbids assistance in tired-of-life cases. We may hope that the changed stance of the Royal Dutch Medical Association will help to clarify the legal position, but some physicians may resist surrendering the security of this belief. Another common reason for refusing is that the patient has no lethal illness, or is not in the terminal stage of that illness. Some of the doctors involved are conscious of using their own norms instead of the legal ones, some others aren't (Van de Vathorst et al. 2011). Yet another fairly common reason is that in the view of the doctor the requirement of unbearable suffering has not been met. In some of those cases these doctors may believe that, even if there is unbearable suffering in some sense, it is not of the right kind, being merely existential suffering. Research has shown that still 17 % of the doctors interpret 'unbearable suffering' to require serious physical symptoms (Van Tol et al. 2010).[21] In other cases doctors have difficulty to empathize with the patient, for example because he seems not to be prepared to accept any decline of abilities he once had, or because he simply insists on the authority of his own personal view without being ready to discuss or even explain it. Doctors commonly think that the choice for suicide or euthanasia should be the outcome of a process of intensive interaction between doctor and patient. An important consideration may also be that in the case of old age afflictions which are not life-threatening, it is more difficult than in many cancer cases to establish that there is no alternative way to protect the valuable elements of the patient's life which are threatened by these afflictions and by the loss of functional abilities resulting from them. People who believe they cannot cope with their losses can sometimes be helped to find a more or less precarious balance again. Finally for most doctors the number of cases in which they are able to handle a case of assistance emotionally is limited, and there are signs that many of them have the feeling they are at that ceiling already.

---

[18] The Belgian law requires "a serious and incurable ailment caused by accident or illness".

[19] According to Rurup 2005, Chap. 2, see note 7, 1 of the 6 cases in which a doctor refuses a request concerns a tired-of-life case. According to Chabot (personal communication) this is true of 27 % of the cases identified in Chabot 2007 in which no "severe illness" existed and in which the doctor refused a request.

[20] On reasons for refusal see Pasman et al; Chabot 2007, 128; and in particular Van de Vathorst et al. 2011.

[21] See also Jansen-van der Weide et al. 2005; Pasman et al. 2009; Van de Vathorst et al. 2011, Chap. 9.

Some of these reasons for refusal fully fit in with the framework of the legal requirements of due care, but others clearly don't. Therefore it should be recognized that the Citizen's Initiative is pointing to a real problem, even if it is mistaken in its interpretation of the euthanasia law, and in its exclusive reliance on the principle of respect for self-determination. What about the solution proposed, to create a new class of care providers? A prior question should be whether elderly people who want to end their lives, always need any kind of assistance at all, and, if so, what kind of assistance. If people demand the right to make their own decisions 'and stand the consequences', shouldn't it be preferable for them to implement those decisions themselves, without burdening others? That is a question which should have a special appeal to people who give such a high priority to respect for personal autonomy.

On this point the Citizen's Initiative uses a false rhetoric which is well-known from right-to-die advocacy all over the world. If people don't get assistance, it is suggested, they have only one option left: To kill themselves by hanging or drowning or shooting themselves, or by throwing themselves before the train or from a high building. However, a well-considered choice for death is almost never executed by such horrible means. The suicides which we know from the official suicide statistics are almost invariably impulsive actions of despairing people.[22] People who make a sober decision, preferably in dialogue with friends and relatives, after due consideration of the alternatives, by now know that other means are available.

In its recent decision *Haas v. Switzerland* the European Human Rights Court has stated that the state has a legitimate interest in barring people's access to deadly pharmaca, and in principle I agree.[23] However, it is questionable how realistic that ambition still is. Mr Haas requested his psychiatrist, and 170 other psychiatrists, to provide him with pentobarbital, without examining him on any other condition but his competency. But if he had not spend years of his life in fighting legal battles which he could not win, he could have arranged to buy pentobarbital in Mexico, where it is freely available in animals' shops. By now one can even order it on telephone and several other deadly drugs can be bought online. From the point of the legislator, however, this is not a relevant alternative. The state cannot say to Mr Haas: We don't need to lift the barrier to such drugs, because you are smart enough to know how to get around it.[24]

But precisely for very old and frail people there is another alternative which certainly is relevant: They can stop eating and drinking. The common belief that this leads to an inhumane death is mistaken (Chabot 2007, 2008; Ganzini 2003). Recent research by Chabot has shown that in the Netherlands the number of people making that choice is of a magnitude comparable to the number of euthanasia cases. (To my

[22] Because of its reliance on such statistics most present research on suicide gives a false picture of the phenomenon, Chabot 2007.

[23] See note 17.

[24] It is, however, an urgent task for governments to reconsider the doctor's monopoly on the prescription of deadly drugs. Either the inaccessibility of such drugs to others should be enforced, if it is still possible to do that, or the monopoly should be lifted. Otherwise the law will quickly become a dead letter.

knowledge we have no data for other countries.) In 75 % of the cases the relatives report that this has been a humane route to the self-chosen death, and Chabot argues plausibly that the problems which have arisen in the other 25 % of the cases could have been prevented by better preparation and adequate palliative care.

This raises the question whether it should be allowable, or even obligatory for doctors to assist in providing that care, or in supervising its provision. In its recent position statement the Royal Dutch Medical Association says it is obligatory. When a person has made that decision in a voluntary and well-considered way, and has shown that he is in earnest about it, the physician need take no stance concerning the validity of that decision in order to be prepared to alleviate that person's suffering. That is what doctors are for.

In saying that this is a humane way I do not claim it is an easy one. (Although the weaker a person's physical condition, the easier it is.) In the beginning there is a feeling of hunger which, however, disappears after a few days when the person is on a diet of zero calories. Afterwards the most important care consists of preventing a dry mouth. In a few cases sedation may be necessary in order to prevent delirium. If the person persists, that seems to me the best possible test of the resoluteness of her choice.

This view has an interesting philosophical pedigree. Arthur Schopenhauer had a critical view of suicide because it did not square with his ideal of the self-denial of the will. But he made an exception for "the free choice of starvation deriving from the highest degree of asceticism".[25]

Beyond the recognition of the extension of the doctor's palliative task we need no new institutional arrangements for providing elderly people with Drion's pill. For in a sense most elderly people already have that pill at their disposal.

Acknowledgement Thanks to Boudewijn Chabot, Mette Rurup, Dorothea Touwen and Henri Wijsbek for helpful comments and/or suggestions.

References

Birnbacher, D. 2010. Geleitwordt. In *Ausweg am Lebensende, Selbstbestimmtes Sterben durch Freiwilligen Verzicht auf Essen und Trinken*, ed. Boudewijn Chabot, and Christian Walther, 9–11. Munchen: Erst Reinhardt Verlag.

Cassell, E. 1991. *The nature of suffering*. New York: Oxford University Press.

Chabot, B. 2007. *Auto-euthanasie: Verborgen stervenswegen in gesprek met naasten*. Amsterdam: Bert Bakker.

Chabot, B. 2008. *A hastened death by self-denial of food and drink*. Amsterdam.

Drion, H. 1992. *Het zelfgewilde einde van oude mensen, met reacties van Ch. J. Enschedé, H.M. Kuitert e.a.* Amsterdam: Balans.

Dworkin, R., T. Nagel, R. Nozick, J. Rawls, and J. Jarvis Thomson. 1997. Assisted suicide: The philosopher's brief. *New York Review of Books*, March 27.

Edwards, S.D. 2003. Three Concepts of Suffering. *Medicine, Healthcare and Philosophy* 6: 59–66.

Feinberg, J. 1986. *Harm to self.* Oxford: Oxford University Press.

[25] Quoted from Birnbacher 2010.

Fried, L.P., C.M. Tangen, J. Walston, et al. 2001. Frailty in older adults: evidence for a phenotype. *Journals of Gerontology, Series A* 56: M146–M156.
Ganzini, L., E.R. Goy, L.L. Miller, et al. 2003. Nurses experiences with hospice patients who refuse food and fluids to hasten death. *New England Journal of Medicine* 349: 359–365.
Gobbens, R., M.A.L.M. van Assen, K.G. Luijkx, et al. 2010. Determinants of frailty. *Journal of the American Medical Directors Association* 11(5): 356–364.
Griffiths, J., H. Weyers, M. Adams. 2008. *Euthanasia and law in Europe*. Oxford: Hart Publ.
Griffiths, J., A. Bood, H. Weyers. 1998. *Euthanasia and law in the Netherlands*. Amsterdam: Amsterdam University Press.
Den Hartogh, G. 2011. Haas tegen Zwitserland. Inleiding en noot bij uitspraak Europese Hof voor de Rechten van de Mens. *European Human Rights Cases* 12(4): 645–657.
Den Hartogh, G. (2012) The regulation of euthanasia: How successful is the dutch system?. In *Physician-assisted death in the Netherlands: A view from inside*, ed. S. Youngner and G Kimsma. Cambridge: Cambridge University Press.
Den Hartogh, G. (2013) Is Consent of the Donor Enough to Justify the Removal of Living Organs? *Cambridge Quarterly of Health Care Ethics* 22(1).
Jaarverslag. 2009. van de Regionale Toetsingscommissies Euthanasie, toegankelijk op: www.euthanasiecommissie.nl/Toetsingscommissie/jaarverslag/default.asp.
Jansen-van der Weide, M.C., B.D. Onwuteaka-Philipsen, and G. van der Wal. 2005. Undecided, withdrawn, and refused requests for euthanasia and physician-assisted suicide. *Archives of Internal Medicine* 165(15): 1698–704.
Nys, H. 2007. Euthanasie: de toekomst van het Belgische model. *Tijdschrift voor Gezondheidsrecht*. 31: 199–203.
Pans, E. 2006. *De normatieve grondslagen van het Nederlandse euthanasierecht*. Nijmegen: Wolf Legal Publishers.
Pasman, H.R.W., M.L. Rurup, D. Willems, and B.D. Onwuteaka-Philipsen. 2009. Concept of unbearable suffering in context of ungranted requests for euthanasia: qualitative interviews with patients and physicians. *British Medical Journal* 339:b4362doi:10.1136/bmj.b4362.
Raz, J. 1986. *The morality of freedom*. Oxford: Clarendon Press.
Rockwood, K., D.B. Hogan, and C. MacKnight. 2000. Conceptualisation and measurement of frailty in elderly people. *Drugs Aging* 17: 295–302.
Rurup, M.L. 2005. *Setting the stage for death: New themes in the euthanasia debate*. Amsterdam.
Rurup, M.L., H.W.R. Pasman, et al. 2010. Wishes to die in older people: a quantitative study of prevalence and associated factors. *Crisis: the Journal of Crisis Intervention and Suicide Prevention*. DOI 10.1027/0227–5910/a000079.
Rurup, M.L., D.J.H. Deeg, et al. 2011. Understanding why older people develop a wish to die: a qualitative interview study. *Crisis: The Journal of Crisis Intervention and Suicide Prevention*. DOI 10.1027/0227–5910/a000078.
Scocco, P. and D. De Leo. 2002. One-year prevalence of death thoughts, suicide ideation and behaviours in an elderly population. *International Journal of Geriatric Psychiatry* 17: 842–846.
Van Tol, D., J. Rietjens, and A. van der Heide. 2010. Judgment of unbearable suffering and willingness to grant a euthanasia request by Dutch general practitioners. *Health Policy* 97: 166–172.
Van de Vathorst, S., M. Schermer, and M. Trappenburg. 2011. Niet-ingewilligde verzoeken om levensbeëindiging. In *Kennis en opvattingen van publiek en professionals over medische besluitvgorming en behandeling rond het einde van het leven*, ed. J.J.M. van Delden, A. Van der Heide, S Van de Vathorst, W. Weyers, and D.G. van Tol, pp. 170–183. Den Haag: ZONMw.
Van Der Wal, G. et al. 2003. *Medische Besluitvorming aan het einde van het leven: De praktijk en de toetsingsprocedure euthanasie*. Utrecht: de Tijdstroom.
Wijsbek, H. (2012). The Subjectivity of suffering and the normativity of unbearableness. In *Physician-assisted death in the Netherlands: A view from Inside*, ed. S. Youngner and G. Kimsma. Cambridge: Cambridge University Press.

Chapter 11
Imagining Good Aging

Frits de Lange

11.1 Introduction

Good aging seems to be a *contradictio in terminis*. Is old age not the lesser part of life, a final phase of loss and inevitably decay? Is it not made bearable only with a whiff of humour à la Maurice Chevalier, who said: "Old age isn't so bad when you consider the alternative"? The combination 'good aging' seems to reveal a cynical ethical attitude: Good aging is not a subject for ethics.

I think it is, however. Reflection on good aging is, in the first place, a variant of deliberation on what makes life good, ethics' core concern. Although old age is the best part of life for only a few people, it is not per definition the worst. Good aging is, as I understand it, first of all simply making the best of growing old or making old age a good part of life as well. But I want to push the matter a bit further because a more challenging question is not only what makes life good while aging but also what—if anything—makes aging a good thing. Are there goods to aging in addition to the goods *in* aging?

To explain my position, I will first (Sect. 2) make some remarks on my understanding of ethics and its relationship to aging. I will then (Sect. 3) construct an inventory of ingredients that a theoretical approach to good aging needs to include. The question if there is any good in aging is subsequently dealt with in what I would call an existential gerontology. In addition to being constructive, ethical reflection on good aging also needs to be critical. Therefore, I will (Sect. 4) conclude with questioning some master narratives and gerontological paradigms of 'normal' aging.

F. de Lange (✉)
Protestant Theological University, Groningen, The Netherlands
e-mail: fdelange@pthu.nl

M. Schermer, W. Pinxten (eds.), *Ethics, Health Policy and (Anti-) Aging: Mixed Blessings*, Ethics and Health Policy 1, DOI 10.1007/978-94-007-3870-6_11,

11.2 What Kind of Ethics? Which Aging Discourse?

With Agnes Heller I contend that the task of ethics as reflection on the good life may focus on, but is not restricted to, *normative deliberation* on practical problems (*What is good to do?* Let us call this Ethics I). But I also contend that ever since classical times moral philosophy could not fulfil that task without an *interpretation* of the human condition with all its ontological and metaphysical perplexities (*What does it mean to live a good life?* This can be called Ethics II). Through its expertise in these normative and interpretative matters, ethics could also fulfil an *educational/therapeutic* function as *paideia* (*How should I live?*—Ethics III) (Heller 1988). I think, therefore, that it is also ethics' public task to develop tentative *views of good aging* that can help older persons make more sense of their old age. In that sense, ethics takes an advocatory stance, choosing the side of the elderly. Thus far ethics. But what about aging? We can distinguish three different discourses with respect to old age in our advanced liberal society (Baars 2010). There is: An *elderly discourse*, focusing on strategic policy decisions about societal consequences and the economic effects of the demographic transition towards a 'greyer' society (pensions, citizenship and participation, access to and distribution of healthcare; cf., e.g., Moody 1992); an *aging discourse* on the physical and biomedical aspects of aging (cf., e.g. Harris 2007); and a *growing older discourse*, where the meaning of old age as a distinctive phase in life and the question of good aging is central (cf., e.g., Cole 2010). The role of ethics as practical guidance in morals is evident in the first two discourses: Intergenerational and distributive justice in access to healthcare and welfare, good care in residential settings, civil participation and human rights for the elderly, life extension, and human perfectibility through genetic enhancement. As someone involved in training professionals in spiritual care who work with elderly in healthcare and social welfare contexts, however, I am especially interested in the third discourse: What does it mean to grow and to be old (cf. De Lange 2007, 2010a)?

Heidegger and others used a phenomenological analysis of 'being there' (*Daseinsanalyse*) to draw far-reaching ontological conclusions about life's meaninglessness or—as Albert Camus called it—its absurdity. An existentialist like Maurice Merleau-Ponty, in my opinion, would suit the ethical job better: His analysis of embodiment and temporality did not lead him beyond the recognition of a fundamental ontological *ambiguity*: A human life is a struggle between light and darkness, sense and non-sense (Merleau-Ponty 1945). I think this also applies to an old human life.

One experiences meaninglessness, according to Merleau-Ponty, when the embodied dialogue between oneself and one's surroundings comes to a halt. There is nothing that 'speaks' to us anymore and appeals to our ability to understand. To express it in terms of an embodied language: We can no longer find an appropriate *attitude* in space and time. 'Meaning' in this phenomenological sense is not a cognitive-reflective concept but refers to the primordial experience that the world and I somehow fit together. This experience makes me trust life and surrender myself to it day by day.

Good aging in such an existential sense is about reclaiming meaning, thus understood, from meaninglessness.

An ethics committed to good aging, I contend, is normative in the sense that it seeks to understand how long a meaningful dialogue can be continued between old, vulnerable and decaying human bodies and minds, and their environment. Life is worth living as long as it makes sense to the individual person, by imbuing him or her with an *appetite for life* and willing them to go on with living.

This therapeutic ethic (Heller's Ethics III) can fulfil its task only if it is derived from sound interpretation, *thick descriptions*of good and evil in the experienced reality of aging (Ethics II). Interviews, ego documents in art and literature, either expressive or reflective, supply rich sources. Not only will a theory of good aging that foregoes the rich and complex experience of growing old not help the elderly—it will be bad ethics as well.

I think that ethics as public philosophy also should contribute to the development of 'social imaginaries' (Taylor 2004) of good aging: A set of self understandings and common expectations that, even without being explicitly articulated, give people a sense of shared meaning. Views that help older people make the best of growing older, allowing as much human flourishing as possible. In doing so, ethics has both a constructive and a critical task.

11.3 A Constructive Task: Views of Good Aging

11.3.1 Normative Conditions for a Good Old Age

An ethics that deals deliberately with the question of good aging needs, first of all, to describe general normative conditions for a good old age. In this regard, 'good aging' is just a more specific variant of ethical theories of the good life. With respect to goodness, at first sight there is nothing special about old age. There are several options for theorizing in this field. Living a good life, one can say, requires at least that access to the basic natural and social primary goods (rights and liberties, powers and opportunities, income and wealth, self-respect, health and vigour, intelligence and imagination—'things that every rational man is presumed to want' (Rawls 1971, p. 62)) are optimized for older persons in order to provide them with a quality of life and well-being that corresponds to the minimal requirements of a decent life and of social justice. Despite a deep plurality of views of the good life in liberal democracies about the good life, a thin description of the good will probably result in an overlapping consensus about those basic goods.

From a more hermeneutical perspective, one may become even more substantial perhaps, exploring the moral traditions that formed liberal democracies in the Western world. As I argued elsewhere, a comprehensive approach to good aging then requires a balanced account of at least five dimensions. People age well, if and when (1) their needs and desires are fulfilled (resulting in their physical, psychological, and social well-being); (2) they are respected as full members of the community; (3) they can live according to their own moral convictions and commitments; (4) they are capable of finding meaning in their existence; and (5) they are able to attain a reasonable balance (*phronesis*) between all these dimensions of life.[1] There are different theoretical options. But whatever the ethical position one chooses, there is nothing special about aging. Old people are just like normal people, irrespective of their chronological age.

But what counts as 'normal' and what is good for normal people? For those who are suspicious of cultural biases in thick descriptions of the good but who still want to opt for a general normative perspective, the *capability* approach (A. Sen, M. Nussbaum) offers a promising candidate for ethical assessments of the quality of life of elderly people. It does not want to fix the list of primary social goods ('functionings') to be provided beforehand but identifies the capabilities people ought to have to live the life they have reason to value under their given conditions. The capability approach permits a more specific account of the lifespan, recognizing the fact that older people, in their particular stage of the lifespan, might now value other things more than they did in their younger years, respecting their substantive freedom of agency (cf. Kuh 2007; Gilroy 2006; Lloyd-Sherlock 2002). A good family network, security/ safety, or a religious life, for instance, may become more important than competitive jobs, education, power, and social prestige.

But the more chronological age becomes a discriminating factor in ethical theory, the more the picture of a good old age seems to darken. The elderly entered the ethical debate in the last decades almost exclusively as a social and financial burden. Growing old then seems to equal only an increase in dependence and decay. An understandable perspective, as long as ethics is confined to normative deliberation on practical problem solving (Ethics I). Being old represents 'a new social threat' because the elderly demand a disproportionate share in limited healthcare resources. Simply by the fact of their being old, they cause a "demographic, economic, and medical avalanche... one that could ultimately (and perhaps already) do great harm" (Callahan 1987, p. 20).

In its focus on healthcare, old age as decline, and its apocalyptic picture of a greying population, this debate hardly offers any contribution to views on good aging (Ethics II and III). Although all take a different stance in ethical theory, Norman Daniels' Kantian 'prudential lifespan account' (Daniels 1988), John Harris' utilitarian 'fair innings argument' (Harris 1985), and Daniel Callahan's more communitarian 'natural lifespan' approach (Callahan 1987) are all influenced by the tacit 'paradigm of decline' that so easily dominates the narrative on aging (cf. Gulette 2004). They all argue that people have equal rights to basic healthcare resources but contend that most older people have already had their share because they are on their way out. Age obviously makes some people more equal than others. The young should have priority, not only because it is better for society as a whole to invest in their future (at least according to Harris' utilitarianism) but also to safeguard the right of every

[1] Cf. De Lange 2007, 2010b. As I understand it, the good life is a complex, layered, and open concept clarified only by thick hermeneutic description and interpretation. Far from being coherent and logically consistent, it is an assemblage derived from a variety of classical ethical or philosophical theories embodied in a plurality of traditions of practical wisdom. It is articulated in cognitive discourses but narrated in religious stories and symbols as well.

younger person to live the life that the elderly have already lived (this is Daniels' Kantian and Rawlsian line).

Elements of *ageism*—defined by Robert N. Butler as: "The systematic stereotyping of and discrimination against people simply because of their age" (Butler 1969)—easily slip into debates about scarce resources. This becomes even clearer when, within a communitarian 'natural lifespan' approach (Callahan 1987) that tries to fix the thresholds in the human lifespan chronologically, Callahan uses "age as a specific criterion for the allocation and limitation of healthcare" by denying publicly financed, life-extending healthcare to persons who have lived out a natural lifespan. Although no precise chronological age can be set for determining when a natural lifespan has been achieved, it will normally be expected "by the late 70s or early 80s" (Callahan 1987, p. 171).

A capability approach seems to offer a better opening for a non-biased perspective on good aging than the utilitarian, Kantian, or communitarian theories indicated above. But here also one needs to be on guard for ageism. "Being able to live to the end of a human life of normal length; not dying prematurely, or before one's life is so reduced as to be not worth living" is the first of Nussbaum's central capabilities in her neo-Aristotelian approach (Nussbaum 2006, p. 76). Indeed, what makes a life worth living is up to individual persons themselves, irrespective of their chronological age. One can be over 100 and still (but beware of the hidden prejudice in this 'still') have an insatiable appetite for life. But the trap of ageism starts to open up again, also in Nussbaum's theory (although it is first of all meant as an effective instrument against infant death), as soon as implicit collective images of what 'normal length' and 'prematurely' stand for pop up.

The concept of an *ageless self* may, as a rebound, be taken as an antidote against ageism in issues of intergenerational justice and healthcare allocation. Indeed, the expected quality of life offers a better, more just criterion for the allocation of medical treatment than chronological age alone. In addition, that the self is ageless is an understanding with much psychological and phenomenological evidence (cf. Kaufmann 1986; De Beauvoir 1972). But I am afraid that the concept of a timeless identity does not take us any further in developing views of good aging. Banning watches, calendars, and diaries from one's life may be an understandable act of revolt against society the moment one retires. It may quite effectively overcome the hegemony of the social clock for a while. But biographical time will move on, continuing to affect one's body and one's relationships, like Cronus devouring his own children (Baars 2006). The presumption of a timeless, disembodied, isolated transcendental subject may be a controversial option in rational theories of justice (Sandel 1982), but it misses the point in developing views of good aging.

11.3.2 Existential Gerontology

Views of good aging should not only be concerned with the optimal general life conditions of the elderly but also with how to make the best of growing old biographically, in terms of personal human flourishing. What are the goods *in*—perhaps even more daringly: What are the goods *of*—aging? Here ethics does not need to start from scratch. A vast body of gerontological knowledge has become available since Elie Metchnikoff proposed in 1903 the establishment of this new scientific discipline, researching the biological, social, and psychological aspects of old age, and developing strategies and therapies to improve an old person's life. The hard data of empirical research have taken over the well-intended advice of centuries of wisdom, self-help, and consolation literature, starting with Cicero's *De Senectute* (cf. Cole 1992). As an applied science, gerontology provides us with a treasure trove of 'best practices', even if 'best' and 'good' in this context do not often mean more than temporary evasions of the hardships of old age.

Here the disciplinary frontiers between gerontology and ethics become blurred. Empirical knowledge gained from the objectifying position of the gerontological spectator can be reinterpreted in a normative mode. I think therapeutic ethics turns into *existential gerontology* here: It does not speak about 'old age' as an objective phenomenon but reflects on the lived experience of older persons, bringing them as much human flourishing as possible by helping them find an optimal relation to their own aging. I use the term 'existential' here broadly, without reference to any particular existentialist philosophy (cf. Cole 1992, p. xxv). As a philosophical perspective, 'existential', as I understand it, has to do with the recognition that we are situated subjects, embodied and living in time, having a pre-reflective relationship to the perception of our bodies in space and in time. An existential gerontology critically articulates and ethically evaluates concepts, stories, images, models, and methods that match old people's first-person experiences.

I would like to mention two examples of gerontological knowledge that can help the elderly live a better life. The Berlin gerontologists Paul B. and Margret M. Baltes defined a successful lifespan development as the conjoint maximization of gains (desirable goals or outcomes) and the minimization of losses (undesirable goals or outcomes) (Baltes 1997). They discovered a pattern in how elderly people, confronted with increasing physical and mental limitations, develop adaptive strategies to reach a new balance enabling them to optimize their social functioning and personal well-being. This strategy of Selective Optimization with Compensation (SOC) entails that, after an impairment due to age, an aging person will respond with a combination of three strategies: *Selection* (a different priority in goals or abandoning one goal for another); *Optimization* (realizing the restricted goal(s) by optimal and intensive efforts to enhance one's use of means and resources); and *Compensation* (putting other skills and resources in effort as a response to the loss of means). They presented the concert pianist Arthur Rubinstein as an example. As an 80-year-old, he was asked in an interview how he managed to maintain such a high level of performance when playing piano, he explained his coordination of three strategies. First, he played fewer pieces by concentrating on Chopin (selection), he practiced these pieces more often (optimization), and to counteract his loss in speed he now used a kind of impression management, playing more slowly before fast segments to make the latter appear faster (compensation). Why should we not reread Baltes' description of this often unconscious practice in a reflective and prescriptive manner? Why should we not rephrase it into a hypothetical imperative: "If you want to go on living a good life in old age, develop a reflective attitude towards your aging in selective optimization with compensation." In public welfare policy, this imperative could be creatively worked out in prevention campaigns and training programs for the elderly.

The same obtains for another gerontological insight, also employable in empowerment projects, that aims at an increase in self-management competencies in older persons. This example touches closely on one's identity and self-esteem in high age. The gerontologists Brandstädter and Greve (1994) distinguished three dynamic psychological processes in aging people that enable them to relate existentially to their changing bodies and minds. From an ethical point of view, they can be considered strategies of dignity: They allow people to keep their sense of self-respect and self-esteem into high old age. Experiences of decline and loss may have a negative impact on one's identity. A person is no longer the one he or she used to be and accepted as being. One can no longer be the person one intends and wants to be. The first option is to try to undo the loss by actively changing one's condition, in order to let it fit one's familiar self-image (*assimilation*) again. Loss of memory is fought by notes, myopia by glasses. One's self-image appears undamaged; one's identity does not need to be adjusted. A second strategy is called *accommodation*: It pops up when the first option reaches its limits. Projects are abandoned, plans are changed, goals are reformulated. One needs simply to admit and accept that one is getting older and integrate that acceptance in the story of who one is. Accommodation aims at the ability to 'reinvent' oneself and presupposes that people have not foreclosed the narrative of their life already. One's life plan—defined by Bernard Williams as "a nexus of projects which largely give meaning to life"—can still be adapted or even rewritten. Self-acceptance may not only imply the acceptance of the person one has become but also of becoming someone other than the person one has been. Older persons capable of this plasticity in their aging show—as gerontologists say—a 'resilience' that contributes to the optimization of their well-being in every respect.

Those unable or unwilling to reinvent themselves may opt for an *immunization* of their identity. They are and continue to be, in their own perception, the individuals they once were and do not want to reconsider themselves the person they have become now. They freeze their identity and isolate themselves from any feedback from their environment making a renewal of their self-definition impossible. The thermostat in their identity machine broke down. Mark Freeman speaks of 'narrative foreclosure'. The book of one's life is closed; no new chapter can be written.

Many older persons will more or less consciously use a mix of these three 'techniques' to keep up their sense of self-worth. A therapeutic ethics, however, will extract a normative maxim from them: *If one wants to make the most of one's aging, one needs to continue to re-invent oneself by adjusting one's self-image time and again to one's biographical age.*

A gerontological ethics developed in this vein cannot be anti-ageist in the sense that it tries to deny or stop the arrow of time from flying. It has to follow that arrow and stay as close to it as possible. Gerontological ethics is a *pro-ageism* product: Making the best of old age cannot do without the *recognition* (*Anerkennung*) of the inevitability of aging, as Hans-Martin Rieger rightly argues (Rieger 2008). But his German *anerkennen* has many layers, with far-reaching normative, anthropological connotations. Should the *Anerkennung* of aging also mean its acceptance, acknowledgment, approval, even acclamation and appreciation? Is aging *as such* a good thing?

Here the empirical positivism of gerontology reaches its limits. Within theories of active, healthy, or 'successful aging', the question what—if any—is the meaning of growing old is not dealt with, let alone answered in the positive. But I think ethics cannot refrain from the ontological. An existential gerontology has to deal not only with what are the goods *in* old age are but also with what the goods *of* aging are.

The founders of gerontology and developmental theory were less hesitant in putting these questions on their agenda than their empiricist successors. They deliberately touched on the existential meaning of the final part of life. In their theories of growing old as a period of 'disengagement' (E. Cumming and W.E. Henry), 'ego integrity' (Charlotte Bühler, Erik E. Erikson), 'mature wisdom' (Carl Gustav Jung), they all presupposed that old age represents a stage of life with specific qualities, not simply a bleak and outdated version of adult life (Daatland 2005). They thus provided old religious questions with new, secular answers. The meaning of old age in modernity, however, has shifted in their view: Its central task is no longer the preparation for eternal life but the development of a balanced life review. Old age, in addition, is not simply undergone as one's lot but constructed as a reflective project.

Perhaps these theories are not only outdated from an empirical point of view but also unsatisfactory from a more philosophical perspective. They might be too permeated by the pre-understandings of the European, white, male, well-educated bourgeois milieu of its authors. So the philosophical homework still needs to be done. Because the positivism of current gerontology prevents it from being an ally in these matters, an existential gerontology should look for support in sound philosophy and theology.

Without being able to explore this further, I will present some questions that should be dealt with. Should the human lifespan be regarded as a being-toward-death (*Sein-zum-Tode*, Heidegger) or as a being-against-death (Emmanuel Levinas)? Are we born to die or, on the contrary, born to start something new in the world (Hannah Arendt)? Does aging inevitably bring with it a diminishment of the appetite for life (Spinoza's *conatus essendi*), or does it lead to a different—although tragically final—taste of its never-ending flavour? Does death, coming closer in old age, 'belong' to life, or is death, up until the last breath of a living person, the ontological denial and opposite of life? As a philosopher and theologian, I would like to opt for the latter, without be able to argue my case here.[2]

These questions, however, are not just interesting for lovers of metaphysics: They have relevance for the general public because they influence vital policy decisions concerning life and death. In the Dutch discussion, for example, on assisted suicide after a 'completed life' (cf. Chap. 10), for example, they run through the entire debate,

[2] Impressed as I am by the forceful fragments from the legacy of Paul Ricoeur, written in his last years: *Lebendig bis in den Tod/ Vivant jusqu'à la mort*, Französisch- Deutsch, Felix Meiner Verlag: Hamburg 2007.

often without being made explicit. Indeed, it is not just a debate about autonomy, self-determination, the individual right to die, and the moral claim on others to assist actively; it is also a debate on the question what it means to claim one's life is 'completed'. Does the concept, as I shorten the issue now impermissibly, stand for the physical and/or mental exhaustion of an older person or for his or her biographical achievement? I personally have not yet reached the age to decide the question first hand. But the closing words with which Paul Ricoeur, then 87, signed his final great work—'Inachèvement, Paul Ricoeur' (Incompletion, Paul Ricoeur)—makes me feel less at ease with a concept like 'completion' (Ricoeur 2006 p. 506).

11.4 A Critical Task: Questioning Master Narratives and Paradigms of 'Successful' Aging

A gerontological ethics will be extremely sensitive to the often *implicit* normativity in, what I above called the elderly discourse and the aging discourse. *One-dimensional conceptions of good aging* may dominate the master narratives of aging, designed to guide the 'biopolitics' (state policy, biomedical and gerontological research) in our advanced liberal societies. (Rose 2001, 2007). I think an ethics that pretends to develop views of the good for aged people, needs to take sides with the elderly themselves, and must remain suspicious of becoming incorporated into elderly and aging *discourses*about them.

At first sight, collective master narratives of aging no longer seem available in late modern society. We have to construct our own individualized, reflective identities in old age as well, with no normative models available (Giddens 1991; Bauman 2001; Beck 2002). There is no 'normal' process of growing old. Even the choice of aging 'naturally', admitting no medical or technical end-of-life interventions is an intervention, a deliberate choice in the creation of one's own autobiographical myth (Rose 2001, p. 16, cf. McAdams 1993).

But individual constructions of meaning are not invented out of the blue, since they are mediated by public discourses with a strong power of definition. In mediaeval times, old age counted as the preparation for eternal life. Good aging consisted in public withdrawal and religious introspection (*memento mori*). Modernity replaced religion with work, and considered old age as the *dolce far niente* of not being forced to work any longer. Good aging then is: Enjoying one's retirement. Late modernity now comes with a third candidate master narrative: Third-Age hedonism: The vital, active, autonomous, and healthy senior as role model, and young-old age as the prime time for self-realization (Laslett 1989).[3] As master narratives create winners and losers, in Third Age hedonism, the new losers are the lower class, chronically

[3] Cf. *The Economist* (December 18–31, 2010), cover article: 'The joy of growing old (or why life begins at 46)': 'Stiffening joints, weakening muscles, fading eyesight and the clouding of memory, coupled with the modern world's careless contempt for the old, seem a fearful prospect—better than death, perhaps, but not much. Yet mankind is wrong to dread aging. Life is not a long slow decline from sunlit uplands towards the valley of death. It is, rather, a U-bend. When people start out on

ill, dependent, and—in general—the oldest old. Fourth Age is considered not as a new chapter in the book of life, but as a superfluous, senseless epilogue, hopefully shortened by the compression of morbidity (cf., for a more extensive critique, De Lange 2008 and Chap. 8, this volume).

One would expect a more critical stance from gerontology, especially when it comes to the meaning and specific qualities of the fragile and vulnerable Fourth Age. But gerontology as a discipline rarely functions as a visionary guide for society's outlook and policy on aging. More often, it tends to be a willing follower of prevailing biopolitics (Katz 1996). The dominant paradigm of the so-called 'new gerontology' of healthy and active aging offers an optimistic and activist view of what a 'good' old age can be, informing people that *whoever they are*, successful aging—defined almost exclusively in terms of vitality—can be attained through individual choice and effort. The policy advantages are obvious: Old bodies will be more controllable and less expensive if we keep them healthy, socially active, in charge of themselves, and feeling well. Accordingly, all emphasis in elderly policy will be laid on prevention, empowerment, and self-management in order to stimulate the resilience of old people's subjectivity (Holstein and Minkler 2003, quoting Rowe and Kane 1998). Although legitimate in its aim to slow down the inevitable physical and mental decay of old age—as I argued earlier in my contribution—the successful aging model has a strong *anti-aging* tendency in monopolizing one aspect of the aging process by neglecting an important other one. Aging is not only a matter of loss but a changing balance of gains and losses throughout a life course, as Paul and Margaret Baltes stated convincingly in their lifespan development model. Metaphorically speaking, while our physical and mental *hardware* may deteriorate, our *software* (but Baltes also speaks, out of a long, venerable tradition, of the 'wisdom' of old age) may compensate, even transcend the losses (Baltes 1987, 1997, cf. also the work of Tornstam 1997 on 'gerotranscendence'). Therefore, it should be a matter of public policy to assist older persons explicitly in finding this balance, not only by preventing the losses of old age but also by discovering, exploring, and exploiting its gains. Especially the oldest old among them who are struggling to find the meaning in and of their advanced age, are a matter of concern here but neglected by a policy that is predominantly focused on keeping the vital senior vital.[4] "[H]ow can we respect age if we do everything in our power to deny it?" the critical gerontologists Holstein and Minkler ask. "What most assume as a matter of course in youth and middle age—that is, health and activity," they rightly object, "cannot be the critical measure of success in old age" (2003, p. 795).

---

adult life, they are, on average, pretty cheerful. Things go downhill from youth to middle age until they reach a nadir commonly known as the mid-life crisis. So far, so familiar. The surprising part happens after that. Although as people move towards old age they lose things they treasure—vitality, mental sharpness and looks—they also gain what people spend their lives pursuing: Happiness.' (p. 33).

[4] I refer to the 'Expertise Network Elderly and Life Questions' in the Netherlands that stimulates organizations and workers in the care and welfare area to develop instruments for making questions of meaning explicit, especially among the oldest old. (http://www.netwerklevensvragen.nl).

What will be then the critical 'measure of success' in old age, in positive terms? I think the question should be on the agenda for ethics, but the answer postponed until further notice—for the first and last ethical expert in reflecting the good in and of aging will not be an ethical theory or a strategic agenda but the old person herself.

References

Baars, J. 2006. Het nieuwe ouder worden. Paradoxen en perspectieven van leven in de tijd. Amsterdam: Uitgeverij SWP.

Baars, J. 2010. Goed ouder worden: Een verkenning. *Tijdschrift Voor Humanistiek* 11(41): 6–15.

Baltes, P.B. 1987. Theoretical propositions of life-span developmental psychology: On the dynamics between growth and decline. *Developmental Psychology* 23(5): 611–626.

Baltes, P.B. 1997. On the incomplete architecture of human ontogeny. Selection, optimization and compensation as foundation of developmental theory. *American Psychologist* 52(4): 366–380.

Bauman, Z. 2001. *The individualised society.* Cambridge: Polity Press.

Beauvoir, S.De. 1972. *The Coming of Age.* New York: G.P. Putnam.

Beck, U., and Beck-Gernsheim. 2002. *Individualization.* London: Sage publications.

Brandstädter, J., and Greve, W. 1994. The aging self: stabilizing and protective processes. *Developmental Review* 14(1): 52–80.

Butler, R.N. 1969. Age-ism: Another form of bigotry. *The Gerontologist* 9: 243–246.

Callahan, D. 1987. *Setting limits. Medical goals in an aging society.* New York: Simon and Schuster.

Cole, T.R. 1992. The journey of life. A cultural history of aging in America. New York: Cambridge University Press.

Cole, T.R., R.E. Ray, R. Kastenbaum. eds. 2010. *A guide to humanistic studies in aging. What does it mean to grow old?* Baltimore: The Joihn Hopkins University Press.

Daatland, S.O. 2005. Quality of life and aging. In *The cambridge handbook of age and aging*, ed. Malcolm L. Johnson 371–378. Cambridge: Cambridge University Press.

Daniels, N. 1988. Am i my parents' keeper? An essay on justice between the young and the old. New York: Oxford University Press.

Giddens, A. 1991. Modernity and self-identity. Self and society in the late modern age. Cambridge: Polity Press.

Gilroy, R. 2006. Taking a capabilities approach to evaluating supportive environments for older people. *Applied Research in Quality of Life*, 1(3–4): 343–356.

Gulette, M.M. 2004. *Aged by culture.* Chicago: University of Chicago Press.

Harris, J. 1985. The value of life: An introduction to medical ethics. Boston: Routledge and Kegan Paul.

Harris, J. 2007. *Enhancing evolution. The ethical case for making better people.* Oxford: Princeton University Press.

Heller, A. 1988. *General ethics.* Oxford: Basil Blackwell Ltd.

Holstein, M., and M. Minkler. (2003). Self society, and the 'New Gerontology'. *The Gerontologist* 43(6): 687–796.

Katz, S. 1996. *Disciplining old age. The formation of gerontological knowledge.* Charlottesville: University Press of Virginia.

Kaufman, S.R. 1986. *The ageless self. Sources of meaning in late* life. New York: Meridian.

Kuh, D.K. 2007. A life course approach to healthy aging, frailty, and capability. *Journal of Gerontology Series A Biological Science*, 62(7): 717–721.

Lange, F.De. 2007. De mythe van het voltooide leven. Over de oude dag van morgen. Zoetermeer: Meinema.

Lange, F.De. 2008. De armoede van het zwitserlevengevoel. Pleidooi voor beter oud worden. Zoetermeer: Meinema.

Lange, F.De. 2010a. *Waardigheid. Voor wie oud wil worden.* Amsterdam: SWP.

Lange, F.De. 2010b. Good aging. A comprehensive approach. In *Successful aging, spirituality, and meaning. Interdisciplinary perspectives. Studies in spirituality*, ed. J. Bouwer Supplement 20. Titus Brandsma Institute, Peeters Leuven 2010, 171–189.

Laslett, P. 1989. *A fresh map of life. The emergence of the third age.* London: Weidenfeld and Nicolson.

Lloyd-Sherlock, P. 2002. Nussbaum, capabilities and older people. *Journal of International Development*, 14(8): 1163–1173.

McAdams, D.P. 1993. The stories we live by: Personal myths and the making of the self. New York etc.: Guildford Press.

Merleau-Ponty, M. 1945. *Phénoménologie de la perception.* Paris: Gallimard.

Moody, H.R. 1992. *Ethics in an aging society.*London: The John Hopkins University Press.

Nussbaum, M.C. 2006. *Frontiers of justice: Disability, nationality, species membership.* London: The Belknap Press of Harvard University Press.

Rawls, J. 1971. *A theory of justice.* Harvard University Press.

Ricoeur, P. 2006. *History, memory, forgetting.* Chicago: University of Chicago Press.

Ricoeur, P. 2007. Lebendig bis in den tod/ vivant jusqu' à la mort, französisch- deutsch. Hamburg: Felix Meiner Verlag.

Rieger, H-M. 2008. Altern anerkennen und gestalten. Ein beitrag zu einer gerontologischen ethik. Leipzig: Evangelische Verlagsanstalt.

Rose, N. 2001. The politics of life itself. *Theory, Culture, and Society*, 18(6): 1–30.

Rose, N. 2007. The politics of life itself. Biomedicine, power, and subjectivity in the twenty-first century. Princeton and Oxford: Princeton University Press.

Rowe, J. and Kane, R. 1998. *Successful aging.*New York: Random House.

Sandel, M.J. 1982. *Liberalism and the limits of justice.* Cambridge: Cambridge University Press.

Taylor, C. 2004. *Modern social imaginaries.* Durham: Duke University Press.

Tornstam, L. 1997. Gerotranscendence. The contemplative dimension of aging. *Journal of Aging Studies*, 11(2): 143–154.

Chapter 12
Justice and the Elderly

Anders Schinkel

> In the eye of nature, it would seem, a child is a more important object than an old man; and excites a much more lively, as well as a much more universal sympathy. It ought to do so. Every thing may be expected, or at least hoped, from the child. In ordinary cases, very little can be expected or hoped from the old man. The weakness of childhood interests the affections of the most brutal and hard-hearted. It is only to the virtuous and humane, that the infirmities of old age are not the objects of contempt and aversion. In ordinary cases, an old man dies without being much regretted by any body. Scarce a child can die without rending asunder the heart of somebody. (Adam Smith, The Theory of Moral Sentiments, 219)

> [C]are for children, elderly people, and people with mental and physical disabilities is a major part of the work that needs to be done in any society, and in most societies it is a source of great injustice. Any theory of justice needs to think about the problem from the beginning, in the design of the basic institutional structure, and particularly in its theory of the primary goods. (Martha Nussbaum, Frontiers of Justice, 127)

12.1 Introduction

Analytical philosophy has not shied away from such a big question as what a just society would look like, how it would be organized.[1] Sometimes a particular segment of society, or a specific dimension is singled out—often in response to a perceived lacuna—as in work on gender justice or justice and the family (e.g. Kirp et al. 1986; Robeyns 2007). When it comes to the elderly, however, broad perspectives and wider visions are nowhere to be seen. A thorough search of the academic literature reveals that the conjunction of 'justice' and 'elderly' has resulted in just a few topics for philosophical and ethical debate.

[1] Rawls (2003a, b, 2005) is the paradigmatic example. A recent critique of theorists' preoccupation with this question is Sen (2009).

A. Schinkel (✉)
Faculty of Psychology of Education, University of Amsterdam, The Netherlands
e-mail: a.schinkel@vu.nl

M. Schermer, W. Pinxten (eds.), *Ethics, Health Policy and (Anti-) Aging: Mixed Blessings*, 147
Ethics and Health Policy 1, DOI 10.1007/978-94-007-3870-6_12,

Healthcare rationing, or age rationing in healthcare, is one of the main topics (Daniels 1988, 2008, p. 488 ff; McKerlie 1989, 1992, 2001; Brock 1989; Wicclair 1993, Chap. 3; Williams 1997b). The question here is whether or not it is just to use age as a criterion to decide how healthcare resources are to be distributed among patients. A related but separate debate concerns the 'fair innings argument'. Is there such a thing as a 'normal lifespan', a 'fair share' of what life has to offer; and if so, does(n't) that mean that one's claim to healthcare resources is weaker? (E.g. Callahan 1995; Williams 1997a; Rivlin 2000). A second topic is that of (state) pensions: What does a just pension system look like? What challenges does an aging population pose in this respect? Such questions are central here (Johnson et al. 1990; Laslett 1992; Myles 2002, 2003; Schokkaert and Van Parijs 2003; Howse 2004; Willmore 2004). An important segment of the literature is not primarily concerned with these fairly concrete normative issues themselves, but rather with the question how to conceive of justice (a just distribution of resources or social goods) between age groups or generations; e.g. should we compare whole lives, or should we also make comparisons between age groups, or within the lifespan?[2]

Note that none of these debates takes as its starting-point the question concerning the elderly's claims to justice. Rather, they concern justice between age groups (or generations—this itself is, or has been, under debate), justice between the young and the old, rather than justice for the elderly.[3] The focus of these debates is also rather narrow, concerning distributive justice only, and more particularly the fair distribution of financial resources in healthcare or social security.

The third main topic is that of filial morality (see, for instance, English 1979; Callahan 1985; Post 1989; Li 1997; Keller 2006; Schinkel 2012). This topic does, to some extent, concern justice for the elderly.[4] Yet it is about a fairly limited—although very important—set of questions, such as whether children have special duties (including obligations of justice) to their parents, and if so, which. Moreover, the debate about filial obligations has largely neglected the relations between such obligations and questions of distributive justice or social justice in general.[5]

The general point made in this chapter is that the philosophical perspective on issues of justice related to the elderly should be widened considerably. Many important questions in this area are now neglected. In order to avoid this, we should ask what a just society would look like in its elderly-related aspects. What would be

---

[2] See especially the work of Daniels and McKerlie.

[3] Of course, when a just distribution of (financial) resources *between* the young and the old is reached, this must be just both to the young and the old, so that in this respect justice *for* the elderly has also been done. But the starting-point is very different from the one I suggest in this chapter. (In fact, these debates are often sparked and fuelled by a concern that the elderly are taking *too much*, leaving too little for younger generations.)

[4] Most contributors to this debate do not explicate whether the duties of adult children towards their parents are *moral* duties or, more specifically, duties of justice. It is *my* contention that some of the issues at stake here are matters of justice, but I believe that some of the contributors to this debate would agree (e.g. Sommers 1986; Post 1989).

[5] Wang (1998) and Thompson (2003) do address these connections, as do I in Schinkel (2012).

the place of the elderly in a just society? What would be their role, how would they be treated, what would be the attitude towards old age and the elderly, as expressed in the behaviour of members of younger generations, as well as in institutional arrangements?[6]

To ask this big question, or set of questions, presupposes that it is possible to make meaningful general statements about the place of the elderly in a just society, and this in turn presupposes certain answers to two other preliminary questions, namely whether it would be ageist to even ask this question, and whether it makes sense to speak of 'the elderly'. My view is that one can legitimately speak of 'the elderly'—as long as one is carefully aware of the dangers involved in using such a general term—and that to ask about the place of the elderly in a just society does not presuppose an ageist perspective. I offer some arguments for this in 3.1.

So, if we can meaningfully speak of 'the elderly' without being ageist, what are we after when we ask about the place of the elderly in a just society? Here is a brief answer. Justice does not dictate one form of life, one single way of organizing society. I cannot argue for this here, but I can explain what I mean. I do not believe that all sound thinking about justice must result in one and the same picture of a just society. It is much more plausible to think that the principle of justice can—in theory, at least—play a regulatory role in various types of societies as they arise and develop in different geographical and historical circumstances. This will be easier to accept by people who use a threshold-conception of justice—so that a just society would be one, for instance, in which no-one is allowed to fall below this threshold—or those who take 'justice' to apply only to the basic institutional structure of a society, than, for instance, by perfectionists, whom thresholds cannot satisfy, and who would like to see the whole of society modelled after a particular substantive conception of justice. Even perfectionist conceptions, however, would have to allow for the possibility of a diversity of just societies, since the formal concept of justice will always legitimately derive part of its content from the social world as it is. For instance, respect and disrespect may have different forms in different societies, and certainly not all of this variety can be ruled out a priori as being incompatible with justice.

This also means that different roles for elderly people, and different attitudes towards the elderly, are compatible with justice. For instance, in some societies the elderly may continue to play an active role in the economy—and may have to do so for the sake of the survival of the community—whereas in other, more affluent societies something called 'retirement' may exist, a pension age and a pension system

[6] For the purpose of this chapter, I do not consider it necessary to define what I mean by a 'just society' with great precision. A just society would contain no serious structural injustices, either in the (basic or nonbasic) institutional structure, or in societal or generational attitudes towards specific groups. But it would not have to be perfectly just, if that were to mean the total absence of any form of injustice. It would still be possible for some injustice to occur, for instance as an unintended side-effect of certain policies; but in a just society action would be taken to remedy this, of course. A just society, then, is an ideal – but it should be formulated as a realistic ideal. An assumption underlying this chapter is that even present-day affluent Western societies fall far short of this ideal when it comes to the place and treatment of the elderly in these societies.

that allows elderly people to retire from their working lives. Neither is in itself just or unjust; the judgement in each case depends on many contextual variables.[7]

The answer to the highly general question as to the place of the elderly in a just society, then, can only be equally general and abstract—initially, at least, for it can be 'operationalized' in a number of less abstract questions. What is always owed, to everyone, is recognition. As will become clear in Sect. 2, I do not use this term as recognition theorists like Axel Honneth (1995) and Nancy Fraser (1996, 2000) do. They are much more interested in societal groups' emancipatory struggles for recognition, in marginalized group's efforts to acquire a fuller social status. While this could perhaps provide an interesting perspective with regard to the elderly as well, it is not the perspective I take here. Nor do I treat recognition, as Fraser does, as a dimension of justice alongside (re)distribution and participation. Rather, I see recognition as the primary act of justice, and its basic form. A just society gives the elderly the recognition they are due—not collectively, not as a group, but simply as individual people. This is the central claim of this chapter. The formal conception of justice on which it relies will be explained further in Sect. 2 and fleshed out in terms of more concrete issues of justice in Sect. 3, showing how philosophical debate on justice and the elderly can and should be widened. This section also offers examples of what positions the perspective of justice as recognition urges us to take or reject on various issues. Together these sections constitute the main body of this chapter. The concluding Sect. 4 sketches some of the policy implications of justice as recognition.

12.2 Justice as Recognition

12.2.1 Justice and Recognition

Justice may be defined as the fulfilment of legitimate claims. That is, to practice justice is to fulfil the legitimate claims that others have upon one; the principle of justice is the general moral requirement that legitimate claims be fulfilled. The virtue of justice is the disposition to fulfil legitimate claims. A just institution is one that tends (within the limits of the function for which it was designed) to fulfil and promote the fulfilment of legitimate claims. Finally, a just society is a society that is organized such as to fulfil and promote the fulfilment of legitimate claims; one that is characterized by just institutions, and the tendency to promote the virtue of justice.

There can only be a legitimate claim to something that could, in principle, be provided; I cannot be entitled to what is simply not there, and cannot be brought about by (reasonable) human efforts. Legitimate claims of the kind with which we are concerned here may be based on a variety of grounds, either substantial (e.g. need, desert, effort, contribution, reciprocity) or formal (equality, the principle of treating

[7] The distinction between ideal and non-ideal justice is also relevant here. Some policy or course of action may be just in non-ideal circumstances, even if it is unjust from an ideal-theoretical perspective. For this distinction see, for instance, Robeyns and Swift (2008).

like cases alike). Normally, then, my sudden craving for an ice-cream generates no legitimate claim on anyone, no obligation on the part of anyone to (help me) satisfy my craving. But my need for sustenance does generate a legitimate claim on others—at the least a claim to their non-interference with my legitimate efforts to fulfil my own need.[8] In short, for it to be a matter of justice that a claim be addressed, the claim must concern an important good.[9]

On one level, recognition may simply concern legitimate claims and their grounds (e.g., recognition of need, merit, contribution, or some other substantial ground of justice). But this—to recognize a person's claim as legitimate in the relevant sense—presupposes a more fundamental kind of recognition, namely recognition of the person in question as a source of legitimate claims.[10] To recognize a claim as a legitimate claim in the sense that one recognizes it as engendering an obligation on one's own or someone else's (possibly some organisation's or institution's) part, is not a merely intellectual exercise. Rather, it implies that a disposition to act on the obligation is created, or a desire that the responsible third party do something to meet its obligation. Similarly, recognition of another as a source of legitimate claims is not only—not even primarily—an intellectual matter, but also an affective and volitional one, bound up with a readiness to act in certain ways.

All efforts to establish justice, whether it be distributive, compensatory or even retributive, ultimately presuppose recognition in the fundamental sense, recognition of each person's 'intrinsic claim for justice'.[11] The term 'recognition' expresses the idea that there is something about those within the scope of justice that makes them the proper object of considerations of justice, something that forms the basis for respect and recognition—something that is, indeed, recognized. Whether this is expressed in terms of (human) dignity (of the 'inalienable' kind), 'infinite preciousness', or 'inherent value' is not of great importance.[12] The latter term is perhaps most to the point in identifying what is at stake here, namely the idea that each person is a bearer of value, apart from its instrumental value to others, the moral qualities of its action

[8] I will say more about the kinds of claim needs may generate in 3.3.

[9] Two popular ways in which these goods have been defined are as 'primary goods' ("things that every rational man is presumed to want" because they "normally have a use whatever a person's rational plan of life", according to Rawls (2003a, p. 54); or "various social conditions and all-purpose means that are generally necessary to enable citizens adequately to develop and fully exercise their two moral powers, and to pursue their determinate conceptions of the good", according to Rawls (2003b, p. 57)), and as 'capabilities' (and/or 'functionings'), which may be defined as actual opportunities and capacities for worthwhile ways of functioning; see, for instance, Sen (1999) and Nussbaum (2006). For a comparison between these two 'metrics' of justice see Brighouse and Robeyns (2010).

[10] This is *not* a phenomenological description of the experience of recognition, but an interpretation of it in terms of its relevance for considerations of justice.

[11] Tillich (1954, p. 63) speaks of "the intrinsic claim for justice of everything that has being". I will not be concerned here with the question whether we should indeed include 'everything that has being' in the scope of justice.

[12] The term 'human dignity' is often used by those who wish to provide a basis for universal human rights; e.g. Griffin (2008). Gaita (2004: esp. the preface) speaks of 'infinite preciousness'. Regan (1983) uses the term 'inherent value'.

and character, or some other individuating characteristic. The important point is that when we act from an awareness of such value, we respond to something—we are not merely attributing value to someone—and when we do so this behaviour can also be said to reveal something about the other; Raimond Gaita calls this 'infinite preciousness', a less sentimental term is 'inherent value'.

12.2.2 Recognition and Equality

Whenever justice is based on a particular ground, such as need, merit, or effort—whenever we use one of these as a criterion of justice—we implicitly introduce the formal principle of equality. This formal principle of justice, that like cases ought to be treated alike, is implicit in, or presupposed by, application of the aforementioned substantial criteria.

For example, when we recognize the importance of desert or merit as a ground of claims people may have on each other, and we use desert as a criterion for, say, assigning awards, we are bound to assign the same award to equal merit, regardless of whose merit it is. For to say, for instance, that John deserves X on the basis of his fantastic Y_j, while denying that Robert deserves X on the basis of his equally fantastic Y_r, does not make sense. That is, it would no longer be true that John deserves X on the basis of Y_j alone. John's deserving X would seem to be related, somehow, to his being John—which, most likely, is no mean feat, but the same could be said about Robert. Applied this way, however, the concept of desert would become completely particularistic and uncontrollable. If Robert's merits are the same as John's, but they are treated unequally, either some other criterion intervened or was joined with that of desert, or an element of arbitrariness was introduced as a result of which the criterion of desert was suspended. It is fair to say, then, that desert normally presupposes equality. A similar argument could be made for the criteria of need and reciprocity. What all these criteria presuppose is a formal criterion of equality: that like cases should be treated alike.

But we can go one step further, and argue that (formal) equality only matters because of a fundamental substantial equality, namely the 'possession' of inherent value (or whatever name one prefers); not to treat like cases alike is a denial of inherent value (most obviously, but not only, of the one who receives less than she is owed), implying that fundamentally, one of the (human) beings concerned is more important than the other. Recognition of inherent value, then, underlies the formal principle of equality.

This formal principle, although implicit in the application of substantial criteria of justice, is often taken to have an independent force and plausibility as well, as a principle of fairness. From this perspective the order of considerations would be different: Justice requires us to treat like cases alike; and what like cases are is determined by one or more substantial criteria. These specify the meaning of 'like cases'; they impose a perspective on the group of potential recipients of justice, on those within the scope of justice. But this presupposes that we have determined the scope of justice—who or what falls within the scope and who does not. Again, recognition of others as sources of legitimate claims logically precedes application of the formal principle of equality.

Justice as recognition is bound up with the notion of substantial equality. Recognition of inherent value is the primary act of justice, and no-one has more inherent value than another.[13] This is why we object to arbitrary discrimination, including age discrimination. If people are taken less seriously, or their needs are neglected, simply because they are old—perhaps because of some prejudice against old people—this is discriminatory, because it implies a denial of their equal inherent value, a false suggestion that they are less important than others (see Sect. 3.1).

12.3 Some Dimensions and Domains of Justice in Relation to the Elderly

The obligation to give every human being the recognition they are due as human beings is at once highly abstract and (in some contexts) very concrete. It is abstract in the sense that this principle does not tell us how to behave in particular situations, except that it has to be in accordance with the value we find inherent in those involved in the situation. Yet at the same time it can be very concrete in the sense that it applies to all our dealings with concrete others (as well as more abstract others—those further removed from us in space and time), and that recognition or the lack of it are (literally) embodied in our attitude, posture, words, and actions. In face-to-face interaction, the obligation to recognize the other as a source of legitimate claims—or as we might put in everyday life: as a human being (like oneself)—becomes highly concrete. Recognition is to some extent something one can do, and do immediately, i.e. unmediated by goods distributed or institutional arrangements adapted to meet criteria of justice.

Yet the latter are also important instruments of recognition, and sometimes the best way to show one cares, and to show that one recognizes others' inherent value is to manipulate resource flows or legal and institutional arrangements. These, too, may exhibit an 'attitude' of (dis)respect to (groups of) people. A healthcare system biased towards particular age groups would constitute a good example.

Below I will briefly discuss three dimensions of justice as recognition and two domains in which recognition is of particular importance.[14] The three dimensions distinguished here are: non-discrimination (Sect. 3.1), just distribution (Sect. 3.2), and just care (Sect. 3.3). The two domains I will discuss here are the family (Sect. 3.4) and society in general (Sect. 3.5).

[13] There are different ways of dealing with intuitions about so-called 'marginal cases' or differences between human beings and animals, but I cannot go into this here.

[14] The distinction between dimensions and domains is intuitive. A dimension of justice constitutes a specific sense in which people may be treated justly or unjustly; each dimension offers one coordinate of justice in any concrete situation. Domains are relational settings in which justice may or may not be done. But some dimensions may have a connection with a particular domain and vice versa; I treat 'just care' as a dimension of justice as recognition, for instance, but 'care' could also be seen as a domain of justice.

12.3.1 Non-Discrimination

Adam Smith observes that people tend to care more about young children than about old people, and there is a clear suggestion in the quotation I used as a motto that this has to do with our response to (seeing) the 'infirmities of old age'. Most young and healthy people probably dislike the sight of flecked and wrinkled skin, bent and trembling bodies, especially when the old people whose bodies they are, are no longer in complete control of them.[15] (Apart from the sight, there may also be sounds and smells perceived as disagreeable.) There is a real danger that the aversive reaction to such 'infirmities' leads to a diminished concern about the people suffering from them, to some form of what in the psychological literature is called 'moral exclusion'. As Susan Opotow (1990, p. 2) writes: "Milder forms of moral exclusion occur when we fail to recognize and deal with undeserved suffering and deprivation. The other is perceived as nonexistent or as a nonentity. In this case, harm doing results from unconcern or unawareness of others' needs or entitlements to basic resources (. . .)." In the case of the kind of aversion mentioned above, people may not want to see the elderly people in question, if only due to some primitive mechanism that makes people unable to cope with their own frailty and mortality, and therefore may choose to block out these elderly's needs, interests, et cetera.[16] But there may also be a more direct link between perceived weakness and a tendency to disregard the other's interests.

People's aversion of the 'infirmities of old age' may be one cause of (passive) discrimination of elderly people, i.e. of a lack of recognition of elderly people as sources of legitimate claims, equal to any other person.[17] But the stereotypical association of old age with such infirmities may itself be a root cause of ageist discrimination.[18] For instance, it may lead people to assume that elderly people in general will benefit less from medical treatment than younger people, which could then be used as an argument for age rationing in healthcare. But, as Rivlin (1995, p. 1180) notes, research shows that "[i]n many cases (. . .) elderly people's response to treatment is as good as young people's".

---

[15] See Keizer's contribution to this volume.

[16] Greenberg, Schimel, and Mertens (2002) relate prejudice against elderly people to fear and anxiety about one's own mortality.

[17] 'Active' discrimination expresses more than a lack of recognition. It shows positive disrespect rather than a 'mere' lack of respect; the person discriminated against is explicitly perceived as worth less than others. This is probably relatively rare when it comes to the elderly. Levy and Banaji (2002) note that ageism differs from prejudice against religious, racial or ethnic groups as well as gender prejudice in that it is much more implicit: there are no hate groups that target the elderly, as there are for the other groups; however, also unlike with religious, racial, ethic or gender prejudice and discrimination, "social sanctions against expressions of negative attitudes and beliefs about the elderly are almost completely absent" (50).

[18] Wicclair (1993, p. 82) defines policies, practices, beliefs, et cetera, as ageist "if they are based on false or unfounded universal or statistical generalizations that attribute negative characteristics to the elderly or old age", such as that all elderly are forgetful or depressed.

One might wonder whether asking, as I do, about the place of the elderly in a just society is itself an expression of ageism. Anita Silvers "questions the common assumption that social justice authorizes disparate treatment of people who are at different stages of the lifespan" (Silvers 2000, p. 204). She is suspicious of the idea that the mere fact of being of a certain age and at a certain stage in one's life would justify different treatment (for instance in a medical context). She is rightly wary of the danger of ageism: to discriminate on the basis of the criterion of age, where this criterion is irrelevant[19]—and one might argue that the criterion of age as such is never relevant, except perhaps where law has made it relevant as a proxy for some other, ultimately relevant criterion. In this light it might seem ageist to ask about the role and treatment of the elderly in a just society, for this might be thought to assume that the elderly require a special treatment simply because they are old (i.e. above a certain age).

In reply I would say that to belong to the age group of the (young, middle-aged, old, or very old) old is never a mere fact.[20] One's age is never just a number, but it means that one is at a certain, to an important extent societally and to some extent biologically defined, stage of one's life. For instance, Western societies have created a group of people (above 60–65, usually) who have completed their working-life, and whose place in society is clearly distinct from that of people of or below working age. As long as we remember that 'the elderly' do not exist, in the sense that they are not a homogeneous group, such facts as the above warrant our speaking of 'the elderly' in certain contexts.[21] It does not presuppose that (old) age in itself must entail differential treatment, but draws attention to the fact that people above a certain age (above 60 or 65, say) tend to be in a markedly different life situation from people below that age. For many purposes it will be necessary to be more precise, to say which elderly people exactly we are talking about—institutionalized elderly, for instance, or those suffering from dementia. But in Western societies as they are today, to speak of 'the elderly' is often useful and justifiable—even if it is to criticize socially constructed age categories.

I now turn to one of the main philosophical debates concerning justice and the elderly, that about age rationing in healthcare. The central question in this debate is whether it is acceptable to use age (in particular: old age) as a criterion for the distribution of resources or the offering or withholding of treatments in healthcare. Is it ageist, discriminatory, to use age as a criterion for the distribution of scarce healthcare resources, or not? The question clearly pertains to issues of ageism as well as distributive justice. I will discuss Daniels' 'prudential lifespan account', which supports age rationing, in the subsection on distributive justice. In the present subsection I will look at some other arguments for age rationing, paying most attention to Callahan's 'natural lifespan argument'.

---

[19] This is a short statement of the 'broader' definition of ageism; the 'narrow' definition says that "ageism is discrimination against *older* people on grounds of age" (Bytheway 2005, p. 338).

[20] This is the categorization used by Burnside, Ebersole, and Monea; see Bentley (2007, Chap. 4).

[21] For the heterogeneity of 'the elderly', see Gilleard and Higgs (2000, 2002).

One thing does need to be said here about Daniels' prudential lifespan account. Daniels (2008, p. 478) argues that "treating people differently by age will not invariably or even generally create the objectionable inequalities between persons that race or gender bias produces, as long as the differential treatment operates uniformly over the lifespan of different people as they age".[22] The question is, however, whether it would not still be discriminatory to favour the young over the old (by age rationing of healthcare resources, for instance). Imagine that all human beings were white all of their life, except for one period (between the ages of 20 and 30, say) in which they all turned black; imagine further that black people—that is, everyone during the time they were black—were looked down upon, socially excluded (until they turned white again) and so on. The fact that everyone (reaching that age) would face this treatment would not make it any less discriminatory. Of course, Daniels provides a justification for differential treatment of age groups that is lacking in this (extremely hypothetical) example; but the point here is merely that the mere fact that such differential treatment does not involve discrimination between persons is not enough to show it is not ageist.

Common arguments in favour of age rationing are that elderly people are less likely to benefit from medical treatment than younger people, or that age rationing (in favour of the young) would maximize health benefits. Wicclair deals convincingly with these arguments, noting that, even if they do not reflect ageism (i.e. even if they are not 'based on false or unfounded universal or statistical generalizations that attribute negative characteristics to the elderly or old age'), they do show age bias, since they unjustly favour a particular age group (Wicclair 1993, p. 82). Why? Because these arguments use a suboptimal criterion (age) to maximize or increase the likelihood of health benefits; if this is the objective, Wicclair argues, overall health status is a much more appropriate criterion (ibid., 86 and 87).

Daniel Callahan famously offered another argument, centred round the notion of a 'natural lifespan': "one in which life's possibilities have on the whole been achieved and after which death may be understood as a sad, but nonetheless relatively acceptable event" (Callahan 1995, p. 66, also quoted in Wicclair 1993, p. 93).[23] A natural lifespan, according to Callahan, tends to be reached between people's early seventies and early eighties (Callahan 1995, p. 148). Healthcare resources should be used to help people attain a natural lifespan and to relieve suffering, also beyond that age, but not to extend the lives of those who have already attained a natural lifespan (ibid., p. 53). At some point, elderly people have had 'their fair share of resources' (ibid., p. 140). Although Callahan's proposal may have some intuitive appeal, it is problematic for a number of reasons. It is unclear why elderly people should care less about pursuing their interests because they have already achieved 'life's possibilities', or—and this is particularly problematic from the perspective of justice as recognition—why others should care less about their pursuit of these

[22] His main treatment of this issue is in Daniels (1988).

[23] The idea of 'fair innings' expresses a similar idea somewhat differently, suggesting that those who attain what Callahan calls a natural lifespan have had their 'fair share' of life; for a critique of the fair innings argument see Rivlin (2000).

interests. Some older people may not develop new interests anymore, but the same can be said for many younger people (Wicclair 1993). And though it is true that we tend to react differently to older people's deaths than those of young people, this may depend much more on the manner of their death than on the age at which they die as such (Wicclair 1993). Moreover, our different responses may simply reflect ageist attitudes of the kind observed rather dryly by Adam Smith in the passage used as a motto to this chapter.

Imagine the following scenario, however: a 15-year-old girl and a 75-year-old woman arrive at the same hospital at almost the same time; the old woman comes in just before the girl. They have both been involved in a traffic accident (or separate accidents, for which neither was to blame), as a result of which both require a liver transplant. Otherwise, they are both fit and healthy, and the old woman has led a satisfying life so far. Yet only one donor liver is available, so only one of them can be saved. Now imagine the old woman knows all this, is conscious and—assuming the hospital has a first come first served policy—claims the donor liver for herself. How would we evaluate her act? It is in cases such as these—imagined choice-situations where we are forced to choose between the lives of older and younger people—that Callahan's argument is intuitively appealing. We look differently upon the older woman's claim than we would have upon the same claim, if it were made by someone much younger, a 30-year old for instance. If the old woman had 'given' the liver to the girl, our moral sense would have been satisfied; it would also have been a more 'beautiful' decision.

Even if there is something to Callahan's ideas, however, it does not follow that the healthcare system should employ age rationing. For perhaps justice requires that more resources be channelled to healthcare (including campaigns to convince people to donate their organs), to prevent trade-offs between the old and the young from occurring.[24] As Rivlin (2000) argues, to determine whether someone has had a fair share of resources, we first need to know how many resources there are in total (as well as how many people there are to divide it amongst), and this is the result of political decision making.

Age rationing in healthcare is just one issue relevant to the problem of discrimination of the elderly. Compared to it, other issues that might come under that heading have received very little attention in the philosophical literature. Discrimination can occur on an interpersonal, societal or some intermediate level. The debate about age rationing moves between an intermediate (institutional) and a societal level. More attention should go to the lack of recognition (and therefore the injustice) that may be embodied in personal and societal attitudes towards elderly people. This brings into view a variety of important questions, for instance regarding the problem of paternalism in the care sector, the legitimacy of forced retirement, and present-day Western society's glorification of youth and health and its implications for the elderly and their ability to experience a meaningful old age.

[24] Wicclair (1993, p. 89): "Justice can require greater expenditures."

12.3.2 Just Distribution

Distributive issues are well-represented in the literature: both the age rationing debate and the debate about the pension system concern the distribution of financial resources. Concern about the aging of Western societies and the costs (in healthcare and social security) that might bring also sparked a debate of a more abstract nature, on how we should think about justice between generations or age groups. Norman Daniels (1988), expanding and modifying the Rawlsian approach to intergenerational justice, develops a prudential lifespan account, which reduces the question of justice between age groups (i.c. 'the young' and 'the old') to one of prudence within a single individual's lifespan. After all, the young will become old; so why think of them as distinct groups? Daniels' basic idea is that if rational people have to decide about the distribution of resources for (health) care across the lifespan, having no knowledge of their own age or life plan (or conception of the good), and knowing they will have to live their whole life with the choice they make now, they will favour a healthcare system that includes age rationing—and this system, according to Daniels, would therefore be just. They would choose such a system because they would rather increase their chances of reaching a normal lifespan by allocating more resources to earlier periods in life, than gamble that they would reach such a lifespan without those resources, so as to be able to use them to prolong their life beyond the normal lifespan.[25]

Dennis McKerlie, on the other hand, holds that the lifetime perspective needs to be supplemented by a comparison of life-stages (e.g. McKerlie 2001, p. 163 ff.). In some cases, priority should be given to people simply because they are badly off; McKerlie calls this the 'time-specific priority view' (ibid., p. 165). McKerlie believes that "prudence would save very little for extreme old age, which is why we should hesitate to use it as the test of justice for the elderly" (ibid., p. 161). He does note, however, that when elderly people's claims derive from their being badly off, they 'do not depend on their being in a distinctive stage of life, old age'. "[T]heir claims must compete against the claims of younger people who are also badly off." (ibid., p. 166)

From the perspective taken in this chapter, in which justice as recognition is central, it seems problematic to focus on whole lifetimes alone. There are some things or conditions we cannot allow people to suffer—if we can prevent it—even if it happens to everyone if they live long enough to reach that stage. Equality between people (between their whole lives) is not all that matters; no allocation scheme should allow dependent elderly people to go without the necessary nursing care. That we would all do without it is neither here nor there. At any stage in life, people should be able to live with such dignity as is available to them in their condition. It should be noted that, although this argument counts against the idea that equality between lifetimes is all that matters, my conclusion here is not at odds with Daniels' view, since his argument for age rationing pertains only to 'very expensive or very scarce life-extending services', not to nursing care and pain relief (Daniels 1988, p. 87).

---

[25] There is obviously some affinity here between Daniels and Callahan.

It seems, then, that implicit in Daniels' prudential lifespan account—and more explicit in McKerlie's view—is an acknowledgement of the importance of a fundamental noncomparative dimension of justice, the dimension of recognition. The fact that Daniels excludes the possibility that age rationing would extend to basic forms of care suggests that he would see some possible distributive schemes as incompatible with respect for people's inherent value or human dignity. It seems likely that he would in fact accept what Nancy Jecker has called a 'decent minimum view', which "describes a level of services that should be available to all without hardship, and regardless of ability to pay or geographical location" (Jecker 1989, p. 87).[26] Justice as recognition demands the same, asking for at least the minimum level of services compatible with recognition of people as sources of legitimate claims.

So, the existing philosophical literature is so strongly focussed on comparative issues that the most fundamental noncomparative dimension of justice, i.e. recognition, has largely been overlooked, even if it is implicitly recognized.[27] Another problem with the literature is that, in relation to the elderly at least, there is very little debate about the currency of distribution: resources, primary goods, opportunities, capabilities? One reason why this is important—apart from the many reasons one may have for favouring one of these options over the others—is that it has important political implications. A focus on capabilities, for instance, would probably not have generated the intergenerational antagonism that the current focus on financial resources sparked and continues to fuel.[28]

Finally, the philosophical literature ought to reflect, much more than it does now, the fact that intergenerational (distributive) justice is inextricably linked with the issue of filial obligations and with questions of intragenerational justice and gender justice. These issues are interlinked in many ways: women constitute the majority of the old and very old old, and tend to be overrepresented among the worst-off pensioners (Arber and Ginn 2005). Inequalities during working-life persist after retirement, and so do gender inequalities. Moreover, inequalities of income that do not result in significant inequalities in capabilities and opportunities may lead to such inequalities after the drop of income associated (for most people) with retirement. In other words: intragenerational inequalities do not just persist, but may become more pronounced in old age. The design of the pension system influences both intra- and intergenerational distributions, as well as distributions between men and women. Government reliance on filial obligations may be used to diminish the need for redistribution, but is likely to enhance gender inequalities, because daughters tend to do most of the caring for

---

[26] Jecker (1989, p. 87).

[27] The distinction between comparative and noncomparative justice derives from Feinberg (1974). Comparative issues are issues where what one person is due can only be determined by reference to what others are due; in the noncomparative case, a person's due can be determined without reference to others.

[28] Irwin (1996) argues that there is little evidence in society of the antagonism between generations that many authors anticipate, and suggests, in effect, that it may be a theoretical construct, resulting from particular (flawed) approaches to demographic changes and social inequality.

dependent parents (Lye 1996). A recognition perspective highlights such inequalities as problematic and requiring justification (if there can be any).

12.3.3 *Just Care*

Mainstream philosophical writing about justice rarely connects it with issues of care. The important work of philosophers like Iris Marion Young, Susan Moller Okin, and Joan Tronto has not been able (so far) to change this.[29] Remarkably, the potential of the concept of recognition to bridge the gap between care and justice has remained unexplored. Recognition theory (Honneth, Fraser, Margalit, and others) ignores the ethics of care, and the ethics of care (in the English language area) has not really thematized the concept of recognition.[30] The concept of recognition is central in Van Heijst's ethics of care (Van Heijst 2008).

Recognition is not only the primary act of justice, it is also at the core of any caring relationship. Care begins with, and always implies, the acknowledgement that the other person matters, that his or her needs, desires, interest—moreover, his or her identity—is taken seriously and taken into account. In formal care settings, it implies recognition of the other as an equal source of legitimate claims, as a bearer of equal inherent value. Importantly, it is the other as a whole, as the particular person (s)he is, to whom such recognition is due. Good care is a response to the specific needs of particular individuals. In certain informal care settings, partiality towards the person cared for plays a more important role; this is not logically at odds with recognition of equal inherent value, but the person cared for will often be more important (more 'valuable') to the carers than someone with whom they have no special relation.

Apart from the concept of recognition, the concept of needs also constitutes an important connection between justice and care (Gheaus 2005, 2009). Care is often defined as a certain kind of response to needs; for instance as "an activity or practice aimed at the meeting of needs in others" (Bubeck 2004, p. 9). And in many theories of justice 'need' is seen as an important substantial ground or criterion of justice (e.g. Rescher 1966; Perelman 1970; Deutsch 1975; Campbell 1988, p. 180; Baker 1990, Chap. 2; Schmidtz 2006, Part 5).[31] Justice requires that certain needs, certain kinds of need, are responded to; in other words: certain kinds of care are required

[29] For some discussion of feminist thought on justice and care see, for instance, the contributions by Deveaux, Held, and Tronto to the symposium on care and justice in *Hypatia*, Vol. 10, No. 2 (1995). For more extensive discussion see Gheaus (2005).

[30] Kittay (1999, p. 71) does speak of "the recognition of equality inherent in the understanding that we are all some mother's child" (which she calls 'connection-based equality'), and others also occasionally *use* the term; yet they do not develop the concept.

[31] Sen (2009, 251–252, 260–261) acknowledges the importance of needs, but emphasizes the relative importance of freedom "(including the freedom to meet our needs)" (251). Engster (2007) makes care central to his moral and political theory: "Because human beings universally depend upon one another for care, we all have moral obligations to care for others in need." (2) Thus, for Engster, a caring society is one in which individuals are enabled, first of all, "to meet their basic

by justice, and in such cases care that falls short of certain standards of good care constitutes an injustice.

For example: when nursing home residents who are not incontinent have to wear diapers all day because there are not enough nurses to take them to the toilet when they need to go, this violates many important needs: the simple need to relieve oneself, as well as needs for hygiene, cleanliness, dignity and decorum, and autonomy (they even become dependent on diapers after a while), and the need to be able to retain one's self-respect.[32] People who are fortunate enough not to be in this situation need only ask themselves whether they would ever consider this an acceptable treatment for themselves to know that this is not compatible with the recognition and respect due to every person. As 'small' a thing as this constitutes an injustice, and one that ought to make one think about the place assigned to (dependent) elderly in our society.

Those needs that engender legitimate claims, needs that give rise to obligations of justice, may require three different responses, according to the kind of claim they engender. They may be claims to non-frustration (or non-interference), accommodation, or fulfilment. Any need that gives rise to obligations of justice engenders a claim to its non-frustration. This applies to the need for food as much as to the need for sex (a need that many elderly people still have, whether young people want to know it or not). Some needs engender legitimate claims to their fulfilment. An old person incapable of feeding himself, for instance, should be fed by others (assuming he wishes to be fed, there is enough food to go round, etc.). When people are capable of looking after themselves, their needs normally only give rise to claims to non-frustration and accommodation, the latter meaning that circumstances favourable to their fulfilling their own needs are maintained or, where necessary, created. (I am assuming a societal context, like ours, in which the creation and maintenance of such conditions is not or only minimally within the power of individuals.)

Although sexual activity declines with age, a significant number of old people have a need for sex.[33] This can be considered a basic and important need, which means that recognition of inherent value is incompatible with neglect of this need, where old people are unable to provide for its satisfaction themselves. Elderly people in care homes, for instance, have a legitimate claim to accommodation of their need for sex, meaning (at minimum) that it should be possible for them to receive sexual partners in a private place. Further issues, such as whether paid sex should be arranged for them, and possibly even be paid for them if they lack the funds, are not settled by the recognition perspective as such. To decide such matters requires specification of the approach, which may proceed in different directions. The need for sex can never

---

needs" (76) – and 'care theory', in this book, provides nothing other than "an alternative vision of a just society" (5).

[32] See, for instance, Tonkens (2007) for comments on this use of diapers.

[33] See, for instance, the study by Lindau et al. in the United States (2007), which showed that 26 % of respondents between 75 and 85 years of age were still sexually active. Significantly more men than women were sexually active at that age.

give rise to a claim to its fulfilment, however, since justice cannot require anyone to perform the necessary actions.

Justice can only require more than accommodation of needs under very specific conditions: (1) when people are unable (i.e. incapacitated or hindered by circumstances) to fulfil their own basic and important needs, and not culpably responsible for this inability themselves;[34] or (2) in those cases where paternalism is justified to prevent serious harm. When these conditions are not met, attempts to go beyond accommodation actually signify a lack of recognition (as in cases of unjustified paternalism). Force-feeding a nursing home resident who has decided, lucidly and competently, to stop eating because she has had enough of life would constitute an example of this.

12.3.4 Recognition in the Family

The recognition that underlies both politeness and decency on the one hand, and the equal respect accorded to people (one hopes) by laws and constitutions on the other, relates to people's personhood in a very thin and abstract sense. The bus driver does not need to recognize me, as the particular individual I am; it is enough that she treats me like a person, a human being like herself.[35] The personhood that needs to be recognized becomes thicker in a care setting, and even more among acquaintances, friends, and family. In such settings, we expect people to acknowledge not merely our personhood in the thinnest and most abstract sense, but, to varying degrees, our individuality. By all kinds of small acts or displays of attention, a nurse can show his or her awareness that one patient is not the same as another, that they are different individuals, with different needs. But especially among intimates—though often also outside this circle—we need more than this: we need to be valued for who we are and what we do or create, we have a need for (deserved) esteem.[36]

This makes it especially important to think about the place of the elderly in family relations—relations that are undergoing important changes (see Lowenstein 2005). With increasing longevity, it is more common for there to be four generations of one family alive at the same time; sometimes two older generations are to some extent dependent on (grand)children. The trend in government policies in developed countries is to rely increasingly on informal care, and people's sense of filial obligation (e.g. Simonazzi 2011; Aronson and Neysmith 1997). At the same time,

---

[34] Culpability does not always matter, especially when it comes to basic needs. Even if it is someone's own fault that he lacks food, justice may (in some cases) still require others to provide it.

[35] If I get on the same bus (that she drives) every day, however, I do after some time expect her to recognize me in the most familiar sense of the word, and if she does not, or does not acknowledge it at all, this amounts to a lack of respect.

[36] I cannot go into the relations between recognition as the primary act of justice and people's *need* for recognition in any detail here. Failure of recognition, in the sense in which I use the term, is not only problematic because it frustrates people's need for recognition.

socio-economic reality is such that family care is under pressure. Working parents also increasingly depend on their elderly parents for part-time childcare. Some grandparents may be happy to oblige, but one may wonder whether a society could still be called just if it would be organized such as to make such a practice virtually inevitable. Similarly, it is questionable whether a just society would both undermine the conditions for family care and increasingly rely on it. The implications for gender justice have already been mentioned in 3.2.

What level of informal care a government can reasonably expect people to provide is a complex issue. The answer depends on whether we are asking about adult children caring for parents, elderly people minding their grandchildren, people caring for friends, or yet other arrangements. It also depends on what resources are available on the collective level, on the institutional context, and so on. But recognition of people's equal inherent value does imply that people cannot be used as means to reduce the costs of healthcare. It is unacceptable for grandparents to be pressured into minding their grandchildren. Nor is it acceptable for a government to place the responsibility for eldercare in the hands of adult children. People should have the room to fulfil their moral obligations, but children cannot simply be assumed to be under any obligation towards their parents, for the mere fact of being someone's child is not a ground for obligation. Another problem with policies that shift responsibilities to the informal sphere is that it seems to signify a lack of recognition of those depending on care, for it is likely that their care is now no longer guaranteed; that it comes to fall beyond the sphere of government supervision. The cared-for are left to the goodwill of family and friends—assuming they have such relations. In other words, their care is at risk of moving from the sphere of justice to that of charity. Even if, against this, it could successfully be argued that family are required, as a matter of justice, to provide certain types of care (which would be a very difficult argument to make), the care would only remain solidly within the sphere of justice if the government would make sure that family members met their obligations. A third problem with these policies is that they would further entrench and possibly even increase socio-economic inequalities which, from a recognition perspective, is unacceptable (see Sect. 12.3.2). Fourthly, although close friends or relatives may be best placed to provide a very personal form of recognition, being forced into a caring relationship may also wreak havoc on the original relationship by introducing a one-sided dependency and thereby a power differential.

Problems of recognition in family relations, as in all interpersonal relations, are also related to societal attitudes towards elderly people. The private sphere is not immune from societal influences in this (or any other) respect. Recognition in the family is not separable from recognition in society, to which issue I now turn.

12.3.5 Recognition in Society

Two of the main dimensions of justice with particular relevance in the societal domain have already been discussed: non-discrimination and distribution. With regard to the first issue, it is important to consider, apart from clear and concrete instances of discrimination, the dominant attitude(s) towards the elderly in current (Western) societies. Cuddy and Fiske, for instance, write that "[in the United States] today's elders are seen as low status, which elicits perceptions of incompetence, and passivity, leading to perceptions of warmth"—the latter showing the ambivalence in attitudes towards the elderly. The elderly 'are pitied but not respected' (Cuddy and Fiske 2002, p. 17). They also report findings of other researchers, such as that "[t]elevision portrays only 1.5 per cent of its characters as elderly, and most of them in minor roles", and that the elderly are "more likely than any other age group to appear in television and film as conduits for comic relief, exploiting stereotypes of physical, cognitive, and sexual ineffectiveness" (ibid., p. 3).[37] Unlike in 'traditional' societies, in which the tradition (i.e. passing-on) of knowledge and ways of life, and the imparting of wisdom from the old to the young, are important aspects of intergenerational relationships, in modern Western societies the knowledge of the old is seen as obsolete, their experience as of little use in our fast-changing society; and the elderly are not regarded as wiser than anybody else, but more often as senile. At the same time, "[b]eliefs about the elderly as unable to contribute to society, and hence as dispensable members of a community, and attitudes towards them of dislike and distancing are prevalent" (Levy and Banaji 2002, p. 49).[38] Elderly people seem to have internalized such attitudes: "To our knowledge, the elderly is the only group that shows as strong negative implicit attitudes toward their own group as does the out-group (the young)." (ibid., p. 67)

Such widespread attitudes and implicit ageism do not fit comfortably in the picture of a just society. The examples given above suggest that elderly people are not seen as full members of the (moral) community, that they are not given the recognition they are due as persons, let alone as the particular individuals they are. The existence of these attitudes creates the danger of epistemic injustice against elderly people, of both kinds distinguished by Miranda Fricker (2010). Testimonial injustice occurs when certain people's testimony is systematically given less credibility than other people's, because the former are seen as members of a particular group, where this group is the object of certain negative prejudices (ibid., Chap. 1). For instance, elderly people's opinions might be given a lesser hearing because old people are seen as out of touch with reality. That elderly people seem to have internalized negative attitudes towards their own group suggests that hermeneutical injustice also occurs. Hermeneutical injustice entails that certain people or groups are at a hermeneutical disadvantage, in the sense that they are (relatively) deprived of the means to understand significant aspects of their own experience, or (wrongfully) harmed by a general social lack of resources for such understanding (ibid., Chap. 7). The pace at which society changes may place (some) elderly at a hermeneutical disadvantage, without this constituting

---

[37] They refer to Zebrowitz, L.A., and J.M. Montepare. 2000. "Too young, too old: Stigmatizing adolescents and elders". In *The social psychology of stigma*, ed. T.F. Heatherton, R.E. Kleck, M.R. Hebl, and J.G. Hull, 334–373. New York: Guilford Press.

[38] They refer to Kite, M., and B. Johnson. 1988. "Attitudes toward older and younger adults: A meta-analysis". *Psychology and Aging* 3: 233–244.

an injustice, but insofar as their disadvantage is the result of implicit ageism, it must be considered as hermeneutical injustice. Internalized ageist attitudes may render elderly people unable to make positive sense of their place in society, as well as actively prevent them from seeking and engaging in meaningful activities (for instance because they consider themselves too old, and therefore incompetent).

The other side of the coin is that with the aging of Western societies, the 'grey' lobby has grown more powerful, and the elderly electorate significantly larger, so that some means of making themselves heard—and thereby especially their political influence—have grown stronger. This has sparked concern about possible injustices inflicted on future and younger generations (Van Parijs 1998).

And there are further aspects to the question as to the place of the elderly in a just society. For instance: what opportunities for meaningful participation are available, what opportunities and (conceptual) resources for leading a meaningful life and experiencing a dignified death? These things are influenced as much by prevalent attitudes as the basic institutional structure of a society, and therefore need to be thought about from the beginning.

12.4 Policy Implications and Concluding Remarks

Justice is not the whole of ethics—not by a long stretch. Many ethical issues regarding the elderly are not (primarily) issues of justice, and have therefore remained undiscussed here. There are interesting questions concerning the ethics of aging as such—*what is a good, meaningful, graceful way to age? How should we evaluate techniques of and research into life-extension? Is there a connection between age and wisdom?* And then there are traditional topics like euthanasia and assisted suicide. Some of these are related to issues of justice and recognition, and form part of the context for inquiry into such issues. The field of justice and the elderly, I hope to have shown, is much more encompassing—and complex and interesting—than one would gather from the existing philosophical literature that explicitly connects justice and the elderly.

To broaden the perspective, I began by suggesting that we should ask what a just society would look like in its elderly-related aspects; what the place of the elderly would be in a just society. I suggested that any just society, whatever form it takes, owes the elderly recognition—not as a group, but as individual human beings. I defined recognition of every individual as a source of legitimate claims, of people's equal inherent value, as the primary act of justice. Recognition and nonrecognition can occur in various forms, in different dimensions—of which I discussed three: non-discrimination, distributive justice, and just care—and different domains, such as that of the family, and the societal domain. The recognition perspective suggests certain issues to be issues of justice that are not commonly discussed under that heading, such as the many forms of implicit ageism, and care for dependent elderly. It is naturally allied to approaches that define a decent minimum of services, healthcare, et cetera, and opposed to views that focus exclusively on comparisons between benefits received across whole lifetimes. Justice as recognition urges attention to the specific needs of particular individuals, needs which may engender claims to their non-frustration, accommodation, or fulfilment. By way of example, I suggested that elderly people's sexual needs require accommodation, where necessary, and that the need for food may engender claims to fulfilment of that need. Finally, I looked at issues of (non-)recognition in two domains, the family and society, paying some attention to government policy with regard to the first domain. To conclude this chapter I will sketch some further policy implications of the approach suggested here.

The starting-point, and the point to which we should keep returning, is recognition. The question that should, ultimately, guide policy and policy development is: are elderly individuals given the recognition they deserve as human beings; in other words: is their treatment (at minimum) compatible with recognition of their inherent value, and does it express such recognition? What does this mean for the distribution of financial resources? As long as we can improve, through reallocation of such resources, the overall 'rate of recognition' (i.e. reduce—in ways respectful of people's inherent value—the number of cases in which people are treated as of lesser importance) we should do so. To give an example from the Netherlands: as long as nursing homes are understaffed, so that residents have to wear diapers, there is no justification for building an enormously expensive stretch of road (the A4 between Delft and Schiedam) that will at best only temporarily relieve the problem of traffic jams. What is compatible with recognition of inherent value or not will always depend greatly on the available means—means that must be distributed over many important goals. The perspective of justice as recognition emphasizes that which goal is at any point to be prioritized depends on what, at that point, will best express recognition of the importance of all those influenced by the decision.

Concretely, the recognition perspective favours government support of autonomy, independence, and abilities and opportunities for social participation. Hence, the recent decision by the Dutch government to largely abandon the 'personal budget' that allowed people to arrange their own care, must be regarded as regrettable.

The now current detailed quantification of care and the concomitant language of care 'products', 'units of care', care 'consumers' and 'producers', et cetera, are antithetical to the perspective of justice as recognition. The current care system creates a fictional (numerical) reality that functions as a smokescreen, obscuring the practical reality of (elder)care and defending it even where it is indefensible. That a certain amount of time and money, invested in a prescribed number of units of care, has found its way from a care producer to one of their clients tells us nothing about whether this person has been given the recognition (s)he is due.

In areas like that of healthcare (the rationing debate) and the pension system (intergenerational justice) the recognition perspective does not obviously point to one particular policy. Rather, it provides a test for any policy we do choose, requiring us, for instance, to investigate it for the presence of implicit ageism. This may also be the most the government can do to counter ageism in society at large; anti-ageism campaigns, such as the recent Scottish campaign, are another option, but their effectiveness is hard to predict.

Elderly people themselves also bear some of the responsibility for the image of old age. For some, the post-retirement period seems to constitute a second age of irresponsibility, a new childhood, in a sense, a time of consumptive enjoyment—or at least it may look like that to people of working age (who, often enough, can't wait to get there themselves). But on the other hand, even if some elderly people were to see it that way, who could blame them? What place do Western societies have for the elderly, what serious role can they offer them? Societies that pin their hope for survival on innovation and change are more likely to overestimate the importance of youth than to take elderly people seriously. In this area as in many others—ecological sustainability, economic policy, labour and immigration policy, to name a few—to realize justice seems to require that we row against the stream.

Acknowledgments In writing this chapter I have benefited greatly from various people's helpful comments on earlier drafts; thanks are due especially to Martijn Boot, Patrick Delaere, Anca Gheaus, Elke Müller, Gijs van Oenen, Ingrid Robeyns, Maureen Sie, Christian van der Veeke, Nicole van Voorst Vader-Bours and Arno Wouters at Erasmus University Rotterdam, as well as to the editors of this volume. The research for this paper was done in the context of Ingrid Robeyns VICI project Social Justice and the New Welfare State, financed by the Netherlands Organisation for Scientific Research (NWO).

References

Arber, S. and J. Ginn. 2005. *Gender dimensions of the age shift*. In Johnson (ed.), pp. 527–537.

Aronson, J. and S. M. Neysmith. 1997. The retreat of the state and long-term care provision: Implications for frail elderly people, unpaid family carers and paid home care workers. *Studies in Political Economy* 53: 37–66.

Baker, J. 1990. Arguing for equality. London: Verso.

Bentley, E. 2007. Adulthood. London: Routledge.

Brighouse, H., and I. Robeyns, eds. 2010. *Measuring justice: Primary goods and capabilities*. Cambridge: Cambridge University Press.

Brock, D. W. 1989. Review: Justice, healthcare, and the elderly. *Philosophy and Public Affairs* 18(3): 297–312.

Bubeck, D. 2004. Care, gender, and justice. Oxford: Clarendon Press.

Bytheway, B. 2005. *Ageism*. In Johnson (ed.), pp. 338–345.

Callahan, D. 1985. What do children owe elderly parents? The Hastings Center Report 15(2): 32–37.

Callahan, D. 1995. *Setting limits: Medical goals in an aging society, with 'a response to my critics'*. Washington, DC: Georgetown University Press.

Campbell, T. 1988. Justice. Houndmills: MacMillan Education Ltd.

Cuddy, A. J. C., and S. T. Fiske. 2002. *Dod dering but dear: process, content, and function in stereotyping of older persons*. In Nelson (ed.), pp. 3–26.

Daniels, N. 1988. Am I my parents' keeper? *An essay on justice between the young and the old*. Oxford: Oxford University Press.

Daniels, N. 2008. Justice between adjacent generations: Further thoughts. *The Journal of Political Philosophy* 16(4): 475–494.

Deutsch, M. 1975. Equity, equality, and need: What determines which value will be used as the basis of distributive justice? *Journal of Social Issues* 31(3): 137–149.

English, J. 1979. What do grown children owe their parents? In *Having children: Philosophical and legal reflections on parenthood*, ed. O'Neill and William Ruddick, pp. 351–356. New York: Oxford University Press.
Engster, D. 2007. The heart of justice: Care ethics and political theory. Oxford: Oxford University Press.
Feinberg, J. 1974. Noncomparative Justice. The Philosophical Review 83(3): 297–338.
Fraser, N. 1996. Social justice in the age of identity politics: Redistribution, recognition, and participation, the Tanner Lectures on Human Values Stanford University. Tanner Humanities Center, Lecture Library. http://www.tannerlectures.utah.edu/lectures/documents/Fraser98.pdf. (Accessed 3 April 2012).
Fraser, N. 2000. Rethinking recognition. New Left Review 3: 107–120.
Fricker, M. 2010. Epistemic injustice: power & the ethics of knowing. Oxford: Oxford University Press.
Gaita, R. 2006. *Good and evil: An absolute conception*. Abingdon: Routledge.
Gheaus, A. 2005. *Care and justice: Why they cannot go together all the way*. Ph.D. thesis. Budapest: Central European University.
Gheaus. 2009. How much of what matters can we redistribute? love, justice, and luck. Hypatia 24(4): 63–83.
Gilleard, C., and P. Higgs. 2000. *Cultures of aging: Self, citizen and the body*. Prentice Hall Harlow: (Pearson Education Limited).
Gilleard, C., and P. Higgs. 2002. The third age: class, cohort or generation? *Aging and Society* 22(3): 369–382.
Greenberg, J., J. Schimel, and A. Mertens. 2002. *Ageism: Denying the Face of the Future. In Ageism: stereotyping and prejudice against older persons*, ed. T. D. Nelson, 27–48. Cambridge, MA: The MIT Press.
Griffin, J. 2008. *On human rights*. Oxford: Oxford University Press.
Irwin, S. 1996. Age related distributive justice and claims on resources. *The British Journal of Sociology* 47(1): 68–92.
Held, V. 1995. The meshing of care and justice. *Hypatia* 10(2): 128–132.
Honneth, A. 1995. *The struggle for recognition*. Cambridge: Polity Press.
Howse, K. 2004. What has fairness got to do with it? Social justice and pension reform. *Aging Horizons* 1: 3–14.
Jecker, N. 1989. Should we ration health-care? *The Journal of Medical Humanities* 10(2): 77–90.
Johnson, P., C. Conrad, and D. Thomson, eds. 1990. *Workers versus pensioners: Intergenerational justice in an aging world*. Manchester: Manchester University Press.
Keller, S. 2006. Four theories of filial duty. The Philosophical Quarterly 56(223): 254–274.
Kirp, D. L., M. G. Yudof, and M. S. Franks. 1986. Gender justice. Chicago: The University of Chicago Press.
Kittay, E. F. 1999. Love's labor: Essays on women, equality, and dependency. New York: Routledge.
Laslett, P. 1992. Is there a generational contract? In *Justice between age groups and generations*, ed. P. Laslett and J. S. Fishkin. New Haven: Yale University Press.
Levy, B. R., and M. R. Banaji. 2002. Implicit ageism.
Li, C. 1997. Shifting perspectives: Filial morality revisited. *Philosophy East and West* 47(2): 211–232.
Lindau, S. T., et al. 2007. A study of sexuality and health among older adults in the United States. New England Journal of Medicine 357(8): 762–774.
Lowenstein, A. 2005. Global aging and challenges to families. In *The cambridge handbook of age and aging*, ed. M. L. Johnson, 403–412. Cambridge: Cambridge University Press.
Lye, D. N. 1996. Adult child-parent relationships. *Annual Review of Sociology* 22: 79–102.
McKerlie, D. 1989. Equality and time. Ethics 99(3): 475–491.
McKerlie, D. 1992. Equality between age-groups. Philosophy and Public Affairs 21(3): 275–295.
McKerlie, D. 2001. Justice between the young and the old. Philosophy and Public Affairs 30(2): 152–177.

Myles, J. 2002. A new social contract for the elderly? In *Why we need a new welfare state*, ed. G. Esping-Andersen, 130–172. Oxford: Oxford University Press.

Myles, J. 2003. What justice requires: Pension reform in aging societies. *Journal of European Social Policy* 13(3): 264–269.

Nussbaum, M. 2006. *Frontiers of justice: Disability, nationality, species membership*. Cambridge, MA: The Belknap Press (Harvard University Press).

Opotow, S. 1990. Moral exclusion and injustice: An introduction. *Journal of Social Issues* 46(1): 1–20.

Perelman, C. 1970. *The idea of justice and the problem of argument*. London: Routledge & Kegan Paul.

Post, S. G. 1989. What children owe parents: Ethics in an aging society, Thought 64(255): 315–325.

Rawls, J. 2003a. *A theory of justice (Revised Edition)*. Cambridge, MA: The Belknap Press (Harvard University Press).

Rawls, J. 2003b. *Justice as fairness: A re-statement*. Cambridge, MA: The Belknap Press (Harvard University Press).

Rawls, J. 2005. Political liberalism (Expanded Edition). New York: Columbia University Press.

Regan, T. 1983. *The case for animal rights*. Berkeley: University of California Press.

Rescher, N. 1966. Distributive justice. Indianapolis: Bobbs-Merrill.

Rivlin, M. M. 1995. Protecting elderly people: Flaws in ageist arguments. British Medical Journal 310: 1179–1182.

Rivlin, M. M. 2000. Why the fair innings argument is not persuasive. *BMC Medical Ethics* 1(1).

Robeyns, I. 2007. When will society be gender just? In *The future of gender*, ed. J. Brown. Cambridge: Cambridge University Press.

Robeyns, I., and A. Swift, eds. 2008. Social justice: Ideal theory and nonideal circum-stances. *Special Issue of Social Theory and Practice* 34(3).

Schmidtz, D. 2006. Elements of justice. New York: Cambridge University Press.

Schokkaert, E. and P. Van Parijs. 2003. Debate on social justice and pension reform: Social justice and the reform of Europe's pension systems. *Journal of European Social Policy* 13(3): 245–279.

Sen, A. 1999. *Development as freedom*. Oxford: Clarendon Press.

Sen, A. 2009. *The idea of justice*. London: Allen Lane.

Silvers, A. 2000. Aging fairly: Feminist and disability perspectives on intergenerational justice. In *Mother time: women, aging, and ethics*, ed. Margaret Urban Walker, 203–226. Inc, Lanham: Rowman & Littlefield Publishers.

Simonazzi, A. 2011. Home care and cash transfers: The search for a sustainable elderly care model. In *Gender and well-being: The role of institutions*, ed. E. Addis, P. de Villota, F. Degavre and J. Eriksen, 127–146. Farnham: Ashgate.

Tillich, P. 1954. *Love, power and justice*. London: Oxford University Press.

Tonkens, E. 2007. "Het best bewaarde geheim". De Volkskrant 19-12-2007.

Tronto, J. C. 1995. Care as a basis for radical political judgments. *Hypatia* 10(2): 141–149.

Van Heijst, A. 2008. Iemand zien staan: Zorgethiek over erkenning. Kampen: Klement.

Van Parijs, P. 1998. The disfranchisement of the elderly and other attempts to secure intergenerational justice. *Philosophy and Public Affairs* 27(4): 292–333.

Wicclair, M. R. 1993. Ethics and the elderly. New York: Oxford University Press.

Williams, A. 1997a. Intergenerational equity: An exploration of the 'Fair Innings' argument. Health Economics 6: 117–132.

Williams, A. 1997b. The rationing debate: rationing healthcare by age—the case for. *British Medical Journal* 314: 820–825.

Willmore, L. 2004. Population aging and pay-as-you-go pensions. *Aging Horizons* 1: 23–29.

Part III
Looking Ahead: Ethics and Prolongevity

Chapter 13
Ethical Perspectives in Biogerontology

Sebastian Sethe and João Pedro de Magalhães

13.1 Introduction

From the perspective of scientists involved in biogerontology, the branch of science focused on the biology of aging, ethical themes can be classified as either belonging to an 'inner sphere' where the conduct of the aging research itself is under ethical scrutiny; or, secondly, an 'outer sphere' where questions are raised about the philosophical and social implications of 'curing aging'. There is of course some overlap between these spheres (most evidently when communicating between them) but generally commentators tend to focus on the 'outer sphere'. Here, we shall focus on the 'inside' perspective of moral agents involved in biogerontology. We make no claim to be representing 'the biogerontology perspective' nor do we aim to chart, let alone consider in depth all the ethical issues that might arise from this perspective. However, in a debate where commentators—including biogerontologists—tend to discuss abstract positions, we suggest that considering these issues from the particular vantage point of a research protagonist provides a useful further angle to enrich the discussions.

There are some 'established' ethical issues that arise in biogerontology as they do in other fields, albeit with some particular characteristics: Research relying on elderly people may be more difficult to conduct since they may be more prone to frailty and regenerate less quickly plus the capacity for giving informed consent may be a greater issue than in other demographic groups. Animal research may involve keeping animals a longer time in confinement than in other fields giving rise to different husbandry requirements. Where research focuses on genetic or medical data privacy may be an issue. This is not necessarily surmountable simply by resorting to anonymization of data—a data subject may resent that her information

J. P. de Magalhães (✉)
Integrative Genomics of Ageing Group, Institute of Integrative Biology, Biosciences Building, University of Liverpool, Crown Street, Liverpool L69 7ZB, United Kingdom
e-mail: jp@senescence.info

S. Sethe
NorthEast England Stem Cell Institute, United Kingdom
e-mail: ssethe@sgul.ac.uk

M. Schermer, W. Pinxten (eds.), *Ethics, Health Policy and (Anti-) Aging: Mixed Blessings,* Ethics and Health Policy 1, DOI 10.1007/978-94-007-3870-6_13, is being used in research to 'cure aging' whether or not personal data protection is an issue. These examples illustrate that even 'standard' ethical challenges may be structured differently in biogerontology, but it is worth remembering that when one considers the most immediate ethical issues that biogerontologists face, these are not very different from those encountered by researchers in other fields. Other such 'standard' issues have to do with a researchers' personal ethics: How one treats one's students and staff, how one behaves ethically as a peer and author and so on. At least one of these 'standard' ethical challenges merits some further discussion.

Herein, we discuss a spectrum of ethical considerations as they present themselves to biogerontology internally, when considering how to communicate the scientific developments and potential technological advances; and externally, when thinking about long-term social consequences of anticipated technological progress in this area, in particular issues derived from the potential for 'curing aging' such as overpopulation and cultural stagnation.

13.2 Communicating the Potential of Biogerontology

A primary ethical challenge to researchers is to communicate truthfully with non-specialists in explaining their research, the development and potential of their field. Aging is a personal, emotive and complex topic and in the battle to secure funding researchers must be mindful of not creating unrealistic hopes. In the context of 'curing aging' especially, one looks at a long history of hope and desire driving science and pseudo-science. So what can ethically be said about the potential of curing aging? Herein, we first discuss the state and potential of the biogerontology field, before comparing the study of aging to that of other diseases and discussing the motivations of researchers in this field.

13.2.1 Present Understanding of Aging and Potential Interventions

The underlying cellular and molecular mechanisms of aging and the process(es) driving the aging process are still poorly understood. Conceptually, biogerontologists have proposed two broad explanations for aging: Damage-based theories of aging that posit that aging results from random or stochastic damage and programmed theories of aging that suggest that aging results from predetermined mechanisms, usually with an element of genetic regulation (de Magalhaes 2011). Damage-based theories include the free radical theory of aging, which suggests that a gradual build-up of oxidative damage with age drives the aging process, and the idea that DNA damage accumulation with age causes the physiological and functional decline we call aging.

To understand modern 'programmed' theories it is necessary to appreciate that evolutionary origins of aging differ from those of other traits: Contrary to other processes, such as development which is orchestrated by our genome (with environmental inputs playing a role) and is the product of millions of years of evolution, human aging is not thought to have evolved 'for a purpose'. Instead, aging can be considered a by-product of evolution. Because in the wild only a small percentage of organisms survive long enough to reach 'old age' the force of natural selection declines with age and so evolutionarily there is little pressures to, for example, favour genes that are only beneficial late in life (Kirkwood 2005). Therefore, it is widely accepted among evolutionary biologists that human aging is a process that was not 'selected for' by evolution but rather escaped the force of natural selection.

Because of the evolutionary theory of aging, modern 'programmed' theories tend to focus on programmatic aspects of aging, such as gene expression changes or genetically regulated chains of events. The developmental theory of aging, for instance, argues that developmental mechanisms can regulate aging as an indirect consequence of developmental processes optimized to maximize physiological function for reproduction which then become detrimental in adulthood (de Magalhaes and Church 2005). A combination of theories of aging cannot also be excluded and more than one may turn out to be correct, but this distinction between whether aging is caused by stochastic processes or by the genetic program has implications in terms of developing interventions and in the philosophical interpretation of those interventions: Are we repairing damage or are we trying to change our genetic program?

In spite of our incomplete understanding of aging, progress in biogerontology in the past few decades has been impressive, in particular at the genetic level. Researchers have identified hundreds of genes that when manipulated change the aging process in model organisms, from simple yeast to invertebrates and even mammals like mice (Kenyon 2010; de Magalhaes et al. 2009). For example, it is possible to extend the lifespan of mice by up to 50 %, delay the appearance of age-related diseases and increase health by disrupting a single gene (Flurkey et al. 2002). A number of gene variants have also been associated with human longevity and we are starting to understand the gene difference behind why some people live longer than others (Browner et al. 2004). This knowledge of the genetics of aging and longevity gives scientists a blueprint for intervening in the aging process, as discussed below.

The strongest effect from environmental manipulations of aging in mammals is observed from caloric restriction, an intervention that consists of restricting caloric intake without malnutrition, and which in some mouse strains (but not in all) can extend lifespan by up to 50 % (Fontana et al. 2010). It is unlikely such marked effects are applicable to humans, but a delay of human aging by targeting caloric restriction pathways and genes is possible within the coming years (de Magalhaes et al. 2012). Indeed, pharmaceutical or nutraceutical targeting of aging and/or caloric restriction related genes or pathways could lead to the development of new drugs for age-related diseases and potentially retarding the human aging process: The so-called anti-aging pill (Stipp 2010). Numerous companies and research labs are working in this paradigm with potential drugs undergoing clinical trials. For example, the drug rapamycin can extend lifespan by 9–14 % when fed to middle-aged mice (Harrison et al. 2009).

While testing anti-aging drugs and diets for long-term effects is problematic, it is plausible that some have short-term effects on health parameters and on age-related diseases that can be tested in a clinical setting to obtain approval from regulatory agencies. It will then be up to the consumers to decide, based on short-term human clinical trials and animal studies, whether to adopt the anti-aging pill, diet or lifestyle in the long-term. Therefore, to some degree the life extending applications of research on aging are on the horizon. It will be up to individuals whether to adopt them or not and thus providing accurate and intelligible information to the public is a crucial task of biogerontologists.

In summary, although the essence of the basic process of aging remains contentious, there are many technical possibilities for how aging might be slowed. As a basic distinction, we can differentiate between 'genetic engineering' of the yet unborn and dealing with aging in adults. In the long-term reality, there may be a combination of these two possibilities. (Germline interventions are of course ethically contentious for other reasons. Aging researchers working in genetics are challenged to develop a position on this topic. However, as we have pointed out above, genetic research in aging may well pave the way to interventions other than germline alteration, ranging from adult gene therapy, to pharmaco- and nutrigenomics. Therefore, the ethical issues in germline gene therapy are not necessarily implicated in all research on the genetics of aging and are not what we would focus on here.)

13.2.2 The Possibility of a Cure

Most biogerontologists agree that life extending applications of research on aging are plausible (Butler et al. 2004). Contested remains the factor by which life expectancy can be increased (Richel 2003). Assuming we 'cure' aging and thus eliminate (or at least prevent the age-related increase in incidence of) all age-related pathologies, including cancer, heart disease, and neurodegeneration, our average lifespan would increase dramatically. A life expectancy of over a thousand years seems theoretically feasible. (The average lifespan ($t_{0.5}$) of a non-aging population is given by the equation: $t_{0.5} = -\ln 0.5/\text{IMR}$ (Finch 1990). Assuming the initial mortality rate, the IMR, of a typical population in an industrialized nation (0.0005/year) we have $t_{0.5} = 1{,}200$ years. Of course this assumes a constant IMR, which may not be the case if there are wars or pests that increase the IMR or conversely progresses in other areas that decrease the IMR).

Human aging goes beyond the general changes that occur at a molecular or cellular level; it likely has multiple organ-specific determinants (Sames 2005). Consequently, the challenge in aging is to apply both very general and very specific interventions, and to maintain their effectiveness indefinitely—a rather daunting endeavour. Nonetheless, some commentators have boldly suggested have that it may be possible to cure aging within the next few decades (de Grey and Rae 2008; Kurzweil and

Grossman 2004). Consequently, many are very critical of the suggestion that aging can even be cured (Warner et al. 2005).

Strategies to cure aging rely on a complex interface of technological advances in regenerative medicine and gene therapy (de Grey and Rae 2008) or computing and nanotechnology (Kurzweil and Grossman 2004) that may take a long time to materialise. A combination of extraordinary breakthroughs in several areas is required for these predictions to become reality. However, based on what we know about the aging process, there is no scientific reason why a dramatic extension of the mean as well as the highest achieved lifespan should not be possible. Precisely because aging is such a multi-facetted issue, it seems entirely feasible to solve the problem in a piecemeal fashion using a portfolio of medical and technological alternatives. It seems unlikely that a single intervention will suddenly abolish aging. More realistic is a stepwise approach, where life-years are added in small instalments. A paradigm shift would arguably come if this progress were to occur at such a high rate that it outpaces the rate of aging (de Grey 2004a).

However, at this point the ethical issue of responsible communication comes into new focus: If one agrees that aging is, in principle, amenable to drastic alteration, then keeping silent about this possibility is in itself ethically problematic (de Grey 2004b). While there are several ethical problems with unwarranted hype, there is also an ethical obligation not to 'undersell' aging research (Farrelly 2010).

This challenge incidentally arises on 'both sides' of the wider moral debate: The ethical obligation to be clear about the potential of a research field arises regardless of whether one agrees that aging requires a 'cure' in the first place—to be discussed next.

13.2.3 'Simply' Avoiding Frailty?

What can rather be excluded as a possibility is that significant life extension will simply prolong the period of old age. Harking to stories like the classic Tithonus or Gulliver's Struldbrugs, some accounts imagine life extension as the extension of 'old age' and increasing frailty. Fukuyama, for example envisages a geriatric society occupied with the perpetuation of decrepitude (Fukuyama 2002). Many gerontologists would claim that 'compression of morbidity' (Fries 1983) rather than life extension is the practical aim of their studies. This is a very appealing strategy to generate research funding: It is politically palatable by promising to mitigate the pervasive health impact of the 'silver tsunami' while not encroaching on ethically contentious territory. Instinctively, many would agree that a 'quick departure' is desirable.

However, those therapies which will have any noticeable effect on lifespan are very unlikely to act at the stage where system failure is imminent (de Grey 2005). Although there has been some minor progress, there is no evidence that the 'compression of morbidity' approach is effective (Fries 2003; Crimmins and Beltran-Sanchez 2011); in fact, it may detract from effective strategy (de Grey 2006). By making progress against frailty, it has been argued that interventional biogerontology needs to adopt a much more rigorous evidence base (Nadon et al. 2008), but there are also concerns that regimes akin to the current schemes for marketing authorisation may not be appropriate. "A broad spectrum improvement in health is not an outcome that would currently motivate a drug trial and nor is frailty a recognised medical problem" (Partridge et al. 2011, p. 8).

These and similar discussions illustrate yet again the ethical importance of communicating the potential scope and impact of aging research clearly and truthfully: Such research does not fit conveniently into established political and regulatory categories.

13.2.4 Is Aging a Disease?

If there is potential to delay or even stop aging, how does that position a biogerontologist ethically towards the subject? Can aging be regarded as a disease with a view towards 'curing' it? This has significant ethical implications. 'Age' as a cause of death will rarely be found in mortality statistics. Barring accidents, most people in developed nations currently die of vascular complications or cancer—conditions intricately linked with aging, but that fit the common definition of disease. Still, some would object to classifying aging as disease, viewing aging as ubiquitous and natural (cf. Chap. 16).

It could be argued that contrary to popular belief, aging is not universal. A number of complex species, such as lobsters, rockfishes, and tortoises, do not show signs of aging (Finch 1990). Whatever the importance of death in evolution and in ecosystems, aging itself is certainly not a prerequisite to life. As mentioned above, senescence is now thought of as an evolutionary by-product rather than an end in itself. If aging is understood as a stochastic phenomenon, then it has been argued that "this makes aging unnatural and in no way an intrinsic part of human nature. As such, there is no reason why it is intrinsically wrong to try to reverse or cure aging" (Caplan 2005, p. 73).

We would doubt that common occurrence or 'naturalness' can be taken as a serious consideration either for or against treatment. Many of civilisations greatest achievements are a departure from 'nature'. It seems odd if for those people who drive cars, take medicines, wear glasses, receive e-mail, watch television, and do not have to kill their own dinner think life-extension is unnatural. Consequently, the attempts to muster logical arguments distinguishing aging from disease tend to flounder rather helplessly. (Consider: "The difference between a biological cause, such as the mutation in a gene or the malfunctioning of a protein, and a life-process such as aging, is manifest. The gene and the protein are parts of the biological organism, whereas aging is a part of life as a lived process: Life as we live it" (Rehmann-Sutter 2011)). In the process, attempts to divine ethical problems may themselves appear of dubious ethical distinction: A person who is dying of cancer might not care whether the malignancy was triggered, hastened, or caused by a genetic disposition, a mutation, infection, an aberrant stem cell, immune system failure, oxidative stress, pollutants, radiation, or any combination of age-related factors. The question of whether curing cancer in the young is morally desirable whereas curing cancer in the aged is inappropriate shifts the issue of whether aging is a disease from science to politics or indeed to regulation. "The clinical redefinition of aging as a disease state would not only make sense, but it would also foster the development of therapies of benefit to older people" (Gems 2011, p. 109). Thus, in this arena "Should we consider aging a disease?" is essentially asking "Should aging be cured?"

13.2.5 Biogerontology vs. Medicine

For strategic reasons, gerontology is moving away from a juxtaposition of aging and disease towards a discussion of age-associated pathology. It is apparently easier to make the case for curing Alzheimer's, heart disease or stroke, tangible suffering associated with aging than to make a case for tackling the underlying condition.

Underlying conditions are also notoriously more difficult to address than symptomatic treatment. The fact that aging is so complex and pervasive makes it difficult to 'narrow down' findings towards distinguishable treatments (although some insights from biogerontology are more immediately subject to translation (Lim et al. in press)). In fact, in developing 'respectability' biogerontology needed to cast off associations with charlatanry, and "strengthen the boundaries between themselves and anti-aging practitioners and entrepreneurs (Fishman et al. 2008)". Ironically, where biogerontologists have sought this distance, other biological sub-disciplines are challenged and struggling to become more translational. In fact this sort of 'boundary work', vigorously distancing biogerontology from 'anti-aging medicine' (Olshanksky et al. 2002), has not helped to establish biogerontology as part of the moral endeavour of easing suffering. While the field of geriatrics shares with gerontology the 'aging research' umbrella term (and consequently even some conferences, societies, journals and funding pots), the two fields do not always have effective interface—not facilitated by the observation that both fields are inherently interdisciplinary (Clark 1993).

For various historical and conceptual reasons, aging research and medical research remain distinct (de Grey 2007), and this has not helped to clarify the role of aging as a disease.

This has never stopped biogerontology to—rightly—advertise its potential to address epidemiological and medical issues. Lately, aging research has risen to particular prominence as an imperative to tackle otherwise insurmountable demographic challenges (Rae et al. 2010). For this to be effective though, it may be necessary for biogerontology to develop more effective ways of making a difference—and this means going further than offering data interpretation in the field of geriatrics. By aspiring to make a real difference, biogerontology assumes automatically an ethical position towards aging.

13.2.6 Motivation

The motivation for aging research is of course not necessarily linked to wanting to cure aging. Indeed motivations of aging researchers are very heterogeneous (Underwood 2009). 'The project' to cure aging has been linked to various contentious ideological roots, and variously been identified as a political ploy either "aiming to control individuals" or/also as the expression of ultimate narcissistic individualism, as placing a burden on the elderly to perform in a consumerist marketplace, in short as another detriment of shallow capitalist materialism (Jones and Higgs 2010; Lafontaine 2009). This is of course at odds with the 'lived experience' of those suffering from the detriments of aging. Whether we live in a more individualist society or not, the experience of aging is a deeply individual one, not least because the very plasticity of aging explained above manifests at the level of the individual—and it is undeniably the case that for many but not all, the experience of aging is one of personal frustration and suffering.

It is this suffering that presents an ethical motivation to some biogerontologists and justification for supporting research and development in this area. Commentators may be inclined to detect causes for ethical concern in the underlying normative aesthetics, but to cast scientists who support working on aging with a view towards its eventual mitigation and abolition as unwitting cogs in a nefarious capitalist machine offers no useful foundation for different ethical decision making by the accused. Instead it would not seem unreasonable for an ethically motivated biogerontologist to feel a desire that aging research be developed into effective 'anti-aging' treatments. However, we accept that moral agents are challenged to consider not just the ethics of their actions but also the far reaching implications. Consequently, we will turn briefly first to discuss the ethical practicalities of implementing research findings, and then move to discuss some of the potential social consequences in case such implementation should ever come to pass.

13.3 Social Consequences

Every researcher needs to take account of where the research may ultimately lead to. The potential social, economic, psychological and practical implications of drastically increased lifespan have been subject to extensive deliberation. Common to these scenarios is an inevitable element of speculative conjecture. In the following, we will briefly consider the two of the most prominent ethical concerns about the social consequences of successful biogerontology: The 'fairness objection' and the 'overpopulation' argument.

Subsequently, given the 'biogerontology perspective' which we adopt here, we will address two further topics which are arguably of particular concern to scientists active in this field: The suggestion that a long life would lead to stagnant, less fulfilled lives and to diminished social progress and the accusation that research to abolish aging stigmatises and marginalises the elderly.

13.3.1 Cost and Equity

While it is difficult to foresee the costs of a hypothetical cure for aging, it has been assumed that such treatments will be expensive.

The basic premise is not uncontested. Firstly, any economic cost calculation also needs to take account of the economic benefit. Curing aging and extending healthy lifespan would be profitable for nations. Historically, the economic value of increased longevity from 1970–2000 was estimated at $3.2 trillion per year for the US alone through increased productivity and significantly decreased healthcare costs, with economic gains from future health improvements also estimated to be large (Murphy and Topel 2006). In fact, effective anti-aging interventions are likely required to avert significant economic burdens associated with the current demographic situation (Aaron and Schwartz 2004).

Moreover, it could be suggested that even if curing aging is initially expensive, its universal desirability will help to recoup costs of investment relatively quickly, allow for low profit margins, create political and competition pressure to drive prices down, facilitate e.g. mass production and service infrastructures etc. as has happened in similar cases (Lucke et al. 2009).[1]

Others would not wish to wait for such 'trickle down' effect or doubt its efficacy. Nonetheless, some commentators have identified the potential social inequities between those who may be able to afford effective anti-aging treatments and those who cannot pursue human life extension at all (Glannon 2002). To those voices, contemplating further life extension in an unequal world is ethical anathema (cf. Chap. 17). While the argument is usually put more eloquently (Pijnenburg and Leget 2007), it seems subject to distillation into the statement: 'Healthcare in rich countries should not advance until the poorer countries have caught up'. Not only does this argument neglect the considerable burden of age-associated diseases already threatening to crush poor country economies (Smith and Mensah 2003), the underlying ideology would call for the cessation of any number of activities other than those directly aimed at improving life expectancy in poor nations (which are, it has been argued, not predominantly scientific but political barriers).

However, even if we assume for the sake of argument that life extension therapies will always be costly, it still remains questionable whether the notion of inequality is such a moral evil as to require "collective suttee" (Davis 2004). "We do not normally think it an ethical requirement to prevent good being done to some unless and until it can be done to all" (Harris 2002, p. 290). In transplantation medicine, for example, the availability of organs is a very real factor of life extension: Those lucky enough to be allotted an organ, survive much longer. Generally, society has been able to agree on laws that regulate the allocation of these special resources. We have not chosen to destroy all organs as they become available in order to preserve equality in despair. Similarly, if life extension treatments would turn out to be irrevocably scarce, this must not mean that only the ability to pay will be decisive. Treatment could be allotted

[1] For a different view, see chapter 19, this volume.

randomly (Broome 1984), as a social reward (Vance 1956), according to imminent need (Harris 1987), within a general utilitarian framework (Miller and Sethe 2005), or even as "affirmative prolongevitism" (Overall 2003). There is no intrinsic reason why such arrangements must be morally flawed if instituted in a society that can agree on an equitable system of apportionment.

In summary, upon closer examination the spectre of unequal availability cannot present an ethical barrier to biogerontology.

13.3.2 Overpopulation

The other main issue most frequently considered a social implementation objection centres on consequences for population growth.

It is also not always clear why a larger population is considered morally problematic, but most concerns seem to be based on the assumption that an ever greater population leads to a world that is 'not worth living in'. Since Malthus raised his concerns, indeed since the doomsday scenarios were projected in the 1970s, population has increased manifold without any decline in living standards (Trewavas 2002). The question of population pressure is obviously defined by other characteristics than just life expectancy: birth rates, infrastructure, availability of resources, capacity to recycle waste, space management, land use, and concepts of privacy all play a role.

In popular discourse, overpopulation is seen as the root of many environmental problems and societal ills (Ehrlich 1968), but in many respects, moral issues in environmentalism are independent from future developments in population pressure. If we are using finite resources in a non-sustainable manner, then this problem needs to be solved independently of how long people live. Relying on death is not a very creative way to tackle such problems (More 2004) especially considering that population-linked doomsday scenarios have generally been dispelled by human ingenuity (Boserup 1981).

In concentrating on the underlying moral issues at play, we are challenged to question the relevance that these differing visions might have in the first place. If one decided that the vision of a crowed planet is too terrible to permit, what type of intervention should be adopted? Would we decline to invest in medical innovation? Withhold its use? Encourage suicide or sanction killings? In population ethics, one is precariously balancing the real interest of existing people against the hypothetical interests of those projected to be born (Parfit 1983) and potentially also balancing a hypothetical quality of life against the imposition of an early death (Davis 2005; Cutas 2008).

The instinct and desire to procreate is strong in many. This may be due to evolutionary reasons, but also a conscious decision to defy death by trying to perpetuate something of oneself—which indicates that such desires might be less strong in 'immortals' (Perry 2000). Some suggest a scheme where those who have become 'immortalized' could agree not to reproduce (Harris 2000). Apart from enforceability, one could regard this arrangement as troubling where it might lead to social stagnation. Yet, at this point we really feel one enters such a complex area of pure speculation (scientific, economic, political, social, psychological and individual) that we fail to see how one could draw any ethical conclusions that should seriously undermine the moral imperative as we have outlined it to treat aging today. The question of imminence is valid: The most immediately effective way to control population pressures is to stabilise birth rates. Population changes are surprisingly slow in their response even to a dramatic life extension (Gavrilov and Gavrilova 2010). Even if a 'cure' for aging were developed tomorrow, the pressure of population whatever they may be would not amount to a marked increase for many decades. Thus if overpopulation becomes an issue it would be well into the next century or beyond. By then, social, technological and economic parameters are likely to be so different to render all anticipatory ethical concerns moot. Such an array of uncertainties surrounds potential population pressures that it seems presumptuous to preclude today the moral decisions that are the concern, privilege, and responsibility only of future societies. A biogerontologist needs to balance these considerations against the immediacy of alleviating age-associated suffering.

13.3.3 Stagnation

At first, a general fear of boredom may seem like one of the most trivial objections to increasing lifespan. In essence, it assumes that (a) patterns of experience will inevitably repeat and (b) that the ability to derive satisfaction from experience always diminishes in repetition. The somewhat cynical question of boredom alludes to a more serious issue in the longevity of the mind. Part of this argument seems to draw on the (usually unspoken) hypothesis that the human brain is not equipped to deal with vastly extended lifespans. It is common experience that personalities are essentially formed during childhood, whereas adults are often very set in their ways. This could mean that ubiquitous long life leads to an altogether less flexible and dynamic society, where the majority are less willing to change their outlook and convictions, where new culture and technology is stifled, strife and inefficiency perpetuated. There are good reasons to believe that some—if not the majority—of our decreasing ability to learn and adapt as we age is an artefact of brain aging (Lockett 2010) and may thus be subject to remedy of anti-aging treatments (Lynch 2011). In the long term, the 'Markopoulos challenge' is not such a trivial one, but requires a more nuanced approach to the psychology of boredom (Bortolotti and Nagasawa 2009).

Still, those who have spent a 'lifetime' developing a theory, following a creed, or hating an enemy are presumably less likely to change their mind than those younger and less encumbered by their past. As Max Plank suggests "A new scientific truth does not triumph by convincing its opponents and making them see the light, but rather because its opponents eventually die, and a new generation grows up that is familiar with the idea from the beginning" (Planck 1950). Yet the problem of such ossification is not confined to future technology. Similar considerations have inspired legislators to limit the term any one individual can spend in a position of power. To rely on death as a driver of change is to take a very resigned view about our moral responsibility and capacity. Instead, we would agree that rather than dwelling on the ethics of variable aging it is important to address the actual "lived experience of those who engage, refuse to engage, or are unable to engage with these contested domains" (Jones and Higgs 2010, p. 1518).

Ultimately, it is intriguing to speculate that far from having reached its endpoint due to medical technology, evolution by natural selection would come into its own in these futuristic scenarios: Not only will there be strong dispositional selection pressure against those who reject such treatments for ideological reasons, it might well be the case that humanity will undergo a selection where those who can experience the greatest fulfilment from ongoing discovery will choose to live substantially longer, compared to other character types.

13.3.4 Stigmatising the Elderly

Sometimes hidden in some of the more obscure postmodernist critique of efforts to cure aging a particularly worrying allegation emerges that applies directly to the biogerontologist. In efforts to mitigate, abolish, cure, eradicate, defeat aging, are scientists forging a paradigm in which old people must feel as unfortunate failures? Conflating the disease with the sufferer would seem a straightforward fallacy to avoid, but we know from other situations in bioethics and beyond that in fact such issues need a very sensitive and empathetic link between the researcher and the researched. Stigmatization, ghettoization and loneliness of the elderly are a real problem already, but it is true that these could be exacerbated where a narrative of 'successful aging' casts off those who in some way 'fail' to evade it (Vincent 2006). However, this is precisely where biogerontology can make an ethical contribution: What commentators lamenting the impending alienation and stigmatization of the elderly often overlook is that there is in fact no 'golden age' where old people were treated fairly as research subjects or patients. Instead, we know that there is a paucity of research on age-related conditions and that sometimes even the most basic interventions are not applied to the elderly as they are 'likely to die anyway'. Biogerontology can make a real contribution by challenging these assumptions.

A biogerontologist wishing to cure aging will be encountering those who share this aspiration but for whom no feasible treatment exists as well as those who reject the notion and feel shaken by the mere suggestion that what is happening to them may not be 'normal'. Either situation calls for tact, respect, and sympathy. Neither situation will be trivial to navigate ethically. It is this kind of interaction between moral agents that brings us full circle back to what was called the 'inner circle' ethics in the introduction: Away from the flights of fancy that ethicists sometimes posit, the 'social consequences' of biogerontology are being shaped not (only) by the sinister workings of materialist modernity, but by the daily contact between people trying to make sense of their lives and trying to relate to each other with respect and sympathy.

13.4 Conclusion

Here, we have aimed to discuss some ethical issues in biogerontology from the perspective of biogerontology research. We have shown that, by assuming this perspective, certain ethical 'dilemmas' may appear less immediately relevant whereas others come into sharper focus.

On the one hand, research shows that aging is flexible, subject to intervention, amelioration and modulation. No ethical case can be made for denying or suppressing this fact. Aging is associated with evils such as grief, suffering, loss of dignity and loss of freedom. Worries about long term social consequences not only lack evidence, they fail in the face of the immediate ethical challenge. Efforts to hide this fact or to justify its existence on abstract philosophical or ideological grounds are ultimately immoral if they stand in the way of effective action in research or treatment.

However, it is also true that aging and death will remain a 'fact of life'. This opens up a gap, a space for ethical deliberation and concern. The real ethical challenge is to identify involuntary aging as 'the enemy' while not abandoning or belittling the fate of those who are nonetheless subject to aging.

In summary, protagonists in biogerontology are subject to three ethical imperatives: (1) to represent the potential of aging research without hype but also without unwarranted political constraint; (2) to face the fact that aging causes suffering whereas the putative social drawbacks of controlling aging are speculative and contested; (3) to tackle the challenge of 'fighting aging' without fostering ageism.

References

Aaron, HJ and WB Schwartz, eds. 2004. *Coping with methuselah: The impact of molecular biology on medicine and society*. Washington: Brookings Institution Press.

Bortolotti, L., and Y. Nagasawa. 2009. Immortality without boredom. *Ratio* 22(3): 261–277.

Boserup, E. 1981. *Population and technological change: A study of long-term trends*. Chicago: University of Chicago Press.

Broome, J. 1984. Selecting people randomly. *Ethics* 95: 38–55.

Browner, W.S., A.J. Kahn, E. Ziv, A.P. Reiner, J. Oshima, R.M. Cawthon, et al. 2004. The genetics of human longevity. *American Journal of Medicine* 117(11): 851–860.

Butler, R.N., H.R. Warner, T.F. Williams, S.N. Austad, J.A. Brody, J. Campisi, et al. 2004. The aging factor in health and disease: The promise of basic research on aging. *Aging Clinical and Experimental Research* 16(2): 104–11.

Caplan, A.L. 2005. Death as an unnatural process. *EMBO Report* 6(S1): S72–S75.

Clark, P.G. 1993. A typology of interdisciplinary education in gerontology and geriatrics: Are we really doing what we say we are?. *Journal of Interprofessional Care* 7: 217–227.

Crimmins, E.M., H. Beltrán-Sánchez. 2011. Mortality and morbidity trends: Is there compression of morbidity?. *Journals of Gerontology* 66B(1): 75–86.

Cutas, D.E. 2008. Life extension, overpopulation and the right to life: Against lethal ethics. *Journal of Medical Ethics* 34(9): e7.

Davis, J.K. 2004. Collective suttee: Is it unjust to develop life extension if it will not be possible to provide It to everyone?. *Annals of the New York Academy Sciences* 1019: 535–541.

Davis, J.K. 2005. Life-extension and the malthusian objection. *The Journal of Medicine and Philosophy* 30: 27–44.
de Grey, A., and M. Rae. 2008. *Ending aging: The rejuvenation breakthroughs that could reverse human aging in our lifetime*. New York: St. Martinś Press.
de Grey, A.D.N.J. 2004a. Escape velocity: Why the prospect of extreme human life extension matters now. *PLoS Biology* 2(6).
de Grey, A.D.N.J. 2004b. Biogerontologists' duty to discuss timescales publicly. *Annals of the New York Academy of Sciences* 1019: 542–545.
de Grey, A.D.N.J. 2007. Understanding and tackling aging: Two fields communicating (A little) at last. *Rejuvenation Research* 10(4): 637–640.
de Grey, ADNJ. 2006. Compression of morbidity: The hype and the reality, part 2. *Rejuvenation Research* 9(2): 167–8.
de Grey: ADNJ. 2005. Resistance to debate on how to postpone aging is delaying progress and costing lives. *EMBO Reports* 6(S1): S49–S53.
de Magalhaes, J.P. 2011. The biology of aging: A primer. In *An introduction to gerontology*, ed. I. Stuart-Hamilton, 21–47. Cambridge: Cambridge University Press.
de Magalhaes, J.P., and G.M. Church. 2005. Genomes optimize reproduction: aging as a consequence of the developmental program. *Physiology (Bethesda)* 20: 252–259.
de Magalhaes, J.P., A. Budovsky, G. Lehmann, J. Costa, Y. Li, V. Fraifeld, et al. 2009. The human ageing genomic resources: online databases and tools for biogerontologists. *Aging Cell* 8(1): 65–72.
de Magalhaes, J.P., D. Wuttke, S.H. Wood, M. Plank, and C. Vora. 2012. Genome-environment interactions that modulate aging: powerful targets for drug discovery. *Pharmacological Reviews* 64: 88–101.
Ehrlich, P.R. 1968. *The population bomb*. New York: Ballantine.
Farrelly C. 2010. Why aging research? The moral imperative to retard human aging. *Annals of the New York Academi of Science* 1197: 1–8.
Finch, C.E. 1990. *Longevity, senescence, and the genome*. Chicago: The University of Chicago Press.
Fishman, J.R., R.H. Binstock, M.A. Lambrix. 2008. Anti-aging science: The emergence, maintenance, and enhancement of a discipline. *Journal of Aging Studies* 22(4): 295–303.
Flurkey, K., J. Papaconstantinou, and D.E. Harrison. 2002. The snell dwarf mutation Pit1(dw) can increase lifespan in mice. *Mechanisms Ageing and Development* 123(2-3): 121–130.
Fontana, L., L. Partridge, and V.D. Longo. 2010. Extending healthy lifespan–from yeast to humans. *Science* 328(5976): 321–326.
Fries, J.F. 2003. Measuring and monitoring success in compressing morbidity. *Annals of Internal Medicine* 139: 455–459.
Fries, James F. 1983. The compression of morbidity. *Milbank Memorial Fund Quarterly* 61: 397–419.
Fukuyama, F. 2002. Our posthuman future. *Farrar, Straus and Giroux*.
Gavrilov, L.A., N.S. Gavrilova. 2010. Demographic consequences of defeating aging. *Rejuvenation Research* 13(2–3): 329–334.
Gems D. 2011. Tragedy and delight: the ethics of decelerated aging. *Philosophical Transactions Rhode Social* B366: 108–112.
Glannon, W. 2002. Reply to harris. *Bioethics* 16: 292–297.
Harris, J. 1987. QALYfying the value of life. *Journal of Medical Ethics* 13: 117–123.
Harris, J. 2000. Intimations of immortality. *Science* 288: 59.
Harris, J. 2002. Response to glannon. *Bioethics* 16(3): 284–291.
Harrison, D.E., R. Strong, Z.D. Sharp, J.F. Nelson, C.M. Astle, K. Flurkey, et al. 2009. Rapamycin fed late in life extends lifespan in genetically heterogeneous mice. *Nature* 460(7253): 392–395.
Jones IR, Higgs PF. 2010. The natural, the normal and the normative: Contested terrains in aging and old age. *Social Science and Medicine* 71(8): 1513–9.

Jones, I.R., and P.F. Higgs. 2010. The natural, the normal and the normative: Contested terrains in aging and old age. *Social Science and Medicine* 71(8): 1513–1519.
Kenyon, C.J. 2010. The genetics of aging. *Nature* 464(7288): 504–512.
Kirkwood, T.B. 2005. Understanding the odd science of aging. *Cell* 120(4): 437–447.
Kurzweil, R., and T. Grossman. 2004. *Fantastic voyage: Live long enough to live forever*. Emmaus, PA: Rodale.
Lafontaine, C. 2009. Regenerative medicine's immortal body: From the fight against aging to the extension of longevity. *Body and Society* 15(4): 53–71.
Lim, E.L., K.G. Hollingsworth, B.S. Aribisala, M.J. Chen, J.C. Mathers, and R. Taylor. (in press). Reversal of type 2 diabetes: normalisation of beta cell function in association with decreased pancreas and liver triacylglycerol. *Diabetologia*.
Lockett, G.A., F. Wilkes, R. Maleszka. 2010. Brain plasticity, memory and neurological disorders: An epigenetic perspective. *NeuroReport* 21(14): 909–913.
Lucke, J.C., P.C. Diedrichs, B. Partridge, W.D. Hall. 2009. Anticipating the anti-aging pill. Lessons from the history of the oral contraceptive pill and hormone replacement therapy. *EMBO Reports* 10(2): 108–113.
Lynch, G., L.C. Palmer, C.M. Gall. 2011. The likelihood of cognitive enhancement. *Pharmacology Biochemistry and Behavior* 99(2): 116–129.
Miller, J., S. Sethe. 2005. Gods with a limited budget: putting the utility back into utilitarian health politics. Interdisciplinary. *Science Reviews* 30(3): 273–278.
More, M. 2004. Superlongevity without overpopulation. In *The scientific conquest of death—essays on infinite lifespans; librosEnRed*, ed. B.J. Klein, S. Sethe, et al., 169–185.
Murphy, K.M., and R. H. Topel. 2006. The value of health and longevity. *Journal of Political Economy* 114(5): 871–904.
Nadon, N.L., R. Strong, R.A. Miller, J. Nelson, M. Javors, Z.D. Sharp, J.M. Peralba, D.E. Harrison. 2008. Design of aging intervention studies: The NIA interventions testing program 2008. *Age* 30(4): 187–199.
Olshansky, S.J., L. Hayflick, and B.A. Carnes. 2002. No truth to the fountain of youth. *Scientific American* 286(6): 92–95.
Overall, C. 2003. *Aging, death, and human longevity: A philosophical inquiry*. Berkeley: University of California Press.
Parfit, D. 1983. *Reasons and persons*. Arlington: Clarendon.
Partridge, L., J. Thornton, and G. Bates. 2011. The new science of aging. *Philosophical Transactions Rhode Social* B366: 6–8.
Perry, M. 2000. *Forever for all: Moral philosophy, Cryonics, and the Scientific Prospects for Immortality*. Boca Raton, FL: Universal Publishers.
Pijnenburg MA, C. Leget. 2007. Who wants to live forever? Three arguments against extending the human lifespan. *Journal of Medical Ethics* 33(10): 585–7.
Planck M. 1950. *Scientific autobiography and other papers*. London: Williams and Norgate.
Rae MJ, RN Butler, J Campisi, AD de Grey, CE Finch, M Gough, GM Martin, J Vijg, KM Perrott, BJ Logan. 2010. The demographic and biomedical case for late-life interventions in aging. *Science Translation Medecine* 2(40): 40cm21.
Rehmann-Sutter, C. 2011. Aging as disease?. *Swiss Medecine Weekly* 141.
Richel, T. 2003. Will human life expectancy quadruple in the next hundred years? sixty gerontologists say public debate on life extension is necessary. *Journal of Anti-Aging Medicine* 6(4): 309–314.
Sames, K. 2005. Aging as a consequence of organ differentiation. In *Extending the lifespan: Biotechnical, gerontological, and social problems; LIT*, ed. K Sames, S Sethe, A Stolzing.
Smith, S.M., G.A. Mensah. 2003. Population aging and implications for epidemic cardiovascular disease in sub-Saharan Africa. *Ethnicity and Disease*, 13(2 SUPPL. 2): S2–77–S2-80.
Stipp, D. 2010. *The youth pill: Scientists at the brink of an anti-aging revolution*. London: Penguin Books.
Trewavas, A. 2002. Malthus foiled again and again. *Nature* 418: 668–670.

Underwood, M., H.P. Bartlett, W.D. Hall. 2009. Professional and personal attitudes of researchers in aging towards life extension. *Biogerontology* 10(1): 73–81.
Vance, J. 1956. *To live forever*. New York: Ballantine.
Vincent, J.A. 2006. Aging contested: Anti-aging science and the cultural construction of old age. *Sociology* 40(4): 681–98.
Warner, H., J. Anderson, S. Austad, E. Bergamini, D. Bredesen, R. Butler, et al. 2005. Science fact and the SENS agenda. What can we reasonably expect from aging research?. *EMBO Report* 6(11): 1006–1008.

Chapter 14
Immortal Ethics

John Harris

Life-extending therapies and optimistic discussions of their promise and probable effect are an increasing dimension of serious scientific and philosophical discussion.[1] If such therapies ever become reality (Bodnar et al. 1998; Weinrich et al. 1997), and if our bodies could repair damage caused by disease and aging "from within"(McBearty et al. 1998) the effects not only on personal health and survival but also on society and on our conceptions of ourselves and of the sorts of creatures we are would be profound (Thomson et al. 1998; Pedersen 1999; Mooney and Mikos 1999). If we could switch off the aging process(Lanza et al. 1999a; Lanza 1999b) we could then, in Lee Silver's words, "write immortality into the genes of the human race" (these possibilities were rehearsed in the BBC TV *Horizon* program).

14.1 Familiarity with Immortals

Increased longevity and its logical extension, some would say its *reductio ad absurdum*, immortality, have a long history. The human imagination is familiar with the idea of immortals and mortals living alongside one another and interacting. The *Iliad*, the *Odyssey*, the Bible, the Koran, the Ramayama, and Shakespeare's plays all have made such ideas familiar, and even modern classics have taken seriously the possibility of immortality. In his celebrated trilogy in five parts *The Hitchhiker's Guide to the Galaxy*, Douglas Adams imagines a man who had achieved immortality by accident:

This chapter has been published earlier as: Harris, J. (2004). Immortal Ethics. Annals of the New York Academy of Sciences, 1019, 527–534.

[1] I have benefited from the incisive comments of my colleague Søren Holm.

J. Harris (✉)
University of Manchester, United Kingdom
e-mail: john.harris@manchester.ac.uk

M. Schermer, W. Pinxten (eds.), *Ethics, Health Policy and (Anti-) Aging: Mixed Blessings*, Ethics and Health Policy 1, DOI 10.1007/978-94-007-3870-6_14,

> To begin with it was fun, he had a ball, living dangerously, taking risks, cleaning up on high-yield long-term investments, and just generally outliving the hell out of everybody. In the end it was the Sunday afternoons that he couldn't cope with, and that terrible listlessness which starts to set in at about 2.55 when you know that you have had all the baths you can usefully have that day, that however hard you stare at any given paragraph in the newspapers you will never actually read it... and that as you stare at the clock the hands will move relentlessly on to four o'clock, and you will enter the long dark teatime of the soul (Adams 1982).[2]

Despite the apparent pessimism of this passage many people would be prepared to endure "the long dark teatime of the soul" in exchange for immortality.[3] Indeed, there is much evidence both from literature, and in the literature, that suggests that many people are willing to trade off quality of life for longevity (Weiss 2000). From the pact of Faust, celebrated by writers from Marlowe to Goethe, to Bram Stoker's vampires,[4] to choices made by cancer patients with a terminal diagnosis (Slevin et al. 1990), the evidence is strong that people want extra life time even at substantial costs in terms of pain and quality of life, even when outcomes are highly uncertain.

14.2 Immortality is not Invulnerability

Note that immortality is not the same as invulnerability, and even "immortals" could die or be killed. Accidents, infectious diseases, wars, and domestic violence would all take their toll, and although we might hope for progress in combating existing diseases, the development of new threats, such as HIV/AIDS and the emergence of variant Creutzfeldt-Jakob disease have demonstrated, may increase rather than reduce human vulnerability over time. If we add to this the diminishing effect of proven therapies such as antibiotics through the emergence of resistant strains of bacteria, it is difficult to predict the likely levels of 'premature' deaths in a future in which increased life expectancy was developing and spreading through the human population.

14.3 Life Extension Saves Lives

When we save a life, by whatever means, we simply postpone death. Life saving is just death postponement. This is a truth from which it follows that life-extending therapies are, and must always be, life-saving therapies and must share whatever priority life saving has in our morality and in our social values. So long as the life is of acceptable quality (acceptable to the person whose life it is) (Harris 1987), we have a powerful, many would claim an overriding, moral imperative to save the life, because to fail to do so when we can would make us responsible for the resulting death (this claim is defended in detail in Harris 1987, 1980).

---

[2] For the record, the immortal's name was Wowbagger (p. 9).

[3] And we should note that Wowbagger himself did find something meaningful to do through all eternity.

[4] I am grateful to Simon Woods for insights into the un-dead.

Three main sorts of philosophical or ethical objections have been leveled at life extension. It has been claimed that life extension would be unjust; it would be pointless and ultimately unwanted because of the inevitable boredom of indefinite life and would in any event be nugatory or self-defeating because personal identity could not survive long periods of extended existence. I may wish to be immortal but in the end it wouldn't be 'me', so the project fails. Finally it is claimed that life extension would be prohibitively expensive in terms in increased healthcare costs. We will look at all these issues now, but necessarily briefly.

14.4 Global Justice

One thing we do know is that the technology required to produce such results will be expensive. For existing people with multiple interventions probably required, the costs will be substantial. To make modifications to the embryo or even to the gametes before conception, people will have to be determinedly circumspect about procreation and will probably need to use reproductive technologies to have their immortal children. Even in technologically advanced countries therefore, 'immortality', or increased life expectancy is likely to be confined to a minority of the population. In global terms, the divide between high-income and low-income countries will be increased, with low-income countries effectively denied access to the technology that might make some of their citizens immortal. The issue of the citizens of rich countries gaining further advantages over the poor will rightly disturb many. How are we to understand the demands of justice here?

14.4.1 Parallel Populations

A feature of life-extending treatments, which seldom has been thought through, is the fact that as treatments become available we will face the prospect of parallel populations, of 'mortals', and 'immortals', existing alongside one another (Silver 1999). Thus, the problems of global justice will be repeated in those societies able to implement life-extending therapies. Just as there will exist parallel societies, some able to provide immortalizing therapies and some not, so within those societies that have the technology and the resources required there would exist parallel populations of mortals and immortals. This of course is precisely the destiny for which the poetic imagination has prepared us, literally from 'time immemorial.'

Although such parallel populations seem inherently undesirable and even unfair, it is not clear that we could, or even that we should, do anything about such a prospect for reasons of justice. If immortality or increased life expectancy is a good, it is doubtful ethics to deny palpable goods to some people because we cannot provide them for all. And this unfairness is not simply contingent, a function of a regrettable, but, in principle, removable lack of resources. There will always be circumstances in which we cannot prevent harm or do good to everyone, but surely no one thinks that this affords us a reason to decline to prevent harm to anyone in particular. If twins suffer from cancer and one is incurable and the other not, we do not conclude that we should not treat the curable cancer because this would in some sense be unjust to the incurable twin. We don't refuse kidney transplants to some patients unless and until we can provide them for all with renal failure. We do, however, have a clear ethical responsibility to ensure that the question of which of those who could benefit receives the treatment should be decided according to some just principle of distribution. We don't usually regard ourselves as wicked in Europe or North America because we perform many transplants (this claim is defended in detail in my *Violence and Responsibility* and in *The Value of Life*), whereas low-income countries perform few or none at all. The solution, however, is certainly not to say that we will outlaw transplantation unless and until equitable distribution on some agreed principles can be guaranteed. The introduction of any new complex and/or expensive technology raises these problems. The impact on global justice or on justice within societies is important and must be addressed; it is a principled objection, but not an objection in principle to the introduction of life-extending therapies. The principle requires that strenuous and realistic efforts be made to provide the benefits of the technology justly and as widely as possible, not that the benefits be denied because of the impossibility of ensuring adequate justice of provision.

14.4.2 Immortality as the Side Effect of Therapy

Remember that immortality is not unconnected with preventing or curing a whole range of serious diseases. It is one thing to ask the question "Should we make people immortal?", and answer in the negative; it is quite another to ask whether we should make people immune to heart disease, cancer, dementia, and many other diseases and decide that we should not, because a 'side effect' of the treatment would be increase in life expectancy. We are then unlikely ever to face the question: Should we make people immortal, 'yes', or 'no'? We may rather be called upon to decide whether we should treat a particular disease when we know an effective treatment will extend lifespan.

It might then be appropriate to think of immortality as the side effect of treating or preventing a whole range of diseases. Could we really say to people "You must die at the age of thirty or forty or fifty, because the only way we can cure you is to extend your lifespan?" Faced with such a choice, an individual might well say, "Let me have my three score and ten and then let me die." Given the quite pervasive and irrational hostility to euthanasia, whether societies would be willing to allow such bargains to be made is doubtful.

14.5 Longevity is a Rational Good

Given that people want life and fear death, it is difficult not to see longevity, and perhaps immortality, as a palpable good. Many have taken issue with this claim on two main grounds: either that indefinite life eventually would become terminally boring or that over long periods of survival personal identity could not be maintained and so the survival of a particular individual would prove illusory. Elsewhere (see Harris 2002) I have criticized, and I believe decisively refuted both these objections. Suffice it to say that only the terminally boring are in danger of being terminally bored, and perhaps they do not deserve indefinite life. Those who are bored can, thanks to their vulnerability, opt out at any time. But those of us who do not have terminal failure of the imagination should be left to create new ways of enjoying life and doing good. It is easy to see that that personal identity is not required for a coherent desire for indefinite survival. Suppose 'Methuselah' has three identities, A, B, and C, and that C can remember nothing of A's life. But suppose the following is also true: A will want to be B who will remember being A, B will want to become C who will remember being B but possibly not remember being A. It is not irrational for A to want to be B and not irrational for A to want to be B partly because he or she knows that B will be able to look forward to being C, even though by the time she is C she won't remember being A. Thus, even if personal identity in some strict sense fails over time, it is not clear that a sufficiently powerful motivation for physical longevity fails with personal identity. This would remain true however many selves 'Methuselah', turns out to be.

Prominent among recent denigrators of the idea of life extension has been Leon Kass, who identifies the core question as the following: 'Is it really true that longer life for individuals is an unqualified good?' Kass has many arguments against life extension, all of which fail disastrously (Harris 2002). We have space to consider only his main objection: "For to argue that human life would be better without death is, I submit, to argue that human life would be better being something other than human... The new immortals, in the decisive sense, would not be like us at all. If this is true, a human choice for bodily immortality would suffer from the deep confusion of choosing to have some great good only on condition of turning into someone else." (Kass 2001) Insofar as this claim of Kass's relies on claims about psychological continuity, over time it has the problems we have already considered. However, Kass's argument seems to be suggesting a more simple objection: that since the (current) essence of being human is to be mortal, immortals would necessarily be a different type of being and therefore have a different identity. There is a sense in which this is true, but not, I think, any sense in which it would be irrational to want to change identity to the specified extent. Someone who had been profoundly disabled from birth (blind say, or crippled) and for whom a cure became available in his or her mid-forties would become in a sense a different person. They would lead a different type of life in many decisive ways. It does not follow that the blind or crippled individual has no rational motive to be cured. It would be both odd and cruel say to them, as Kass presumably would have us do, 'it is deeply confused to want to cease to be disabled because then you will no longer exist.'

14.6 Population Policy

Many people addressing the question of life extension have assumed that such a possibility will have a disastrous effect on the world's population with the present generation living indefinitely and a procession of subsequent generations adding to the congestion (Glannon 2002). However, this is by no means either a likely or even the most likely scenario. The effect of life extension on population will be a function of several different factors, the outcomes of which are all difficult to predict. The first is the degree of uptake, which itself will be heavily dependent on cost and availability of the therapies. Granting, as we have, that life-extending therapies gradually will become available, cost, risk, and uncertainty will mean that for a very long time the numbers of people availing themselves of such therapies will be a tiny proportion of the world's population. We already have noted a possibly increasing human vulnerability due to new infectious diseases or antibiotic resistant strains of bacteria. Again it is difficult to predict the continuing effect of these on population or how the advent of some immortals would affect the equation. Disease may well continue to be an effective leveler, improving its own technology as we improve ours. And of course immortal but vulnerable people will continue to die in accidents and from injuries received.

14.7 The End of Reproduction

Should we assume the necessity for, or desirability of, the creation of future generations? Is there a moral difference between a future that will contain x billion people succeeded by another x billion different people and so on indefinitely, or x billion people living indefinitely and replacing themselves on the (rare?) occasions when they are killed? Although, as we have noted, this is an unlikely scenario, posing the question in this stark form enables us to ask an important question. That question is whether what matters morally is that life years of reasonable quality exist or that different people with lives of reasonable quality exist. Put in this way the problem assumes a familiar form—should we maximize life years or individual lives? (There is an enormous literature on this. See, e.g. Harris 1987, 1997; McKie et al. 1998) From the life years perspective, it ought not to matter how many new people the world would contain but simply how many life years of acceptable quality it will contain. Those who, like me, find the life years approach unsatisfactory will be inclined to think that individual lives matter. But even so, it could consistently be held that it is the individual lives of existing people that matter, not how many new individual lives there will be.

However, the argument for making sure that there will be new generations is not settled by the outcome of the debate between those who think that future lives count equally with existing lives and those who do not. One group of such reasons has to do with the desire to procreate and the pleasures of having and rearing children (Harris 1998, 1999).

The second set of reasons has to do with the advantages of fresh people, fresh ideas, and the possibility of continued human development. If these reasons are powerful, and I believe they are, and if the generational turnover proved too slow for regeneration of youth and ideas and for the satisfactions of parenting, we might face a future in which the fairest and the most ethical course might be to contemplate a sort of 'generational cleansing'. This would involve deciding collectively how long it is reasonable for people to live in each generation and trying to ensure that as many as possible live healthy lives of that length. We then would have to ensure that, having lived a 'fair innings' they died at the appropriate time to make way for future generations. Achieving this result by voluntary or ethical means might be difficult; attitudes to suicide and euthanasia might change, but probably not overnight.

Christine Overall, in her book *Aging, Death and Human Longevity* (Overall 2003), has found it difficult to be sure of Harris's attitude toward 'generational cleansing'.[5] For the record, I think it would be unjustifiable, and therefore it is difficult to see how we could resist death-postponing therapies.

14.8 Immortality is Cost-Effective

Søren Holm[6] has suggested that immortality so far from increasing health costs per individual actually might dramatically reduce them; there might be in short an economic discounting argument for the public funding of 'immortality', interventions.

Let us assume the following: (1) for both mortals and immortals, there is the same period of old age with increased healthcare costs (say 10 years, but the length does not matter for the argument) and the same costs of treatment during those years (let's say £ 10,000 on average); (2) the mortals will reach this period in 70 years and the immortals in 1,000 years; (3) there is a 1 % per year rate of real economic growth. The present-day discounted costs of treating a person in 70 years will be £ 4,948, whereas the present-day cost of treating the same person in 1,000 years will be 43 pence thus makes economic good sense to invest now and postpone healthcare costs from 70 years into the future to 1,000 years into the future, and as is evident from the figures, it makes sense even if immortals would have a much longer and more costly old age (because of the discounting, even a 10-fold increase in costs would not matter).[7] Add to this the probability that a greater number of immortals would die as the result of accidents rather than long drawn out illnesses and the economic arguments grow stronger still.[8]

14.9 Conclusion

For the first time in human history we face the prospect of a truly open future, involving sequential as well as simultaneous opportunities, and stretching, open-ended before the individual in an unprecedented but truly liberating pathway. We

---

[5] I deliberately choose the term "generational cleansing" for its obvious unpalatable connotations.

[6] In a personal communication. The calculations are those of Søren Holm.

[7] Douglas Adams used a similar argument to show that the costs of traveling in time to eat at "the Restaurant at the end of the universe" would bring the price the price of eating at the most expensive restaurant of all time easliy within the reach of a humble budget. "All you have to do is deposit one penny in a savings account in your own era, and when you arrive at the End of Time the operation of compound interest means that the fabulous cost of your meal has been paid for". See his *The restaurant at the end of the universe.* Pan Books. London. 1980: 81.

[8] See Chap. 19, this volume, for a critique on this view.

should be slow to reject cures for terrible diseases even if the price we have to pay for those cures is increasing life expectancy and even creating immortals. Better surely to accompany the scientific race to achieve immortality with commensurate work in ethics and social policy to ensure that we know how to cope with the transition to parallel populations of mortals and immortals as envisaged in mythology.

References

Adams, D. 1982. Life, the universe and everything. London: Pan Books.
BBC TV horizon programme 2000. Life and death in the 21st century. Broadcast, January 2000.
Bodnar, A.G., M. Ouellette, M. Frolkis et al. 1998. Extension of life-span by introduction of telomerase into normal human cells. *Science* 279: 349–352.
Glannon, W. 2002. Identity, prudential concern, and extended lives. *Bioethics* 16: 266–283.
Harris, J. 1980. Violence and responsibility. London: Routledge, Kegan Paul.
Harris, J. 1987. QALYfying the value of life. *Journal Medicine of Ethics* 13: 117–123.
Harris, J. 1987. *The value of life*. London: Routledge.
Harris, J. 1997. The rationing debate: Maximising the health of the whole community. The case against: what the principal objective of the NHS should really be. *British Medicine of Journal* 314: 669–672.
Harris, J. 1998. Rights and reproductive choice. In *The future of human reproduction: Choice and regulation*, ed. J. Harris and H Søren, 5–37. Oxford: Oxford University Press.
Harris, J. 1999. Clones, genes and human rights. In: *The genetic revolution and human rights, the Amnesty Lectures 1998*, ed. J.C. Burley, 61–95. Oxford: Oxford University Press.l
Harris, J. 2002. Intimations of immortality—the ethics and justice of life extending therapies. In *Current legal problems*, ed. M. Freeman, 65–97. Oxford: Oxford University Press.
Kass, L.R. 2001. L'ehaim and Its limits: Why not immortality?. *First Things* 113: 17–24.
Lanza, R.P, J.B. Cibelli, M.D. West. 1999a. Human therapeutic cloning. *Nature Medicine* 5: 975–977.
Lanza, R.P., J.B. Cibelli, M.D. West. 1999b. Prospects for the use of nuclear transfer in human transplantation. *Nature Biotechnology* 17: 1171–1174.
McBrearty, B.A., L.D. Clark, X.M. Zhang et al. 1998. Genetic analysis of a mammalian wound-healing trait. *Proceedings NationalAcademyof Science USA* 95: 11792–11797.
McKie, J., J. Richardson, P. Singer, H. Kuhse. 1998. *The allocation of healthcare resources—an ethical evaluation of the "QALY" approach.* Aldershot, UK.: Ashgate Pub.
Mooney, D.J., A.G. Mikos. 1999. Growing new organs. *Scientific American* 280: 60–65.
Overall, C. 2003. Aging, death and human longevity. Berkeley, CA: University of California Press.
Pedersen, R.A. 1999 Embryonic stem cells for medicine. *Scientific American* 280: 68–73.
Silver, L.M. 1999. *Remaking eden*. London: Phoenix Giant.
Slevin, M.L., L. Stubbs. and H.J. Plant. 1990. Attitudes to chemotherapy comparing views of cancer patients with those of doctors and the general public. *British Medical Journal* 300: 1458–1460.
Thomson, J.A., J. Itskovit-Eldor, S.S. Shapiro et al. 1998. Embryonic stem cell lines derived from human blastocysts. *Science* 282: 1145–1147.
Weinrich, S.L., R. Pruzan, L.B. Ma et al. 1997. Reconstitution of human telomerase with the template RNA component hTR and the catalytic protein subunit hTRT. *Nature Genetics* 17: 498–502.
Weiss, K.M. 2000. The biology of aging and the quality of later life. In *Aging 2000: Our health care destiny. Vol. 1. Biomedical issues.* ed. Gates & Samorajski.

Chapter 15
Will You... When I'm Sixty Four?

Inez de Beaufort

But tell me yet; this thing, thus daubed and oiled,
Poulticed, plastered, baked by turns, and boiled,
Thus with pomatums, ointments, lacquered o'ver
Is it a face, Usidius, or a sore?
(Iuvenalis)

15.1 The Old and the Beautiful

What does aging have in store for our appearance? Baldness, wrinkles, blubbery arms, potbellies, drooping eyelids, facial hair (for women, and it is worse if you are not used to it), grey hair, varicose veins, breasts (for men, and it is worse if you are not used to it), sagging breasts (women), erection problems and the female equivalent, bottoms that have lost much of their past allure, dentures.... some of these will inevitably hit us all, if we age.

But do not despair: If you do not want to sell your soul to the devil in exchange for the fountain of youth and immortality—and I highly recommend you do not as the endings usually leave much to be desired and the quest is a hard one, we know that from fictional stories such as Oscar Wilde's *Dorian Gray*, and Goethe's *Faust*—there is cantassium glusosamine, cantassium Gingko Biloba, Maximuscle Maxidetox, MRM 5-HTP Griffonio Bean Extract, and a lot more as a simple search on the internet will learn. 'Looking younger' is a huge industry and involves lots of money. "Stop the clock. Whether wise beyond our years or big kids we all want to stay young. Scientific breakthroughs have brought some amazing supplements to us. They are proven to reduce the wear and tear of every day use that our bodies are subject to. Combine these with relaxation, good food and exercise and you will remain forever young." (www.igetfit.com)

I. de Beaufort (✉)
Erasmus MC/University Medical Center Rotterdam,
Department of Medical Ethics and Philosophy of Medicine, The Netherlands
e-mail: i.debeaufort@erasmusmc.nl

M. Schermer, W. Pinxten (eds.), *Ethics, Health Policy and (Anti-) Aging: Mixed Blessings*, Ethics and Health Policy 1, DOI 10.1007/978-94-007-3870-6_15,

Apparently many aging people want to change their appearance and look younger. Because they feel younger than they look or they look older than they feel, because they feel ugly or unattractive, maybe because they are or because others make them feel ugly and unattractive, and because society expects or even demands them to look younger.

Suppose, hypothetically, that there was a pill that could make you look ten years younger, would you take the pill? (Presume, as one can safely do with hypothetical pills, that it has no side effects and that it is affordable.)

Suppose we could change our looks permanently or temporarily in order to look younger, what are moral arguments against and in favour of such changes?

That is the issue I will discuss. It is a kind of 'reconnaissance' mission. I will just raise some questions and discuss some answers. Some might think this is a frivolous subject. Generally speaking I do not consider it beneath my dignity to discuss frivolous themes and even to take frivolous matters seriously but I want to start by emphasizing that there is suffering, unhappiness and that many interests (financial and other) are at stake.

I'm interested in the theme because appearance is important, though it may be politically incorrect to admit this, and of course because I'm getting old myself. I once was consulted about an assisted suicide request of an Aids patient in an advanced stage of his disease. He could have lived for a year or so, but he asked for assisted suicide because he was losing his good looks. For him that was extremely important. He felt he lost his identity and he didn't want to be there for the final 'meltdown' as he called it. My initial reaction, 'you have so much to live for, this is not good enough a reason', made me wonder how crucial appearance is.

That was when I was young myself. Since, a friend once asked me if I had grandchildren. Me, a grandmother.... the very thought of it... I wondered: Do I look like a grandmother? I try not to. Were all my (expensive) efforts in vain? Do I behave like a grandmother? Certainly not! I wouldn't even know how as I try to act as ungrandmotherly as possible. Now why did I care so much? It had to do, I realized, when analysing my discomfort with the innocent question—I could have been a grandmother after all—with my view on the story of my life. I did not want to be a grandmother yet, as being a grandmother did not fit in the story of my life at that particular moment. I resented that others thought I was.

Friends with whom I discussed my concerns and fading looks were divided on the subject. 'Why look older than necessary, dye your hair' (I already did), 'Go be botoxed and lifted.' Others brought forward the '*Sic transit Gloria mundi* argument': One has to accept that one is old or older and looks old or older. We age. It shows. Too bad. This is the time for lace handkerchiefs, demure grey twinsets and blouses with white collars and above all sensible shoes. (Sensible shoes... the horrifying thought.)

15.2 Changing One's Looks: Rejuvenating Enhancements

> As you grow old, you lose your interest in sex, your friends drift away, your children often ignore you. There are many other advantages, of course, but these seem to be the outstanding ones. (Richard Needham)

There is, of course, a wider debate on the morality of changing, preferably improving, one's appearance or looks. Here I will discuss enhancement of appearance in the sense of looking younger: Rejuvenating enhancements, looking younger than one's 'real age'. Of course that is a complex notion itself. What is one's 'real' age—is that how old you feel, is it related to the birth date in your passport, how to combine biological and biographical notions? I will for the sake of argument think of age in years. It is also hard to say: Now you look older than 55 or now you look younger than 61. Age comes in episodes or ranges, say of 5–10 years. It is the ranges one usually wants to change. It would be absurd to say: 'Well, now you look younger than yesterday' or 'you look precisely 68 years and five months'.

15.2.1 How: Camouflage and Vitalizing

There are many different methods: Surgery, beauty treatments, creams, pills, stem cell treatments, training, lifestyle choices, offered by different professionals from hairdressers to doctors. There are different results with regard to the 'lastingness' of results and the way they are 'incorporated'. This is morally relevant. I will distinguish between camouflage strategies with semi-permanent, renewable results, e.g. hair implants, dying of hair, facelifts, as well as temporary, non-permanent outside changes e.g. make-up, clothes. Changing the appearance is the sole goal. It is focused on the outside. It has to do with hiding one's age.

Then there are changes in appearance brought about by lifestyle, exercise, prevention of diseases, and increasing fitness, the vitalizing strategies. Looking younger or anti-aging is aimed at by rejuvenating the whole body (or parts of it). It changes more than appearance only. It actually changes one's bodily condition, metabolisms and functioning. Enhancing one's appearance may be one of the goals, but not necessarily the only goal. To increase fitness, regain energy or not having certain diseases also makes you look younger. One can imagine genetic treatments or stem cell treatments in the future. Dying your hair is wonderful and makes you feel younger, but cannot be compared with real or complete rejuvenating methods inside the body as that has to do with actually influencing the aging process and its consequences.

15.2.2 The Importance of Appearance

A first, preliminary, argument against such enhancing enterprises would again be to argue that appearance does not matter. But old dreams, cold creams, eternal youth, looking beautiful, mankind has always looked for the fountain of youth. (The alternative is to stay young forever by dying young, but of course that is quite a high price to pay given the fact that one cannot profit from it oneself.) Appearance is important. It is closely linked to our identity; we are who we are not only but also because of the way we look. We are not free floating personalities, but flesh and blood people (I have discussed that elsewhere, see De Beaufort and Bolt 2010). Also character and appearance are closely related. We define or show ourselves in a certain way. We can look radiant, or sexy or careless; a chosen sloppy appearance is also a statement about one's appearance, or professorial image (except for sensible shoes); et cetera.

The argument that appearance doesn't matter is separating who we are and how we look in an unrealistic manner. Appearance does matter. It may also be (naïve) wishful thinking, often of those whose appearance leaves something to be desired, or it may be—closely connected—a moral point of view: Appearance should not matter. You (ought to) love or appreciate people because of who they are, not because of how they look based on the underlying moral ideal that it is all about the persons' character (the secret of many marriages, at least after some time). But if you see your favourite aunt who is suffering from dementia in the nursing home smelling of urine and looking not at all like her former well groomed elegant self, you are angry and sad. She may not realize it, but you are sad because this appearance does not fit her personality and her style.

It is politically correct to say that 'little ugly things' can be beautiful in the eyes of the beholder (one's lover or of course one's mother), and I'm sure some are highly appreciated 'warts and all', but still men start drooling when seeing Scarlet Johansson, or Anita Ekberg coming out of the Trevi Fountain. Women start drooling—correction: Women do not drool—they enjoy seeing George Clooney.

In short: There is a complex interplay between our experience of ourselves as bodies and our bodies as ourselves, the way we embody our character and use our appearance, and the way others see us and treat us, e.g. obese persons, tall persons, little persons, beautiful persons et cetera.

I use the term appearance and not beauty. Beauty is of course a complex philosophical notion. I will not go into that. I purposely use the term appearance because it is more encompassing than beauty, it is about style, looks, expression, energy et cetera.

15.3 Some Arguments

I will discuss some arguments against and in favour of enhancement of the aging appearance in the sense of looking younger. They are closely connected in the two sides of the coin way. They are:

1. The proper season argument
2. The pressure from society argument
3. The more important problems in the world argument

15.3.1 The Proper Season Argument

> The course of life is fixed, and nature admits of its being run but in one way, and only once; and to each part of our life there is something seasonable; so that the feebleness of children, as well as the high spirit of youth, the soberness of maturer years, and the ripe wisdom of old age—all have a certain natural advantage which would be secured in its proper season. (Cicero)

How to think of this 'proper season' argument in this context?

My grandmother was an every-inch-a-lady-lady. She looked like a lady should look in her days. Refined, immaculate, elegant, pearls, grey silk blouses, grey hair. She probably would have been shocked by the way I dress at my grandmotherly age. If I would tell her that Madonna was around 50 and Tina Turner around 70, she would most certainly think I was joking or suffering from early onset dementia. If she and I could meet at the same age, which I realize is impossible, I think she would probably ask me why I deny my age, do not look like a lady, borrow my daughter's sneakers (and vice versa). For her it would not be proper given my season.

There are two strains of argument with regard to the proper season argument, the first is the content of proper—the notion of what is proper has changed—the second strain is more complicated and has to do with the notion and length of a proper season.

15.3.1.1 Proper

The debate on the content of proper is on fashion, the relation between age and fashion, on good manners and how to behave, on different cultural developments and on aesthetic judgments. I would tell my grandmother that times have changed, fashion has changed, that she had fewer technologies at her disposal had she wanted to look younger, which she probably would not have thought proper. The elderly now look different and what is proper has simply changed. Akin to the fact like that in her time ladies did not wear trousers. She might regret that change, but it is the way fashion changes.

Sometimes examples are given to demonstrate that a certain change or treatment is not an aesthetic enhancement. That is usually an argument against certain methods of enhancement by showing the sometimes dramatic results, rather than against the 'properness' of enhancement in general. It is an argument against certain methods, not against the underlying idea of looking younger as such. A botched facelift is not an argument against wanting to look younger.

One might also argue that there is another, and right or proper, kind of beauty rather than the attempts to rejuvenate. This might comfort the elderly: 'You now have a different kind of beauty or an interesting appearance.' One can be beautiful when looking old, it is just of a different kind. The story of your life is reflected in your appearance; the history of pain, sorrow, wisdom et cetera. A face that does not age reflects emptiness and hollowness. Old faces are interesting and beautiful to look at, not thanks to the aging, but because of the aging. There is no need for rejuvenating enhancement. It would even be counterproductive. It would be a nightmare if one's age were not at all visible. A seriously underestimated view. There is a need for a different view on beauty and aging. In the words of Hilhorst who has developed this argument:

> If an 80-year-old woman uses all kinds of aesthetic surgical means to look 40, some people will probably not see her as beautiful, but feel sorry for her and perceive her attained 'beauty' as unnatural and distasteful. Her face may look young, but her eyes, her way of speaking and her movements still betray her real age, in the same way that operation on a child with Down's syndrome cannot efface the fact of its anomaly. Even if there were no physical signs

> to give her away, we would feel uncomfortable with her. In the light of what we see as properly human in the context of age, her new face does not fit her as a person. If there is physical beauty in an old woman, as there can be (given the formal norms), her old and beautiful face will include the wrinkles that inherently go with aging. This again signifies that judgements about human physical beauty are relative to their contexts, i.e. relative to age. (Hilhorst 2002, p. 20)

One could add to the argument that it would be socially very confusing if we all looked young and you had no idea whether someone was 64 or 42. I mean: Imagine making a pass at a grandmother.

I would very much like to be convinced by the old-is-beautiful argument, and when I have run out of strategies I will seriously consider being converted, but I do think it holds only to a certain degree. If 'enhancements' try to erase the stories from faces and bodies then they miss the point. We do need to see some differences between generations. But softening the consequences of aging, is not taking away the history. When looking at the photos in an interesting book on elderly sex (Bot 2004) one is touched by the passion, but also shocked by the merciless effects of aging on the bodies (well, at least I was).

15.3.1.2 Proper Season

The notion of proper season raises complex questions. In their kind attempts to comfort you, people bring forward the liberation argument. 'The advantage of being old is that you do not have to stress about your appearance (or many other things for that matter), it provides a kind of freedom and liberation. Relax and enjoy this season! You can let all that worrying about how you look, what men (or women) think of you, et cetera behind you and finally feel comfortable in your own body. You're alive, you're healthy, what more do you want?' This argument may be convincing for some, but not for others. Berlusconi certainly would not be convinced by the argument (not that I think the argument would be much stronger if he was convinced). But what is crucial here is that it very much depends on individual lives and priorities and activities, as well as on beauty ideals catered to the elderly. One may want to postpone the moment of acceptance and liberation as it is not the right time yet.

15.3.1.3 Acceptance

Closely related to the proper season often seems to be the 'acceptance imperative'. One should accept growing old (and decay and looking old and finally death) as a natural process. It is a plea against resisting what is going to happen anyway, the inescapable fact that we are—so far—finite beings, and that we have to suffer the disadvantages and inconveniences of old age or of growing older (not the same by the way!) and might as well get used to it.

The acceptance imperative has a certain appeal as we all recognize that there is a certain sadness or sometimes even tragedy in those trying too hard to escape from it, and not succeeding. It is sad if people try to look younger and fail miserably; it may demonstrate some sort of self depreciation or even self loathing. We tend to feel rather sorry for these people. They seem in a way to linger in the wrong chapter or their story. But who says when you have to leave which chapter or season and move on to the following one? And if they were to succeed in looking younger would it not be different then? Is it the failure of the attempt rather than the attempt itself that saddens us?

Of course the acceptance imperative is problematic if one is really confronted with choices to live longer and/or look younger. To accept what cannot be changed may be a sign of wisdom and resignation may be the best answer to fate, however hard. In the words of the writer Sandor Marai:

> We age slowly. First, our pleasure in life and other people declines, everything gradually becomes so real, we understand the significance of everything, everything repeats itself in a kind of troubling boredom. It's the function of age. We know a glass is only a glass. A man, poor creature, is only a mortal, not matter what he does. Then our bodies age: Not all at once. First, it is the eyes, or the legs, or the heart. We age by installments. And then suddenly our spirits begin to age: The body may have grown old, but our souls still yearn and member and search and celebrate and long for joy. And when the longing for joy disappears, all that are left are memories of vanity, and then, finally, we are truly old. One day we wake up and rub our eyes and do not know why we have woken. We know all too well what the day offers: Spring or winter, the surface of life, the weather, the daily routine. Nothing surprising can ever happen again: Not even the unexpected, the unusual, the dreadful can surprise us, because we know all the probabilities, we anticipate everything, there's nothing we want anymore, either good or bad. That is old age. There's still some spark inside us, a memory, a goal, someone we would like to see again, something we would like to say or learn, and we know the time will come, but then suddenly it is no longer as important to learn the truth and answer to it as we had assumed in all the decades of waiting. Gradually we understand the world and then we die. (Marai 2003)

But does that also hold if we could change it? It seems quite unlikely that if our life-expectancy were to be increased dramatically, most would say: 'No thank you, ever so kind, but dying in my forties or sixties is quite a proper season to die.'

The acceptance imperative is quite popular when it comes to new technologies, and also was brought forward when opponents of reproductive technologies said to infertile couples that they should accept their infertility, and—adding insult to injury in my view—described how wonderful life can be without children.

Would we accept the argument when it comes to Viagra? Being old leads to being impotent, accept it. Of even worse, describe how nice life can be when impotent.

I admit suffering from an allergy with regard to the acceptance imperative: It has been misused too often in the history of mankind—towards women in particular. And I do not think it will solve the moral qualms regarding rejuvenating treatments.

15.3.1.4 Is Normal Proper and Proper Normal?

The story of Benjamin Button by F. Scott Fitzgerald is so confusing because he is born old and his life evolves in the opposite way. Living our story with a beginning, a middle and an end is part of the 'condition humaine'. Therefore, so one might argue, it is wrong to choose rejuvenating enhancements. Why? Because it is normal to age and the proper season is what is normal. Normal is proper. John Harris has dealt with this argument, arguing that many things are normal and that we still do not want them and rightly fight them: "Disease and death are hideously normal" (Harris 2007). Yes, forgetfulness and dementia are normal as well. And surely that is not an argument against developing treatments against Alzheimer's et cetera.

> The normality of the trait in question is clearly doing no work at all in the assessment of its moral acceptability or of the risks it might be worth running to change things. If we subtract the acceptability from the normality we are left with nothing of moral, nor of argumentative force. Traits in short are not acceptable (in the normative rather than of course the simply descriptive sense of 'acceptable') because they are normal, they are acceptable because they are worth having. If they are not worth having, or if they are worth not having, their normalcy seems bereft of interest or force. (Harris 2007, p. 53).

I agree.

15.3.2 It Is All Caused by Social Pressure

'We are forced by societal expectations to try to rejuvenate. It has nothing to do with autonomous choices, with states we want and desire, with our own wants or decisions. Western culture and its merciless pace and idolizing of youth, makes aging difficult. The elderly feel ugly and superfluous and hence they try to do something about it'. This argument is sometimes supported sometimes with (overly?) romantic pictures of the social respect for the elderly, e.g. in Africa. Therefore we succumb to societal pressures. Again particularly women are the victims. It is not bad but sad. It is not the aging and the elderly who actually want the rejuvenating enhancement. It is the social circumstances that create this so-called want and professionals and quacks of course jump on the market. Were we to live in a society that respected the elderly and treated them well, and not like remnants of times gone by who are a logistic nuisance and a financial burden and deplorable cases of early onset of total putrification, then we would not want to change our appearance at all, we would carry our mature ripeness with pride, feel good in our aging bodies and enjoy the attention of the young who want to profit from all our experience and great wisdom.

Ergo: It is the social views on aging and the old that ought to change, then we would not have to stop, slow down or camouflage the aging process. In fact: The options for changing appearances will contribute to the prejudices and prevent changing the view on elderly. It is a vicious circle.

In many ways this argument makes sense. We probably all agree that there is a lot of pressure on elderly to cope with the demands that modern society has in store for them (that may vary from having to master the art of emailing, surfing the internet, to coping with the news, to managing to take care of oneself) and that in Western societies elderly people find themselves in a position that leaves much to be desired. We should also note that the pressure of looking young does not start with and is not exclusively aimed at potential grandmothers and octogenarians being bombarded with possible treatments, already for those over thirty aging and looks are made into an issue.

Indeed: I do feel the sword of superfluousness dangling above my head. There are so many reports on the costs of aging et cetera that one starts to feel guilty for still being around and seriously fears the future.

On the other hand this does not mean that rejuvenating is not also about what individuals want and makes them simply feel better. Not everyone cares. So it is possible to resist. Some care more than others. I do not think that all can be reduced to the external pressure view. It does not imply that individuals have hardly any say in the matter. But the extreme value society places on youth and the consequences of how that affects the elderly and their position in society is certainly to be the subject of further debate.

15.3.3 There are More Important Problems in the World

In India I saw an old paralyzed lady who was begging by moving around her tiny body on a skateboard. The image is fixed in my head. At her age having to live this way.... In fact I don't even know her age. She looked like 80, but she may have been my age. Are our affluent societies' aging 'problems' not luxury problems for which we should feel ashamed?

There are certainly more important problems in the world. Children with their whole life ahead of them starve or die from diseases that could easily be prevented. People die from hunger and deprivation and diseases that could be treated with a part of the money well-to-do elderly spend on rejuvenating their looks.

Also in the rich world there are more pressing problems regarding elderly as this book demonstrates: Decent healthcare, the opportunities to participate in society.

Now what? Such arguments can be complete debatestoppers. And sometimes are meant to ridicule those who discuss these frivolous ideas. It is true: There are more pressing problems. To deny or ignore, however, that appearance can also be a problem is naive. The argument provides a good reason to look critically at societal expectations, priorities in global healthcare et cetera. And it causes specifically thorny moral choices, also for individuals. Should I spend money on elementary food and sensible shoes and give the rest away so that the lady in India can have a decent old age?

15.4 Looking Younger: Arguments in Favour

15.4.1 The Argument of the Story of My Life, the Narrative Fit

Our appearance as I have argued is related to our identity and authenticity, our roles and capacities. There are all kinds of expectations regarding the behaviour and looks of the elderly. There are lots of do's and don'ts. Those who have lived with teenagers probably know all about being instructed on the rules for parents regarding dress and other PFB (parent-fitting-behaviour).

The nice thing about the television programme Benidorm Bastards, an originally Belgian series that now has many international versions as well ("Betty White's Off Their Rockers"), is that it puts those expectations upside down. It is about elderly who in no way behave like they are expected too, which provides quite a refreshing and instructive perspective and different view of the elderly.

The way our life story is connected to our appearance differs; it depends on culture, on career, and on personality. Some may not care and go for a Miss Marple appearance and enjoy it, whereas others find it very difficult and will try to look younger as in their story this is an important part of their identity and their social role. If you are a gardening grandmother that is different from being a former big CEO, or an ethics professor, or a former ballet dancer, or a nurse.

So: For some rejuvenating their appearance is more important than for others. Like it is more difficult for a conductor to become deaf, or for a painter to lose his eyesight et cetera. That does not necessarily change with age. It is no coincidence that Jane Fonda and Joan Collins may resort to cosmetic surgery rather than Marguerite Duras or Madeleine Allbright or Margaret Thatcher, let alone Mother Theresa. It is too simple to see Jane and Joan as victims of the ideal of youthful beauty who sadly cannot say goodbye to former looks and cannot cope with aging or mortality. It is an important part of their life story and their identity, their authenticity. We have to respect that. This is what I mean with the narrative fit. Interestingly, it is closely related to the proper season argument, but emphasizing the individual's own view on the proper colouring of the season.

15.4.2 The Longevity Context: Energy and Vitality

> Ask not what can my appearance do for me, but what can I do for my appearance? (Inez de Beaufort)

If our lifespan is (significantly) expanded, say with 60 more years or so, it is not likely that one would like to look or live or feel like a frail ninety year old for 60 more years (from 90–140). If one were to receive the gift of 60 more years without sex, without the ability to travel, napping in a wheelchair through most of the day, struggling with a failing memory, would one actually want such an extension of life? I doubt it! The motivation for longevity is usually zest for life, wanting to do and experience things, and generally speaking one needs energy and a more or less able body to do that. One does not want the clock to stop at 20, and maybe some think of a kind of prolonged middle age, say a long summer and autumn, not a long winter, but long forties of fifties... Postponing diseases of old age, or even eradicating them, will slow down the aging process, and certainly also be translated in our appearance but it is uncertain how we will look and what we will be able to do. We will age, but differently. We will have to rethink completely what proper seasons are.

I guess that if anywhere, it is in this field of preventing diseases related to aging, that the fountain of youth will be found. This might be the field of serious enhancement. Enhancing by actually feeling younger and having one's whole body involved, is superior to whatever camouflage strategy. Probably the rejuvenating influence on appearance will be a (welcomed) side-effect as the goal of treatments or lifestyles strived for is to postpone death and prevent diseases. But it will certainly influence appearance as well.

15.5 Conclusion

How important one's appearance is, is inextricably linked to the story of one's life. The fitness and energy argument seems to be very convincing to me. I do not see that the elderly should be held captive by a proper season argument. Policymakers should consider how they can cater to the needs of different elderly and to influence the pressure on aging people and the ideal of youthfulness. Also there should be sensible and controlled portals to provide information on what treatments rest on evidence and which are based on pure speculation and wishful thinking. There may be rather dangerous treatments around. Young people should certainly talk to their grandparents in order to try to understand what it means to age. And that there is indeed a value in experience and wisdom that they do lack and that deserves respect. (I guess they are probably more willing to accept this from their grandparents than from their parents.)

And I will let you know when I'm ready to be a grandmother.

References

de Beaufort, I.D., L.L.E. Bolt, and S. Vandamme. 2010. *Ethics and appearance, encyclopaedia of applied ethics.*

Bot, M. 2004. *Timeless love.* Rotterdam: Marrie Bot.

Harris, J. 2007. *Enhancing evolution 52–53.* Princeton: Princeton University Press.

Hilhorst, M. 2002. Physical beauty: Only skin deep?. *Medicine Healthcare and Philosophy* 5: 11–21

Marai, S. 2003. *Embers, 229–230 (original publication 1942).* Hungary: Penguin Books

Chapter 16
Old Age Is An Incurable Disease—or Is It?

Maartje Schermer

16.1 Introduction

For a long time in the existence of the human race, getting old was not such a self-evident part of life as it is today, since many people died young. Nevertheless, dreams of immortality and eternal youth have always been around, and people have always looked for 'a cure for aging' or 'the fountain of youth'. It has been one of the great advances of the 20th century that the average life expectancy, at least in Western countries, has increased tremendously, from as low 25 years in suburbs of industrial towns in the 19th century to around 80 years today (Jones and Whitaker 2009). This advance has been mainly due to improvements in sanitation and the successful combating of infectious disease through vaccination and medication, which have led to a reduction in childhood mortality. Moreover, due to better treatment methods a number of diseases that used to be deadly have now turned into chronic diseases, like certain forms of cancer, or AIDS—people no longer die from them but grow old with them.

What is new in the quest for longer lives, however, is that since the second half of the 20th century aspects of *aging itself* have increasingly become the object of biomedical interventions. A movement of anti-aging medicine has started to emerge. One of the first biomedical interventions that can be placed under the banner of anti-aging medicine is the hormone replacement therapy for postmenopausal women that has been popular since the 1960's (Lucke et al. 2009). Other phenomena that used to be seen as inevitable aspects of aging have also come under medical supervision: brittle bones became osteoporosis (officially a disease since 1994); forgetfulness became mild cognitive impairment (Scully 2004; Whitehouse and Juengst 2005).

A central and hotly debated issue in the context of anti-aging medicine and anti-aging research is whether aging is a normal and natural process, or whether aging is a disease that should be treated or cured (Vincent et al. 2008; Kampf and Botelho

M. Schermer (✉)
Department of Medical Ethics and Philosophy of Medicine,
Erasmus MC/University Medical Center Rotterdam,
Rotterdam, The Netherlands
e-mail: m.schermer@erasmusmc.nl

M. Schermer, W. Pinxten (eds.), *Ethics, Health Policy and (Anti-) Aging: Mixed Blessings,* Ethics and Health Policy 1, DOI 10.1007/978-94-007-3870-6_16,

2009). While some emphasize that aging is a process closely *associated* with certain diseases, others go even further and claim that aging itself is a disease, a pathological and abnormal process—it has even been argued that aging and even death are 'unnatural' and should be overcome (Caplan 2004). This is, by the way, not a completely new view: 2000 years ago already, Seneca declared that "old age is an incurable disease". Others contest this and say that aging is normal and natural. 'Aging' was even voted the number one non-disease in a poll by the British Medical Journal in 2002 (Smith 2002).

The aim of this contribution is to clarify the discussion about the question of whether aging is normal or pathological, a disease or not, by analysing the different perspectives and different concepts involved. I will first show that it is important how the relationship between aging and disease is framed, because this influences ethical and policy debates about aging and about interventions in aging. I will then suggest that the conceptual triad of disease-illness-sickness can help us to better understand the controversies around the question of whether aging is a disease or a natural process, pathological or normal, and whether medicalization of aging is desirable or not. I will conclude that the fundamental issue of whether we *ought to accept* the current biological fact of life (i.e. that our bodies age), or should try to intervene in the aging process, remains a fundamentally normative question, that is not solved by appealing to concepts of health and disease. Rather, this normative question returns within the different conceptualizations of disease, as I will show.

16.2 Is Aging Pathological—and why does that Matter?

Currently, there is a world-wide movement of anti-aging medicine, represented for example by the American Association for Anti-Aging Medicine (A4M), that advocates the use of biomedical knowledge and technologies to intervene in the aging process and to try to slow, stop or even reverse it (Kampf and Bothelo 2009). This anti-aging movement consists of mainly commercially motivated and sometimes questionable businesses of rejuvenation therapies, dietary supplements, anti-wrinkle treatments and the like. Though they claim to base their interventions on the latest scientific discoveries, academic scientists mostly shy away from this 'anti-aging enterprise'. It has been extensively documented how the academic community of biogerontologists, who study the fundamental mechanisms of aging and age-related diseases, have been struggling to draw clear boundaries between these 'quacks' with their this 'snake-oil selling business' and their own 'serious and legitimate' science (Fishman et al. 2008; Vincent 2009; Olhansky et al. 2002; Binstock 2004).

Biogerontologists also have also felt the need to distance themselves from far-reaching aspirations, like considerably increasing the maximal human lifespan or attempting to stop aging altogether. According to popular opinion, as well as the opinion of funding agencies, aging is a natural process and intervening in aging is therefore seen as suspect. Intervening in the aging process is quickly associated with reckless and overconfident attempts to secure eternal youth, deny death, and aim for immortality. As such, it raises concerns that are familiar from the discussion between transhumanists and bioconservatives about the desirability and justifiability of human enhancement. It is often said that enhancements are wrong because they intervene in nature, or because they alter human nature or affect human dignity. Other appeal to the 'playing God' argument, or warn against hubris and the illusion of total control and mastery. In general intervening in a 'natural' process such as aging is not met with much approval and biogerontologists therefore repeatedly stress that their aim is not to fight death, but that they merely hold the legitimate medical aims of healthy aging, compressing morbidity, and increasing the health span (i.e. the period of healthy life). They do not aim for life extension, let alone immortality (Post and Binstock 2004).

One way to make aging itself a legitimate goal for intervention is to stress its association with disease. Even if aging itself is not claimed to be a disease, it is said to be 'characterized by a broad spectrum of disease' (Gems 2011) or 'a process that creates so much susceptibility to disease that it can be approached by researchers with therapeutic intent' (Post 2004). In this endeavour, the biogerontologists encounter opposition from sociogerontologists, who claim that the strong emphasis on the biological processes of aging, its association with disease and the efforts to intervene in the aging process constitute a form of medicalization and biologisation of aging and old age. Such a framing of aging does not do justice to reality, they claim, and reinforced negative stereotypes of aging and the elderly. Therefore sociogerontologists generally resist the framing of aging in terms of pathology and disease.

16.2.1 Treatment or Enhancement

The question of whether aging is a disease or not is therefore, first of all, important because it shapes how interventions in the aging process are looked at. If aging is understood as a disease, interventions to stop or slow down the aging process will be understood as medical treatment. If aging itself is a disease, it is a legitimate target for medical intervention. If, on the other hand, aging is understood as normal and natural, interventions to slow or stop aging will be characterized as enhancements, interventions that go beyond therapy. As mentioned, a significant part of the public holds the assumption that treatment is always good, whereas enhancement is morally suspect. Diseases ought to be treated, but natural states or process should not be meddled with—at least, this is what many people intuitively believe. This distinction may not hold on closer scrutiny, and many ethicists agree that the legitimization of interventions does not depend (exclusively) on their status of treatment or enhancement—enhancements that are not treatments can also be legitimate or even morally obligatory in some cases (e.g. Schermer and Bolt 2011). However, this does not take away from the fact that moral assumptions about the wrongness of intervening in natural and normal states or processes strongly influence public debate and opinion. It does matter how interventions are framed, and what language is used.

Moreover, if interventions in aging can be labelled as treatment, this will have important financial consequences. Researchers will find it easier to get funding, pharmaceutical and other companies will be able to develop and market products that slow, stop or reverse aging and healthcare insurance will have to cover such interventions. Since FDA approval is essential to bring medication on the market and such approval is only possible if the product is indicated for a classified disease, aging must be classified as disease in order for product that directly intervene in aging to count as medication and be sold and reimbursed as medical treatment.

Finally, if aging itself would be considered a disease, it would be drawn further into the medical domain, as something that doctors should intervene in because they have an obligation to treat disease. This is something that is resisted by some, as unjustified or undesirable medicalization of aging.

16.2.2 Medicalization of Aging

Medicalization is a complex term that is often used in a pejorative sense. According to Peter Conrad medicalization can be defined as: "The process in which we are coming to see problems in medical terms, using medical language to describe problems, adopting a medical framework to 'treat' them" (Conrad 1992, p. 209).

Medicalization can refer, first, to the factual increase of medical attention and medical interventions in many domains in life. In this sense, to say that aging and old age are becoming more and more medicalized, means that more medical attention is paid to older people, their health is more closely monitored and more treatments are prescribed to them. This increase in medical attention can have favourable effects on health and well-being of elderly. While it used to be too often the case that elderly peoples symptoms or complaints were ignored because they were thought to just belong naturally to old age, nowadays diseases of old age are taken much more seriously (Ebrahim 2002). Increased medical attention can also have negative effects, however, like side-effects, risks of polypharmacy, and increased iatrogenic damage. Only when the over-all effects of the increased medical attention would be negative, is would be justified to say that medicalization in this sense was bad. Making up a balance between the positive and negative effects of medical interference in aging and old age is important but falls outside the scope of this contribution.

Here, I am primarily concerned with the notion of medicalization that emphasizes the conceptual and linguistic dimensions; as Conrad's definition points out: We are coming to see problems, like aging, in medical terms and increasingly use medical terms like 'disease' to describe them. The critique that has been voiced against this form of medicalization is that a normal human experience, a normal life process, is being turned into something pathological. Especially sociogerontologists have resisted this idea, for several reasons. First, the identification of aging with pathology and disease may have a negative impact on both the public image of elderly people and on their self-image. The more aging itself is seen as disease, the more negative the public view of aging and elderly people will become, is their fear. This may have repercussions for the way in which elderly people are treated, the opportunities they are given et cetera. Moreover, the medicalization of aging promotes a reductionist view on aging, stressing only the negative sides of physical and mental decline and to ignore other aspects of aging. As John Vincent puts it:" The focus on biological failure sets up a cultural construction of old age which generates and prolongs its low esteem [. . . .] Striving for an ever-longer lifespan represents a denial of old age as a valued final part of the life course, and allocates old people to a cultural category characterized by redundancy and despair" (Vincent 2009, pp. 682–683).

So, in summary, the question whether aging is a disease matters because the way in which the relationship between aging and disease is understood influences ethical and policy debates about aging and about interventions in aging. In the following section, I will introduce a conceptual model, the triad disease-illness-sickness, to help clarify the different views on the question of whether aging is pathological or normal and natural.

16.3 The Triad Disease-Illness-Sickness

The triad of disease, illness and sickness stems from medical sociology and is meant to clarify the different perspectives that one can take on the complex phenomenon that 'non-health' is. The concepts of disease, illness and sickness "reflect the professional, personal and social perspectives and concern biological, phenomenological, and behavioural phenomena, respectively" (Hofmann 2002, p. 657). Andrew Twaddle was the first to use this triad in 1967 and has later elaborated it and defended it against critique. He gives the following definitions of the three concepts:

- Disease is a health problem that consists of a physiological malfunction that results in an actual or potential reduction in physical capacities and/or reduced life expectancy.
- Illness is a subjectively interpreted undesirable state of health. It consists of subjective feeling states (e.g., pain, weakness), perceptions of the adequacy of their bodily functioning, and/or feelings of competence
- Sickness is a social identity. It is the poor health or the health problem(s) of an individual defined by others with reference to the social activity of that individual (Twaddle 1994).

These definitions are briefly summarized by Hofmann, who explains that "disease is negative bodily occurrences as conceived of by the medical profession. Illness is negative bodily occurrences as conceived of by the person himself. Correspondingly, sickness is negative bodily occurrences as conceived of by the society and/or its institutions. Occurrences here means process, state or event" (Hofmann 2002, p. 657). For the purpose of this contribution, I am not concerned with the exact definitions or their possible alternative formulations or refinements, nor will I discuss the critique and counter arguments that have been exchanged over this conceptual triad. For present purposes, it is especially important to understand the three different perspectives that are expressed by these different concepts. As Hofmann has argued, the triad has explanatory abilities and can be fruitful for a discussion on difficult and controversial cases, regardless of the strictness of definitions (Hofmann 2002).

The triad can be used as a framework to explain certain controversies concerning the status of specific occurrences of non-health, such as aging. In general, controversies arise when a condition is not neatly covered by all three concepts from the triad, but only two, or even only one of them, apply. The paradigm case in which disease, illness and sickness all apply simultaneously is as follows: A person feels ill, for example because he has a sore throat and aching head, and therefore goes to see a doctor who diagnoses a disease, say, influenza. The person is can then legitimately take a few days off from work and stay in bed: He has been attributed the status of sick by society.

However, the concepts of disease, illness and sickness do not always apply at the same time for a given case or situation. There are cases that are conceived of as both disease and sickness, but not illness, like certain pre-symptomatic conditions. Various forms of predictive screening and testing, like breast cancer screening or neonatal screening, create such cases. The conditions tested for, liked breast cancer or congenital hypothyroidism, are considered to be diseases by the medical profession and constitute physiological malfunctions. Society recognizes these conditions and treats them as sickness by granting resources and setting up institutions for testing and treatment. The persons who are tested do not, however, experience any illness. They are a-symptomatic, as it is called from a medical perspective—they feel normal and healthy, not ill. Another example of a controversial situation is chronic fatigue syndrome (CFS). Here patients clearly feel ill and experience many complaints and symptoms, but from a medical perspective it is difficult to define as a disease, since there is no clear physiological malfunction. Society—at least in the Netherlands[1]—has clearly been confused by this situation and has been wavering between granting CFS patients the status of 'sick' or not.

At this point it is very important to notice that disease, illness and sickness are not static concepts. What counts as disease, illness or sickness may change over time, due to new discoveries and insights, and the borders between the concepts are not sharp but rather blurred. One reason for this is that the spheres of disease, illness and sickness are not independent of each other. The attribution of social status (sickness) is of course not fully independent of the conception of disease as understood by the medical profession. Likewise, the subjective experience of illness is influenced by the way both society and the medical profession understand and respond to ones complaints.[2] Another reason for the dynamics of the triad is that

[1] In the Netherlands, over the years social benefit institutions have on and off considered CFS as a grounds for sickness benefits. At one point in time (2005), the Minister of health officially declared CFS not to be a sickness, but was called back by Parliament decided by a voting that CFS was a sickness. Later on again, in 2010, the Ministry declared CFS a non-disease once again.

[2] For example: infertility did not use to be a sickness, but now qualifies for economic support because it has become treatable as a disease. Moreover, this changes the experience of being infertile form 'bad luck' or 'God's command' to an experience of illness. Another example that the experience of illness can be influences by the social status of an ailment is whiplash. When neck- and back aches

what is understood as disease by the medical profession change over time. This is influenced significantly by new scientific and technological developments, like research into the biological mechanisms of aging. The push towards understanding aging as pathological, a disease, comes mainly from the increased knowledge and understanding of the physiological and functional mechanisms of aging.

Part of the controversies around the status of aging can, I believe, be explained by the fact that aging is a perfect example of a case in which the three concepts of the triad do not easily coincide. The situation is even more complicated though, because even from within each perspective (the biomedical, the subjective and the societal); it is not crystal-clear whether aging is a disease, illness or sickness, respectively, as will be discussed in the following sections.

16.4 Aging as Disease

First we turn to the question whether aging is a normal or a pathological biological process; is it, from a biomedical perspective, a disease? Opinions among biogerontologists are divided on this point. On the one hand are the so-called dichotomists who wish to make a sharp distinction between normal and pathological aging, whereas on the other hand continuists claim that the two are inseparable (Blumenthal 2003).

From a biomedical point of view, which solely looks at biological functioning, there is something to be said for the claim that aging itself is a form of disease. Gerbrand Izaks and Rudi Westendorp, who are both physicians and biomedical researchers, claim that many doctors mistakenly belief that aging is not a disease and that they wrongfully attempt to separate pathological aging from normal aging. They state that normal aging cannot be separated from pathological aging: "Aging is the accumulation of damage to somatic cells, leading to cellular dysfunction, and culminates in organ dysfunction and an increased vulnerability to death" (Izaks and Westendorp 2003, p. 6). They conclude that "a similar process is causing aging and disease in the latter part of life. Therefore, in our opinion, normal aging cannot be separated from pathological processes causing disease later in life. As a consequence, we think that making a distinction between normal aging and pathological aging should be avoided" (Izaks and Westendorp 2003, p. 5).

The authors do not explicate or define their concept or theory of disease, but they implicitly seem to depend upon a Boorse-like idea of disease, one in which normal biological functioning is paramount. However, they reject the idea of statistical normality linked to a reference class, as it is used in Christopher Boorse's well-known biostatistical theory of health. Boorse defines disease as an internal state which reduces one or more species typical functional abilities. What counts as normal functional ability is determined by looking at the typical statistical distribution of

---

after accidents are labeled as 'whiplash' and qualify for reimbursement under an insurance scheme, patients may report and experience more illness than when they do not stand to gain anything from it.

a biological function among a reference class, which means that normal biological functioning must be defined relative to sex and age (Boorse 1977, 1997). Boorse therefore does make a difference between normal and pathological aging. Certain functions may decline with aging, but if they decline in most elderly people, it is statistically normal for the reference class and therefore not pathological. One of Boorse's reasons for holding on to the idea of an age-related reference class is that he wants to maintain a notion of development. He sees the life-cycle of development, maturation and physical decline as biologically normal.[3] The famous biogerontologist Hayflick—also a dichotomist—supports Boorse's view that we should accept the biological lifecycle as the norm. He says: "The goal of arresting the aging process might be viewed in the same light that we view the arrest of our physical or mental development in childhood—as a serious pathology" (Hayflick 2000, p. 269). Eric Juengst has argued that this is not a very convincing argument because arresting childhood development cannot be equated with arresting aging; while most people would agree that developing into a mature human being is desirable, because it opens up possibilities, the same is not true for aging (Juengst 2004).

Interestingly, Izaks and Westendorp reject the Boorsian idea of an age-related reference class, and take young adults to be the reference for all. They state that "it is not appropriate to use old-age-specific normal values. The decision whether a body function of an elderly patient is impaired or not must be based on the same normal values that are used in young adults [. . . .] there is no good reason why the normal values for functions in young adults are not applied in adults at all ages" (Izaks und Westendorp 2003, p. 5). They point out that for elderly people, functional levels below those of young people are often associated with higher mortality and should therefore be considered abnormal.

So here, at the heart of the biological concept of disease, lies a normative controversy: What is the standard of 'normality' that we choose: The optimal functioning of young bodies, or functioning relative to others of the same age? Should we accept it as normal that biological functioning in older people declines (because it statistically does so, and we can give an evolutionary explanation for it), or should we resist it and consider it abnormal (because it leads to higher mortality and is therefore undesirable)? Whether or not aging is considered a disease ultimately depends on such normative choices.

16.4.1 Aging and Disease as Continuous Processes

As mentioned earlier, conceptions of disease are not static and the biomedical understanding of disease is changing with the advances of science. Such changes are also

[3] Since Boorse starts from an evolutionary approach, he sees survival and reproduction as the main goals of all life-forms (including human life). From this point of view survival after the reproductive age may be deemed less important and therefore decline and death after reproductive age are not abnormal and should not be called 'disease'.

important to understand the relation between aging and disease. Current biomedical research in genomics, molecular biology or systems biology is influencing the way in which disease is understood within the biomedical community. More specifically, these types of research are bringing the process character of disease more and more forcefully to our attention. Whereas most traditional theories on the concepts of health and disease appear to understand disease as rather fixed state or entity, genomics- and systems biology-research now contribute to the development of a cascade model of disease, it has been argued. "Molecular changes are supposed to cause further changes, e.g. on the tissue level, then on the level of organs etc., and ultimately lead to symptoms and complaints" (Boenink 2010, p. 17). The cascade model thus implies that "one small step in intracellular processes leads to another, in a stream that with each subsequent step becomes more difficult to stop" (Boenink 2010, p. 17). This view on disease is very similar to the view on aging as it is developing in biogerontology. The biogerontology field is promoting a view of aging as a molecular and genetic process that starts early in life and continues to build up damage to cells and organs (Fishman et al. 2008). The leading idea in biogerontology is that damage and repair are continually present from an early age on, eventually causing an accumulation of damage in cells and tissues, which leads to functional decline over time and eventually to death. The process of aging is thus continuously present throughout life.

Whereas a traditional problem with definitions of disease was to decide how many symptoms should be present to make a diagnosis, or how much deviance from normal values should count as disease, the question now becomes at what point in the process of molecular and intra-cellular changes one can start to talk about disease. The traditional problem was how to draw the line between a state X that is a disease, and a similar state X' that is not. How high must blood pressure be to count as hypertension? How low the bone mineral density to count as osteoporosis? How many memory problems count as MCI or dementia?

However, when we start to look at aging and disease as essentially the same kind of processes, processes that develop over time, as in the cascade model, a new difficulty is added: The question of when—at what moment in time—disease begins. At which moment in time does aging become a disease? The more we know about the start of molecular and intracellular processes early in life, it appears, the more states will become 'disease'.

The developments sketched above make it likely that with further advances in biomedical science the boundaries between aging and disease processes will be blurred further and aging will increasingly be seen not as something associated with old age, but as a disease process that starts early in life and is continuously present. An important consequence of this change in view is that the notion of disease gets disconnected from that of illness. Bio-physiological processes that we call disease no longer necessarily occur simultaneously with symptoms and complaints. Sociologist Nikolas Rose (2007) calls these new states pre- or proto-diseases, a term indicating that they differ from old-fashioned disease partly because they are symptom-free and thus not directly connected to illness. This development does explain part of the unease that underlies accusations of unnecessary medicalization: People are called diseased—and may be eligible for treatment—while they do not feel ill and have no signs or symptoms. This might engender the feeling that treatments are given without good reason, and makes it difficult for people to check whether a treatment is actually working since there are no subjective signs by which this could be measured. Suspicion and distrust of 'the medical complex' may be the result.

16.5 Aging as Illness

We now turn to the next question: Whether aging can be conceived of as an illness. Is aging, from a subjective perspective, something that leads to pain, or weakness, or malfunctioning? I think that in most cases it does, though this may not always be equally serious or disruptive. Even so-called minor ailments associated with age, however, such as mild memory decline, loss of hearing, stiff muscles or bad eyesight, may give rise to subjectively experienced illness.

A very interesting an ethically relevant point here is how the experience of illness is formed and influenced by our expectations of what is normal and natural, and by our ability to adapt. Many people consider certain ailments to be 'normal' at a certain age. When people start to become farsighted at a certain age, and need their first pair of reading glasses, do they consider this an illness, or as a normal sign of aging? When it starts happening more and more often that you cannot remember a name, or find the right words, does that mean you has a memory disorder, or are you just getting old? As discussed by Govert den Hartogh in Chap. 10 in certain weary-of-life cases, where a person requests assisted suicide or euthanasia, it is contested whether the person in question is suffering from disease and illness, or 'merely' from the ailments of old age. Although the illness (not being well) is generally recognized in such cases, it is considered to be caused by normal aging, and not by disease. The distinction between disease and illness has far-reaching consequences here. It is clear, however, that aging whether seen as disease or not, can cause serious illness.

A theory of health in which the subjective perspective is considered to be fundamental, is the holistic theory of health and illness developed by Lennart Nordenfelt (1995, 1998). According to this theory, health means being in a physical and mental state which is such that one in able to realize all one's vital goals. "A is completely healthy if, and only if, A has the ability, given standard circumstances, to reach all his or her vital goals" (Nordenfelt 2007, p. 7). Ill-health in this perspective is not being able to realize one's vital goals, due to physical or mental impairments. The emphasis is on the subjective dimension of illness, on the disability or functional impairments one experiences in relation to one's goals and aspirations. In this sense, getting older will often lead to an experience of ill health, because aging often comes with physical or mental impairments that prohibit the attainment of certain goals. For example, after menopause a woman is no longer able to bear children. If she always wanted to have children, but for some reason has remained childless, it will now be impossible for her to attain this vital goal in her life. If an athlete's goal is to excel in his sport, at a certain age he will find he cannot do so anymore and will have to give up. In short, aging can prohibit the realization of vital goals.

According to Nordenfelt (1995), however, aging is not an illness because as one gets older, one adapts ones vital goals. If you do not expect to see as well, remember as much, or run as fast as you used to, you can adapt your standards and start to pursue different goals. In this way, one can grow older without experiencing much illness. This is what many people actually do: Many elderly people report they are healthy and happy, despite objective disabilities or chronic diseases (Strawbridge et al. 2002). Nordenfelt's notion of health also emphasizes the ability to adapt to circumstances and live a healthy life despite disability or impairment. A recently proposed conception of health as "the ability to adapt and to self-manage" also stresses this element (Huber et al. 2011).

Aging is not experienced as illness by many people and the ailments of old age are apparently quite easily accepted as something that just 'belongs to life'. Now it may be very prudent to adapt ones standards and goals to ones abilities, in order to remain happy and satisfied with life. A prudential reasons, however, is not a normative one. One could challenge the idea of adaptation to aging by asking whether one ought to do so. It may be wise to accept ones ailments and impairments and to lower ones standards, but ought one? Should one adapt ones vital goals to ones age and (reduced) abilities? Here again there is a normative tension at the heart of the theory, for whether aging implies illness is partly dependent upon the question whether we should accept it as normal that certain functions decline with age and adjust our goals and expectations accordingly. The answer to this question may partly depend on our actual capacities for intervening in the aging process; according to Nordenfelt "the major (pragmatic) reason for a distinction between old age and disease seems to be that old age is the inevitable fate of all human beings. In the long run, for reasons of principle, senility cannot be cured" (1995, p. 112). This implies that if senility could be cured, or slowed down—as some biogerontologists belief will become the case—it would make sense to start considering the declining abilities of old age as pathological, as form of illness.[4]

Even if we had these powers to intervene in aging, however, one might still argue that it is more natural or more befitting a good human life, to accept decline and to adapt ones expectations and vital goals. Such an argument would imply a normative claim about human nature, however, that would be difficult to defend (Buchanan 2009).

In sum, whether aging is an illness partly depends upon the factual question of whether people do experience it as such. But for another part it also depends upon a normative question, which comes sharply into focus only once we presume that we can and will develop the technological powers to intervene in aging. That question is, should one accept, and adapt to, the impairments and ailments that come with aging, or may one legitimately reject and resist them?

---

[4] I thank Hans-Jörg Ehni for bringing this point to my attention.

16.6 Aging as Sickness

Finally, we turn to the last component of the triad and ask whether aging is a sickness. It is in the social sphere that labelling aging in terms of pathology encounters most resistance. Socio-gerontologists have claimed that aging should not be understood merely in terms of illness and disease, but as a much broader and much more meaningful aspect of life.

The classical sick-role as described by sociologist Talcott Parsons (1951) implies an entitlement to support, an excuse for inability and an exemption from work or other tasks, and the responsibility to comply with therapy and get well again, in order to be able to resume the normal social role (Hofmann 2002). Sickness counts as an excusing condition: You are not to blame for it, and you are not to blame for being temporarily unable to fulfil certain (social) tasks, roles and requirements, both in public and private life.

There are some similarities between the sick role and the role of elderly people in our society, as Seneca already pointed out: "It may be urged many old man are so feeble that they can perform no function that duty or indeed any position in life demands. True, but that is not peculiar to old age: Generally it is a characteristic of ill health... old age is an incurable disease". In most contemporary Western societies, the incapacities and frailties that come with aging and old age are socially recognized as reasons for financial support, and they are recognized as a reason to be exempted from the duty to work. So, old age gives rise to some of the same social exemptions and privileges as sickness. An important difference is that we do expect sick people to do their best to get well again and comply with the doctor's orders, whereas we do not expect old people to grow young again. We do expect elderly people to live healthy lives and take good care of their health related lifestyle, however (see Chap. 18, this volume). Old age gives rise to some of the same social consequences as sickness.

However, as socio-gerontologists have rightly pointed out aging as a social process encompasses a lot more than just disability, dependence or exemption from work. From a social perspective, aging should not be reduced to biological aging. Getting older and being old have social meanings and are connected with specific social roles, tasks and expectations. From a social perspective, aging is also a matter of accumulated experiences, of shifts in relationships, of changes in responsibilities. Both in work and in private life people take up new roles when they age, like being a mentor for younger employees or becoming a grandparent. In social terms aging, growing older, is a normal or natural process, in the sense that is happens to everyone (at least in the West). Growing older is also embedded in our ideas about the human life cycle and about specific stages of life. As it is phrased in a report of the President's Council on Bioethics: "Aging is not just about old age. It is a crucial part of the nearly lifelong process by which we reach old age .... its product is .... the life cycle itself: The form and contour of our life experienced in time"(President's Council 2003, p. 208).

Over the past decades, social gerontologists and others have promoted the emancipation of elderly and worked to fight ageism. This movement has stood up for the idea that elderly people have their own social roles and social importance; that they are not merely burdensome, frail and demented. The critique on medicalization from this perspective is mainly a critique on reductionism, on seeing aging only in terms of ill health, disease, frailty, dependence and illness, and forgetting other social meanings of aging. One of the characteristics of aging may be that it is accompanied by physical and mental decline, by impairments; but this is not all there is to aging when seen from a social perspective. Elderly people have their own social roles and these should not be reduced to or equated with the sick-role. These are good reasons, I think, to resist turning aging into sickness. Aging may be a disease and may cause illness, but it is not a sickness—aging is aging.

There are reasons to resist turning aging into sickness, and calling aging a disease or emphasizing the related experiences of illness may make this more difficult. However, it is an empirical question whether conceptualizing aging as biological disease will eventually change our social view of the elderly. Moreover, this also depends on whether or not we'll be able to intervene in the aging process. If we call the biological aging process (senescence), a disease it does not necessarily follow that we should confine elderly people to a sick role. Successful medical interventions have already changed social roles of elderly in our society. Elderly people probably have more capacities than they used to have in previous centuries, and fulfil a greater diversity of roles. The difference we can make nowadays between the young old, middle old and oldest old testifies of this. People in their sixties can play different social roles now than they did before, partly because of successes of medicine and healthcare. Intervening in the biological aging process—if we will ever be able to—will probably have an enormous impact on the social role of aging. Instead of pushing elderly people into a sick role, it may as well liberate them and create new roles (the active senior, the post-retirement volunteer et cetera).

16.7 Conclusion

The conceptual triad disease-illness-sickness can help us to better understand the controversies around the question of whether aging is pathological, or normal and natural. By incorporating the biomedical, subjective and social perspectives, the triad brings to light the complex phenomenon that aging is. Part of the controversies around the question of whether aging is a disease can be explained by the incongruence between the three concepts of the triad disease-illness-sickness, and by understanding how changes in conceptualization in one domain affect the others. Especially the critique of medicalization can be explained.

One aspect of medicalization is that new biomedical research into aging mechanisms promotes an understanding of the aging process as 'disease' and thus clears the way for prevention and interventions long before there is an actual experience of illness. This leads to interventions in apparently healthy people and to labelling all elderly people as diseased. This framing of aging as disease also opens the door to research funding and reimbursement of medical costs which may—in the end—be beneficial for elderly.

Another aspect is that once aging comes to be understood as a disease, this tends to turn aging into sickness as well. This may be considered undesirable because it narrows our view on what aging is, makes us lose sight of broader social meanings of aging and old age and stigmatizes elderly people as ill and incapacitated. Framing aging as *essentially* a disease is therefore undesirable, because it has undesirable social consequences.

The underlying normative question remains, however. Ought we to accept the loss of biological function and associated loss of abilities to fulfil our goals? Ought we to adapt our personal aspirations and our social roles to this? Or should we try to further develop our biomedical knowledge and technological powers to avoid or postpone this loss?

We are not going to find the answers to these questions by appealing to notions of health and disease. First, because even if we would agree that aging was or was not pathological, this still would not answer the question of what to do about it in a satisfactory manner. The fact that present practices of research funding or healthcare insurance tend to leap from biomedical labels of 'disease' or 'non-disease' to normative conclusions about obligations and prohibitions does not make for a convincing argument. We need normative reasoning to draw normative conclusions. Second, because in this contribution I hope to have shown that there lies a normative controversy within the different conceptions of disease and illness themselves. The fundamental issue of whether we ought to accept the current biological fact of life (i.e. that our bodies age), or should try to intervene in the aging process, remains a fundamentally normative question, that is not solved by appealing to concepts of health and disease. Rather, this question is reflected in the different conceptualizations of disease and illness. This controversy ultimately depends on a deep seated difference in attitude towards life, a difference in views of 'the good life' and in views about human nature. On one side there is an attitude of acceptance and adaptation to (inevitable) facts of human life, which leads to acceptance of aging as a normal (not pathological) part of life. On the other side there an attitude of rejection of and resistance to the boundaries set by our biology, leading to the view that aging is pathological.[5]

Should we try to cure aging? If biogerontologists continue to label the aging process as a disease process and if they develop effective interventions to slow down that process, this is likely to have effects on our experience of old age as well as on the social appreciation of elderly. Intervening in the biological processes of aging will also affect, and require changes in, the way we are used to looking at and dealing with getting older in the personal as well as the social domain.

[5] There is a parallel here with the two frames that Erik Parens has distinguished in the enhancement debate, the gratitude and the creativity frameworks (Parens 2005).

References

Binstock, R.H. 2004. The search for prolongevity: a contentious pursuit. In *The fountain of youth*, ed. S.G. Post and R.H. Binstock, 11–37. Oxford: Oxford University Press.

Blumenthal, H.T. 2003. The aging-disease dichotomy: true or false?. *Journal of Gerontology: Medical Sciences* 58A(2): 138–145.

Boenink, M. 2010. Molecular medicine and concepts of disease: the ethical value of a conceptual analysis of emerging biomedical technologies. *Medicine Healthcare and Philosophy* 13:11–23.

Boorse, C. 1977. Health as a theoretical concept. *Philosophy of Science* 44(4): 442–473.

Boorse, C. 1997. A rebuttal on health. In *What is disease?*, ed. J.M. Humber and R.F. Almeder, 3–133. Totowa: Humana Press.

Buchanan, A. 2009. Human nature and enhancement. *Bioethics* 23(3): 141–150.

Caplan, A.L. 2004. An unnatural process: why it is not inherently wrong to seek a cure for aging. In *The fountain of youth*, ed. S.G. Post and R.H. Binstock, 271–285. Oxford: Oxford University Press.

Conrad, P. 1992. Medicalization and social control. *Annual Review Sociology* 18: 209–232.

Ebrahim, S. 2002. The medicalisation of old age should be encouraged. *British Medical Journal* 324: 861–863.

Fishman, J.R., R.H. Binstock, M.A. Lambrix. 2008. Anti-aging science: the emergence, maintenance, and enhancement of a discipline. *Journal of Aging Studies* 22: 295–303.

Gems, D. 2011. Tragedy and delight: the ethics of decelerated aging. *Philosophical Transactions of the Royal Society* B366: 108–112

Hayflick, L. 2000. The future of aging. *Nature* 408: 267–269.

Hofmann, B. 2001. The technological invention of disease. *Journal of Medical Ethics: Medical Humanities* 27: 10–19.

Hofmann, B. 2002. On the triad disease, illness and sickness. *Journal of Medicine and Philosophy* 27(6): 651–673.

Huber, M., Knottnerus, A., Green, L., et al. 2011. How should we define health?. *British Medical Journal* 343: d4163.

Izaks, G.J., and R.G.J. Westendorp. 2003. Ill or just old? Towards a conceptual framework of the relation between aging and disease. *BMC Geriatrics*. doi: 10.1186/1471-2318-3-7.

Jones, D.G., and M.I. Whitaker. 2009. Finding a context for discussing human life-extension. *American Journal of Bioethics* 9(12): 77–78.

Juengst, E.T. 2004. Anti-aging research and the limits of medicine. In *The fountain of youth*, ed. S.G. Post and R.H. Binstock, 321–339. Oxford: Oxford University Press.

Kampf, A., and L.A. Bothelo. 2009. Anti-aging and biomedicine: critical studies on the pursuit of maintaining, revitalizing and enhancing aging bodies. *Medicine Studies* 1: 187–195.

Lucke, J.C., P.C. Diedrichs, B. Partridge, and D.W. Hall. 2009. Anticipating the anti-aging pill. *EMBO Reports* 10(2): 108–113.

Nordenfelt, L. 1995. *On the nature of health.* Dordrecht: Kluwer Academic Publishers.

Nordenfelt, L. 1998. On medicine and health enhancement—towards a conceptual framework. *Medicine Healthcare and Philosophy* 1: 5–12.

Nordenfelt, L. 2007. The concepts of health and illness revisited. *Medicine, Healthcare and Philosophy* 10: 5–10.

Olshansky, S.J., L. Hayflick, and B.A. Carnes. 2002. Position statement on human aging. *Journal of Gerontology: Biological Sciences* 57A(8): 292–297.

Parens, E. 2005. Authenticity and ambivalence. *Hastings Center Report* 34–41.

Parens, E. 2011. On good and bad forms of medicalization. *Bioethics*. doi: 10.1111/j.1467–8519.2011.01885.x.

Parsons, T. 1951. *The social system.* London: Routledge

Post, S.G. 2004. Decelerated aging: should I drink from the fountain of youth?. In *The fountain of youth*, ed. S.G. Post and R.H. Binstock, 72–93. Oxford: Oxford University Press.

President's Council on Bioethics. 2003. *Beyond therapy.* New York: Dana Press.

Rose, N. 2007. *The politics of life itself.* Princeton: Princeton University Press
Schermer, M., I. Bolt. 2011. ADHD and the grey area between treatment and enhancement. In *Enhancing human capacities*, ed. J. Savulescu, R. ter Meulen and G Kahane, 179–193. Chichester: Wiley-Blackwell.
Scully, J.L. 2004. What is a disease?. *EMBO reports* 5(7): 650–653.
Smith, R. 2002. In search of “non-disease”. *British Medical Journal* 324: 883–885.
Strawbridge, W.J., M.I. Wallhagen, and R.D. Cohen. 2002. Successful aging and well-being: self-rated compared with Rowe and Kahn. *The Gerontologist* 42(6): 727–733.
Twaddle, A. 1994. Disease, illness and sickness revisited. In *Disease, Illness and Sickness: Three central concepts in the theory of health*, eds. A. Twaddle and L. Nordenfelt, 1–18. Linkoping: Studies on Health and Society.
Vijg, J., J. Campisi. 2008. Puzzles, promises and a cure for aging. *Nature* doi: 10.1038/nature07216.
Vincent, J.A. 2009. Aging, anti-aging, and anti-anti-aging: who are the progressives in the debate on the future of human biological aging?. *Medicine Studies* 1: 197–208.
Vincent, J.A., E. Tulle, and J. Bond. 2008. The anti-aging enterprise: science, knowledge, expertise, rhetoric and values. *Journal of Aging Studies* 22: 221–224.
Whitehouse, P.J., and E.T. Juengst. 2005. Antiaging medicine and mild cognitive impairment: practice and policy issues for geriatrics. *Journal of the American Geriatrics Society* 53: 1417–1422.

Chapter 17
A "Longevity Dividend" for All?

New Interventions into Aging and Justice

Hans-Jörg Ehni

17.1 Introduction

Biogerontology, the scientific discipline that explores the biological foundations of aging, has recently achieved important theoretical advances. Two prominent researchers have stated at various congresses, which united more than 20 % of the scientific community, that biological aging is no longer an unresolved problem (Hayflick 2007; Holliday 2006). This means that not only are the general foundations of biogerontology provided by the theory of evolution becoming more and more elaborated, but also the different mechanisms and processes on the molecular, cellular, and organic level are becoming increasingly known (Kirkwood 2008; Arking 2006). Gaining knowledge about these biological mechanisms opens up the prospect of biomedical interventions that might slow down, prevent or even reverse biological aging. Such prospects have already entered bioethical debates, mainly those related to human enhancement. The desirability of 'ageless bodies' and immortality have been the focus of these outlooks on possible long-term achievements (Kass 2003; Kass 2004; Harris 2004, Chap. 13, this volume, Buchanan 2011).

However, this focus might have diverted the attention from the short-term outlook that biogerontologists provide with regard to the possibilities of developing interventions into aging, and the realistic goals these might be able to achieve in the near future. Of particular interest would be the evaluation of these developments in relation to theories concerning just healthcare. This chapter will attempt to develop a general frame of how such an evaluation could be carried out. As claims regularly resurface that age has to be a criterion for the rationing of healthcare in the context of demographic change, a central question would be whether such new interventions into aging should be made widely available, and whether this is possible or not. What kind of impact this type of medicine will have in the context of existing health inequalities will also depend on availability. For detailed and empirically informed answers to these questions and convincing ethical evaluation, an interdisciplinary

H.-J. Ehni (✉)
Institut für Ethik und Geschichte der Medizin, Universität Tübingen, Germany
e-mail: hans-joerg.ehni@uni-tuebingen.de

M. Schermer, W. Pinxten (eds.), *Ethics, Health Policy and (Anti-) Aging: Mixed Blessings*, Ethics and Health Policy 1, DOI 10.1007/978-94-007-3870-6_17,

dialogue would be necessary involving biogerontologists, geriatricians, medical ethicists, and health policy experts. This contribution is not an attempt to replace such a dialogue, but serves to outline what might be a convincing choice of subjects to be discussed, and in which steps it might proceed. As a basis regarding the biology of aging, I will use some publications that summarize biogerontological knowledge and one statement in particular by leading experts of the field in which they outline technological achievements close at hand and their potential benefits.

On this basis, I will sketch the reasons for the promise of new interventions into aging and speculate what kind of medicine we can expect on this basis. Secondly, I will tackle the question of who will benefit and argue that limited access to this type of medical interventions is a likely outcome. Third, I will try to answer the question of whether we should worry about this from a perspective of justice. I will provide a couple of arguments why we should not, and then I will develop a justice framework that demonstrates reasons for why we in fact should. Finally, I will discuss some options for what could be done to avoid the undesirable consequence of increasing existing health inequities.

17.2 New Interventions into Aging and Medicine

Why should this issue be a matter of concern? One result of the recent research on the biology of aging is that the biological aging process is flexible and can be manipulated. This has been done in a range of laboratory organisms, such as yeast, nematodes, fruit flies and mice with a variety of different methods such as Caloric Restriction, gene manipulation and using the forces of natural selection (Kenyon 2005; Fontana et al. 2010; Partridge and Gems 2006; Masoro 2005; Rose 2008). Other interventions such as anti-oxidants (Golden et al. 2002; Harman 2009), hormesis or the inducement of light stress by different means (Rattan 2008), stem cell treatments in the context of regenerative medicine (Lafontaine 2009), and hormone treatments are discussed, tested, and sometimes even applied in the context of 'anti-aging' medicine, which is still an unproven and dubious endeavour (Butler et al. 2000; Olshansky et al. 2002). As no intervention into the aging process to date has been proven to be safe and effective in humans, the question must be raised whether the results achieved in laboratory organisms are transferrable to humans at all.

One indication that they are is the finding that mechanisms have been conserved across species barriers. Examples for humans are some particular characteristics of centenarians such as increased insulin sensitivity or lowered risk-factors for CVD in persons opting for a calorie-restriction diet. But, aging in humans is a complex interaction of genetic (estimates range from 15–35 %), environmental, behavioural, social and stochastic components. Stochastic components are random events on a molecular or cellular level such as DNA-damage. As a consequence, individual persons differ in their biological aging rates, which are also influenced by a complex interplay of social factors, the environment in which a person lives and individual health decisions (cf Chap. 2).

Based on the acquisition of further knowledge in the biology of aging, on the successful manipulation of the aging process in laboratory animals, on the conservation of aging mechanisms across species borders and on some common traits of centenarians, which might be used to develop interventions, a group of leading biogerontologists (Jay Olshansky, Daniel Perry, Richard A. Miller, Robert N. Butler in Olshansky et al. 2006, 2007) propagate a major investment in aging research by the National Institute of Health. As a realistic return on such an investment, they have claimed that there will be a "longevity dividend", which will be achieved by a prolonged professional life and a decreased burden of chronic diseases in old age. The proponents of the longevity dividend would like to distance themselves from both more optimistic authors such as Aubrey de Grey, who propagates 'ending aging', and the existing movement of anti-aging medicine, which claims to dispose of effective means to influence aging immediately (de Grey and Rae 2007). Olshansky et al. suggest that a seven-year prolongation of the human lifespan and slower aging for the same period of time is a realistic goal that could be achieved within the next 40–50 years.

What about the medical means to achieve this longevity dividend? Some other authors raise the prospect of simple interventions into aging such as CR-mimetics, also with reference to superior regulatory signalling pathways (e.g. Insulin/IGF-1), which will allegedly downregulate many other age-related processes, if successfully influenced. However, as already mentioned, the genetic, social, behavioural, and stochastic components of human aging will probably limit this outlook. Instead, some experts suggest a type of medicine that is adapted to individual aging. Although Olshansky et al. do not discuss what kind of medicine could result from the basic science to which they are referring, their suggestion on how to invest in aging research provides some insight. They ask for a support of basic research into genomics of aging and into regenerative medicine. Further they, suggest fostering clinical trials that investigate the interaction of drugs and lifestyle interventions. Finally, they also ask for investments in preventive medicine and the influence of social conditions. Together with the understanding of human aging as being different from individual to individual and the result of a multifaceted process with many stochastic components, this suggests that influencing aging will probably not be the result of a simple intervention that acts like some kind of magic bullet. Rather, the type of medicine resulting from the suggested research will be a personalized, complex and regular set of services including check-ups, advice, prevention, and treatment. This could start with a personal genome analysis to determine genetic risk and longevity profiles. From this could result particular recommendations on pharmaceuticals and diets, and individual life styles in general, which could lead to slower and healthier aging. Further regular check-ups on the aging of particular tissues and interventions to address the risks and to act according to the findings of these diagnostic procedures could follow. Finally, treatments with different interventions of regenerative medicines based on stem cells or artificial tissue could be used to restore the functions of organs affected by age-related changes. Obviously, it is only possible at this time to speculate about the type of medicine resulting from biogerontological research, and this would best be done—and should be done—in a dialogue between biogerontologists, geriatricians and the specialists in the different relevant fields. But, this sketch is at least one possible outcome that could be reasonably expected considering already existing trends (Hamburg 2005; Micans 2005).

The goals of this kind of medicine, which would be achieved by intervening in the basic process of biological aging, are the following: (1) Slower biological aging; (2) Prevention or postponement of age-associated diseases from which would possibly and hopefully result a compression of morbidity; (3) A longer lifespan. Importantly, the target of this type of medicine would be the aging process itself. It may not be possible to identify a particular age-associated disease as a target indication. Also, it may be difficult to predict whether such diseases would be prevented or merely postponed with the result of compressed morbidity.

17.3 Access to this Type of Medicine will be Limited

A first step in evaluating this type of medicine from the perspective of just healthcare is to try to predict who will have access to it. An attempt to do so could be based on its features as a complex, regular and personalized set of services that combines life style recommendations, diagnostic, preventive and therapeutic measures. Further, the goals and the basic way they are pursued could be considered as well. The question of whether access will be widespread could then be answered in two steps. The first will be to see whether a comprehensive set of measures based on influencing biological aging is likely to be covered by public health insurance in industrialized countries. Limits in access in this respect would be structural limits, primarily by legal structures. A different limit could be e.g. limited access to specialist care. The second step would be to look at individual limits in access to this type of medicine.

In many countries, public health insurance or the public healthcare system will probably not be able to finance many of the relevant interventions and services, at least not if the current legal regulations are not changed. In Germany, for example, the respective law requires the public health insurance to cover the costs for medical interventions if they are necessary for the diagnosis or treatment of a disease (§ 27 SGB—social code book—V). However, as indicated above, the main target of the interventions at stake is biological aging, and there may be no clear indication for the treatment of any disease. It is also not clear to what extent these interventions will be considered as necessary for the prevention of age-related diseases. If the onset of these diseases is merely postponed, it may be difficult to convince policy makers to include the respective measures in universal coverage and to decide from which age to start. Some experts even claim that extensive preventive measures should not be a priority of public health insurance, and consequently not be covered.

Further, the law requires sufficiency, cost effectiveness, appropriateness and necessity (§ 12 SGB V). Due to scarce resources and cost constraints that partially result from the demographic change and technological innovation (Bodenheimer 2005, 2005a), there is some likelihood that new expensive medical services and interventions into aging will not be covered by public health insurance. The appropriateness and necessity of a further lifespan extension might be doubted, as Daniel Callahan has done, in an influential way (Callahan 1977; Stock and Callahan 2004).

Personal limits could hinder access to new interventions into aging as well. If public health insurance will not cover this type of medicine, personal resources for financing access could be insufficient. As stated above, a comprehensive set of medical interventions could be personalized, regular and complex consisting of a combination of repeated diagnostics and therapeutics. If the price of these new technologies fall, as is predicted for whole personal genome scans, it remains that such regular visits and extensive consultations on personal habits and lifestyle choices are time-consuming and therefore also a medical service that could be costly for the less well-off. It is also unlikely the case that everybody will benefit from lifestyle recommendations in the same way. Individual responsibility for bad health choices is limited and also determined by social factors. Social determinants, like living conditions, education, lifestyle, if they remain unaddressed, could prevent some people from benefiting from interventions into aging. Unhealthy diets and their outcomes (e.g. obesity) could present an additional barrier: some risk factors related to obesity could lower the effectiveness of interventions into aging.

In sum, if the assumption is correct that medical interventions into the human biology of aging will be part of a complex and regular medical service, scepticism about whether everybody will benefit from the longevity dividend in the same way or even how widespread such benefits will in fact be seems to be appropriate. Additional barriers might also hinder people from contributing to the longevity dividend, if they lack the job opportunities in their field for a longer working lifetime. Consequently, they may lack the resources for a longer lifespan. Limited access to new interventions into aging and limited potential to contribute to the longevity dividend or enjoy its benefits could increase already existing inequalities in healthy life expectancy. At the same time, the situation of those best off is likely to further improve. A large amount of recent research has provided substantial evidence for a difference in healthy life expectancy according to socio-economic status, e.g. the Marmot report for the UK, and the research by Michael Marmot carried out for the WHO (Marmot 2006). The respective difference in healthy life expectancy in the UK according to the Marmot report is around 18 years; the difference in life expectancy is around 10 years. Due to limited access to medical interventions into aging, this could rise to 25 and 17 years if we use the prediction of Olshansky et al. and if we assume that the worst off are likely not benefiting from the longevity dividend, at least if this is not explicitly addressed by health policy. Compared to the situation of the best off, their position will even worse than it is now.

17.4 Why We Shouldn't Worry

It might seem obvious that this is unjust. In a first step, an analysis of new interventions into aging from a perspective of justice could hold that access to these interventions will create benefits or goods that will be distributed in an unequal way across society, presumably according to the socio-economic status of its members. At least four arguments could be raised against the possible claim that the distribution of these goods is unjust. The first argument is that the distribution itself is just because it results from a principle of justice, merit. Secondly, it could be argued that the distribution is unjust, but that this is only temporary and will in the long run raise everybody's position including that of the worst off. Regarding the goods and not the distribution itself, a third argument could be that these goods are not relevant from the perspective of justice, as they represent no important gains. Along the same line, a fourth argument could be based on the assumption that these goods could have some relevance, but not a very high priority.

The first argument could be considered to be a libertarian one. Being able to afford these technologies and medical interventions and benefit from them would just be another well-deserved reward and incentive for social and economic success and a rational and disciplined lifestyle. In turn, not being able to benefit from new interventions into aging and a shorter less healthy life-span could also be considered the result of bad health decisions and an unhealthy lifestyle accumulating throughout life considered as a whole. However, serious doubts can be raised as to whether this is based on correct assumptions. Responsibility for bad health choices and outcomes play a major role in this line of argument. But, research on influences that date back to early phases of life and on social determinants of health points to the limitations of personal responsibility in this field. As to the other part of the meritocratic argument that access to new interventions into aging should be simply seen as a reward for social and economic success, it remains to be seen what kind of arguments can be put forward that the equal distribution of the benefits of interventions into aging matters.

The second argument is somewhat related to the first. If the most successful people in society will have access to new interventions into aging, it could be argued that everybody will benefit in the long run. Further, trickle-down effects and increased availability because of wider social acceptance could contribute to a wider distribution. First of all, there could be a social benefit for all if a social elite is able to work longer. The benefit from the longevity dividend could be limited to the most successful members of a society, but this will initially have indirect effects on everybody else because of productivity increases through increased longevity. While the technologies involved will get cheaper with time, such increased productivity will create incentives to make them widely available. Allen Buchanan, who uses this argument for enhancement technologies in general, cites a labour productivity increase of 4 % per year increase in life expectancy in a population (Buchanan 2011). But still, it remains questionable whether this will indeed extend to all members of society or not. In some types of low-skilled and manual labour jobs, increases in productivity by working longer could be smaller and the availability of jobs for older workers may decrease with age, as is the case at the present time.

Both arguments are based on the assumption that an equal distribution is not unjust or does not matter in this case. Two other arguments are related to the relevant benefits or goods. These could be pointless. For instance, it could be argued that the achievement of a happy and fulfilling life does not depend on its length beyond a certain period of time. Seen from this perspective, living past the current average life expectancy would not be a substantial gain, and therefore it would not matter if some could not afford it. Indeed, Daniel Callahan has used his concept of a "natural lifespan" in this way, which he has put forward against lifespan extension (Callahan 1977). However, this concept poses many problems. First of all, the current average lifespan of humans in industrialized societies is far from being a result of natural evolution. Then, obviously, this argument is a natural fallacy. Even if a certain length of life is natural, this as such has no normative implications. Other assumptions Callahan makes seem to just beg the question. For instance, he argues that after having reached a certain age people have made all experiences that were useful to a certain character. But even if this were true, which is again based on a very doubtful and thin empirical basis, new interventions into aging might also change this alleged aspect of the human life cycle. Finally, contemporary theories of justice do not focus on well-being or fulfilment but rather on the opportunities to achieve them. Ascetic people, such as monks, could forego certain opportunities because they believe they are irrelevant for them. However, what is important is that they have these opportunities in the first place.

Finally, a fourth argument could refer to just healthcare. From this perspective, access to new interventions into aging may be of importance but of less importance than other aspects of healthcare, either for younger ages or for older people. The longevity dividend could possibly be of less significance and may divert resources from more important goals. Such a goal could be end of life care or high quality care for the oldest old, which may still be needed if chronic diseases are just postponed. Not only could the resources for such purposes be missing, we could also run the risk of suppressing the importance of these issues and judgments could become more biased under the impression of extending a vigorous life-time. As could be argued based on Norman Daniels' prudential lifespan account, reasonable priority setting in healthcare would simply ensure that we reach a certain age (Brauer 2009); a prolongation of the average lifespan would not be a priority of just healthcare facing limited resources (Fleck 2010). But also in this case, many questions remain open. First of all: Would all reasonable people really choose in the same way? Many people are willing to make substantial sacrifices to reach a very old age, e.g. undergoing caloric restriction. Secondly, are the goals of reaching a certain age and prolonging the healthy lifespan by means of interventions into old age really competing for resources? This would again depend on the way basic knowledge in biogerontology can be applied to medical interventions, and if and how the longevity dividend can really be achieved.

17.5 The Relevance of New Medical Interventions into Aging from a Perspective of Justice

The arguments that unequal access to medical interventions into aging would not be unjust were either based on assumptions as to what kind of goods this access creates or the principles for their distribution. I will briefly sketch a framework mainly based on John Rawls' theory of justice regarding these two aspects of distributive justice (Rawls 1971) (other forms of justice such as retributive justice, corrective justice or global justice are beyond the scope of this chapter, although they would be relevant for a comprehensive discussion of this topic). These two aspects are the distributional entities or measures whose distribution is relevant from a justice perspective and the principles according to which they should be distributed.

What kinds of goods are then generated by access to interventions into aging? If agency or positive freedom is considered to be the common denominator of measures for justice, interventions into aging have the potential to produce different categories of such goods or capabilities. There are three basic categories of such goods. The first is improvement of well-being in relation to age-related changes. An example could be macular degeneration such as the cross-linking of crystal proteins in the eye. This might not hinder vision, but can cause discomfort. Thus, it may not be absolutely necessary for agency, but may have a slightly negative impact. The second category is improvement of functional changes affected by age-related decline (cognition, sensual perception, mobility etc.) and improved capabilities as a result. Sarcopenia is a good example. This category would be of obvious importance in the context of agency. Finally, the third category would be life-time. Time as such could be considered a necessary precondition for activities and achievements and thus be a relevant aspect beyond mere physical or mental capabilities or resources.

Considering these categories of goods and capabilities, which result from them age-related interventions are relevant from a justice perspective. This is the case from both a Rawlsian and a capabilities perspective (Nussbaum 1992). It is not necessary to decide here which measure—resourcist or capabilities-oriented—or general approach is more appropriate in this context. The respective interventions would be relevant from the perspective of both. This is not surprising, as the most renowned theories of justice develop a measure on the basis of what they consider to be relevant for agency. In this respect, Rawls puts forward his concept of primary goods, which are preconditions of all kinds of life plans and are necessary for the principle of equality of opportunity. Norman Daniels extends this concept to health insofar as health in the form of 'normal species functioning' is relevant for this equality of opportunity. Furthermore, it is does not matter whether aging is considered as 'normal species functioning' or not—which is denied by some biogerontologists—it certainly is a loss of agency in many respects, and therefore relevant to what Daniels calls an "opportunity range" (Daniels 2008). Interestingly, both Christopher Boorse and Lennart Nordenfelt, two of the most prominent authors of a theory of health and disease, propose that aging could be considered as a condition which should be treated by medicine. Boorse notes that despite the fact that it is not disease, should be retarded or even eliminated if it were possible to do so, and if we decided to do so (Boorse 1997). Nordenfelt writes that prima facie aging could be considered to be a disease, but we do not do so because it is untreatable. Instead he proposes to adapt our vital goals to our age-related capabilities (Nordenfelt 1987). But this could also imply that as soon as aging is treatable, we should indeed try to do so. Tristram Engelhardt points out that age-related changes would obviously have the character of a disease if they were 'premature'. He also stresses that defining aging as a disease should depend on our ability to postpone it. If it is not postponable it leads only to frustration if defined as disease (Engelhardt 1979). On the other hand, aging can be usefully considered as a disease if postponable. What is 'premature' in the context of aging is also not merely determined by nature, but also influenced by society and also by medical possibilities. The aging process is not programmed, but flexible and depends on the social context, and likely in the future also on medical technology. In the same sense, it is unclear what constitutes a 'normal opportunity range' in relation to age and old age. This may not only depend on the social context and the social status of particular person but also on access to new medical interventions into aging prolonging different aspects of agency such as cognition or mobility. After all, one of the consequences of unequal access to intervention into aging could be that such a thing as a 'normal opportunity range' is no longer there for all members of a society.

Based on this assumption of unequal access leading to an unequal opportunity range, it becomes easier to decide what could be the principle of distribution of the goods generated by interventions into aging. Considering that those who are already the worst off will have a position that is actually worse than those with the best position, access to longevity interventions should be equal or the worst off should have priority. Otherwise, the possible impact on existing inequalities in healthy life expectancy will also lead to an increase in social injustice.

If we apply a fair process to the decision about this principle of distribution, such as deciding behind a veil of ignorance, the participants in such a fair process would presumably opt for principles leading to a prioritarian or egalitarian. Would somebody opt for a society in which the situation of those who already live shorter and less healthy lives should not have priority and those best off should have even longer lives? This is unlikely to be the case. Considering the impact that this would have on the 'opportunity range' of the participants in a fair process, they would not opt for a sufficientarian distribution, which would correspond to the arguments one and two put forward on a libertarian basis or on the basis of trickle-down effects. In opposition to the libertarian argument, one could maintain that people who are not aware of their position in society would not opt for an unequal distribution of goods of such relevance for fair equality of opportunity in old age. Even more so if this 'opportunity range' would not solely depend on their own responsibility and decision. The 'trickle down' argument could include the assumption that the position of the worst off would be better if some people had access to interventions into aging. However, the 'trickle down' effect is far from being certain, and there could be plausible alternatives for the improvement of the situation of the worst off. Finally, in opposition to Callahan's' argument of a 'natural' lifespan, it could be argued that only when the impact on the 'opportunity range' is considered, does it become clear whether people appreciate a greater 'opportunity range'. The question would address the reasons why anybody should stop to value aspects of his or her agency simply because a certain amount of time has passed. Even if some of the opportunities no longer seem relevant at second glance, it is unlikely that most people will consider it as irrelevant not to have them at all.

17.6 Conclusion: Setting Priorities in Research

If, from a perspective of justice, the likely outcome of limited access to interventions into the aging process is considered to be unjust, what consequences should be taken? Three basic alternatives shall be discussed here: Prohibition, enabling general access and setting priorities in research. The first option, prohibition of intervention, may seem like the easiest choice from the outset. As there is the potential of social harm, it could be argued that interventions into aging are not permissible, and that it would be legitimate to forbid them. Two reasons, however, can be raised against this. While there is some potential for harm, it is far from clear how high the actual risk is and what the risk-benefit-ratio will ultimately be. Because there could be substantial social and individual benefits as well, as the authors of the longevity dividend convincingly claim, simply prohibiting the related technologies as a precautionary measure will be morally problematic. The second argument is that a prohibition is also difficult, if not impossible, to implement in an appropriate manner, as there will presumably be strong resistance against it. In order to be effective, it would have to take place on a more or less global scale; furthermore, it is hard to imagine liberal countries, the leaders of which are reluctant to prohibit technologies without evidence of concrete risks, following such a path. Guaranteeing universal access, although this would be desirable from a perspective of justice (see also Farrelly 2007), will probably be too expensive if it includes a comprehensive set of technologies and personalized services, which will be developed without further provision.

The third option may appear more promising. This entails setting priorities on publicly funded research, amended by actions on the social determinants of health. Priorities should be set on the research of interventions, which will be affordable and implementable in the context of the public healthcare system. Such priorities should also address important health problems related to chronic diseases. This should be reflected in interdisciplinary cooperation between biogerontologists, geriatricians and health economists. A recent publication on "Rejuvenating Aging research" by the Academy of Medical Sciences has developed an outline of such research priorities, which also reflects such aspects as the importance of health problems, but it does not explicitly address the aspect of justice or access to possible interventions resulting from this research (Academy of the Medical Sciences 2010). Also, the proponents of the 'longevity dividend' are aware that these issues have to be addressed, as they mention cost-effectiveness and social aspects as parts of their suggested program to fund aging research. This can provide a starting point for the conception of a detailed agenda for priorities in aging research. An example of a medical condition that could be addressed in this way is obesity. Obesity increases the risk of different chronic conditions, and it has been stressed recently that it will lead to severe problems for the healthcare systems in many countries (ten Have et al. 2011). Further, researchers have claimed that it could generate a trend that could reverse the gains in life expectancy achieved up to today (Olshansky et al. 2005). There is also a clear connection between obesity and social determinants. All of this indicates that the problem of obesity threatens the benefit that could be generated by the longevity dividend, particularly for those who are already among the worst off, and have the lowest healthy life expectancy.

An example of what a strategy for priority setting could look like is the "3D combined matrix approach", which was proposed by the Global Forum for Health Research, an international organization that dedicates itself to demonstrating the importance of the research of health and health equity (Ghaffar 2009). The three dimensions encompass a public health, an institutional, and an equity dimension. With the help of such a tool, a comprehensive approach to a strategy for the research and development of biomedical interventions into aging could be conceived and subsequently implemented in public funding programs such as the framework program of the European Union. In this context, it should be examined whether the criteria for public health insurance or services should be changed and which health policies are successful in tackling the impact of social determinants of health on healthy life expectancy. We should improve the current social context for a fair distribution of the benefits of new interventions into aging and set priorities in research immediately, instead of just waiting to see what kind of impact they will have, if they become available. That way, there might indeed be a 'longevity dividend' for all.

References

Arking, R. 2006. *The biology of aging: Observations and principles.* New York, NY: Oxford University Press.

Bodenheimer, T. 2005a. High and rising healthcare costs. Part 1: seeking an explanation. *Annals of Internal Medicine* 142: 847–854.

Bodenheimer, T. 2005b. High and rising healthcare costs. Part 2: technologic innovation. *Annals of Internal Medicine* 142: 932–937.

Boorse C. 1997 A rebuttal on health. In *What is disease?*, ed. J.M. Humber, and R.F. Almeder, 1–134. Totowa, NJ: Humana.

Brauer, S. 2009 Age rationing and prudential lifespan account in Norman Daniels' Just health. *Journal of Medical Ethics* 35: 27–31.

Buchanan, A.E. 2011. *Beyond humanity?*. Oxford: Oxford University Press.

Butler, R.N., Fossel, M., Pan, C.X., Rothman, D.J., Rothman, S.M. 2000. Anti-aging medicine. 2. Efficacy and safety of hormones and antioxidants. *Geriatrics* 55: 48–58.

Callahan, D. 1977. On defining a 'natural death'. *The Hastings Center Report* 7: 32–37.

Daniels, N. 2008. *Just health.* Cambridge: Cambridge University Press.

De Grey, A.D.N.J., and M. Rae. 2007. *Ending aging.* New York: St. Martinś Press.

Engelhardt, H.T. 1979. Is Aging a Disease?. In *Lifespan*, ed. R.M. Veatch, 184–194. San Francisco: Harper and Row.

Farrelly, C. 2007. Sufficiency, Justice, and the Pursuit of Health Extension. *Rejuvenation Research* 10: 513–519.

Fleck, L.M. 2010. Just caring: in defense of limited age-based healthcare rationing. *Cambridge Quarterly of Healthcare Ethics* 19: 27–37.

Fontana, L., Partridge, L., Longo, V.D. 2010. Extending healthy lifespan–from yeast to humans. *Science* 328: 321–326.

Ghaffar, A. 2009. Setting research priorities by applying the combined approach matrix. *Indian Journal Medical Research* 129: 368–75.

Golden, T.R., Hinerfeld, D.A., Melov, S. 2002. Oxidative stress and aging: beyond correlation. *Aging Cell* 1: 117–123.

Hamburg, M.A., and F.S. Collins. 2010. The path to personalized medicine. *The New England Journal of Medicine* 363: 301–4.
Harman, D. 2009. Origin and evolution of the free radical theory of aging: a brief personal history, 1954–2009. *Biogerontology* 10: 773–781.
Harris, J. 2004. Immortal Ethics. *Annals of the New York Academy of Sciences* 1019: 527–534.
Hayflick, L. 2007. Biological aging is no longer an unsolved problem. *Annals of the New York Academy of Sciences* 1100: 1–13.
Holliday, R. 2006. Aging is no longer an unsolved problem in biology. *Annals of the New York Academy of Sciences* 1067: 1.
Kass, L.R. 2003. Ageless Bodies, Happy Souls. *The New Atlantis* 1: 9–28.
Kass, L.R. 2004. *Beyond therapy.* Washington, DC: President's Council on Bioethics.
Kenyon, C. 2005. The plasticity of aging: insights from long-lived mutants. *Cell* 120: 449–460.
Kirkwood, T.B.L. 2008. Understanding aging from an evolutionary perspective. *Journal of Internal Medicine* 263: 117–127.
Lafontaine, C. 2009. Regenerative medicine's immortal body: From the fight against aging to the extension of longevity. *Body & Society* 15: 53.
Marmot, M.G., and R.G. Wilkinson. 2006. *Social determinants of health.* Oxford: Oxford University Press.
Masoro, E.J. 2005. Overview of caloric restriction and aging. *Mechanisms of Aging and Development* 126: 913–922.
Micans, P. 2005. The need for anti-aging medicine: the challenges faced to incorporate preventative medicine into the clinic and into society. *Annals of the New York Academy of Sciences* 1057: 545–562.
Nordenfelt, L. 1987. *On the nature of health: an action-theoretic approach.* Dordrecht: Reidel.
Nussbaum, M.C. 1992. Human functioning and social justice: In defense of aristotelian essentalism. *Political Theory* 20: 202–246.
Olshansky, S.J., L. Hayflick, and B.A. Carnes. 2002. No truth to the fountain of youth. *Scientific American* 286: 92–95.
Olshansky, S.J., D.J. Passaro, R.C. Hershow, et al. 2005. A potential decline in life expectancy in the United States in the 21st century. *The New England Journal of Medicine* 352: 1138–1145.
Olshansky, S.J., D. Perry, R.A. Miller, and R.N. Butler. 2006. In pursuit of the longevity dividend.The Scientist 20: 28–36.
Olshansky, S.J., D. Perry, R.A. Miller, R.N. Butler. 2007. Pur-suing the longevity dividend: scientific goals for an ag-ing world. Annals of the New York Academy of Sciences 1114: 11–13.
Partridge, L., Gems, D. 2006. Beyond the evolutionary theory of aging, from functional genomics to evo-gero. *Trends in Ecology & Evolution* 21: 334–340.
Rattan, S.I.S. 2008. Hormesis in aging. *Aging Research Reviews* 7: 63–78.
Rawls, J. 1971. *A theory of justice.* Cambridge: Belknap Press of Harvard Univ. Press.
Rose, M.R. 2008. Making SENSE: strategies for engineering negligible senescence evolutionarily. *Rejuvenation Research* 11: 527–534.
Stock, G., and D. Callahan. 2004. Point-Counterpoint. *Journal of Gerontology* 59A: 554–559.
Ten Have, M., I.D. de Beaufort, P.J. Teixeira, J.P. Mackenbach, and A. van der Heide. 2011. Ethics and prevention of overweight and obesity: an inventory. *Obesity Review* 12: 669–679.

Part IV
Choosing Direction: Healthcare Policy and Ethics

Chapter 18
Healthy Aging and Personal Responsibility

Alies Struijs and Marieke ten Have

18.1 Introduction[1]

In this chapter we explore the question which ethical arguments arise when demanding personal responsibility for lifestyle from the elderly in health policy. Even if the focus on personal responsibility were acceptable for younger citizens (and we believe this to be the case to a certain extent), it is not self-evident that this applies equally to the elderly. We start by describing what we mean by personal responsibility and how the trend of personal responsibility is present in health policy for the elderly. After that, we analyse what arguments pro and contra responsibilisation are mentioned in the literature, and how these general arguments apply to the specific group of the elderly. We conclude by focussing on a specific case of policy, that is, the health check for elder people.

Case
Mr A has a BMI of 28. He follows an unhealthy diet and leads a sedentary life. His family physician advises him to follow a healthy diet and to find an appropriate way of getting more physical exercise. Mr A, however, does not change his lifestyle and his BMI remains too elevated.

[1] We are very grateful to Dick Willems and Maartje Schermer for their comments on earlier drafts of this chapter.

A. Struijs (✉) · M. ten Have
Council for Public Health & Healthcare of the Netherlands
The Hague, The Netherlands
e-mail: a.struijs@rvz.net

M. ten Have
e-mail: m.t.have@rvz.net

M. Schermer, W. Pinxten (eds.), *Ethics, Health Policy and (Anti-) Aging: Mixed Blessings*, Ethics and Health Policy 1, DOI 10.1007/978-94-007-3870-6_18,

Most of us might hold Mr A at least partly responsible for the increased health risks associated with his overweight, and many would think this could be translated in, for instance, a higher insurance premium or higher out-of-pocket payments.

Mr A's age was not mentioned in the fragment. The question of this chapter is: Would it make a difference for the assignment of responsibility if Mr A would be 38 years of age or 73? In other words: Should the arguments that are used in the debate about 'responsibilisation' of citizens for their health and health risks, take age into account?

18.2 The Responsibilisation Trend in Health Policy

In this paragraph we describe what we mean by personal responsibility and how the trend of personal responsibility is present in health policy for the elderly.

18.2.1 Definition Personal Responsibility

Personal responsibility for health means that persons take responsibility for their choices and for the consequences of these choices (De Beaufort 1999).

Responsibility can be divided into prospective and retrospective responsibility (Ten Have et al. 1998; Council for Public Health and Health Care 2002). Prospective responsibility concerns future behaviour. "Prospective responsibility is about what a person should care about and what he should do. It is an action guide" (Verweij 1998). Thus, it may be said that people should care about their future health and act accordingly. Prospective responsibility has a motivating and educative function. When the family physician asks Mr A to adopt a healthier lifestyle, he makes an appeal to Mr A's *prospective* responsibility for his health.

Retrospective responsibility means to hold persons accountable for the consequences of their actions. It implies an evaluation after the action has taken place, to hold someone responsible for risky behaviour that could have been avoided, or for excellent behaviour. According to this interpretation, people can be blamed and held financially accountable, or praised and financially rewarded for their behaviour, for instance by excluding them from care or the compensation of healthcare costs, or by rewarding them for healthy behaviour in the form of a lower insurance premium. Retrospective responsibility has a function of reprisal or blaming. One type of retrospective responsibility are financial measures, for instance to ask higher out-of-pocket payments from Mr A because he did not change his lifestyle. Another type of retrospective responsibility is to exclude people from care, for instance to exclude someone who continues smoking from a second bypass surgery (it may be argued that this is not only because people are held accountable for the consequences of past behaviour, but also because the treatment is less effective when the lifestyle is not changed; this, however, does not change our argument).

Personal responsibility presupposes at least some degree of freedom of choice, which presumes that persons are free to choose (without coercion), that they are able to choose (being competent and well-informed), and that they have options.

18.2.2 Personal Responsibility in Government Health Policy

For some decades now responsibilisation has been a trend in government health policy in the Netherlands and abroad. Citizens are being shouldered with more personal responsibility for their health. Where necessary, care providers are expected to support them in this. Supporting rather than taking over responsibilities is the new catchword. An increasingly guiding and decisive role, and therefore more responsibility, is reserved for health insurers, who buy care on behalf of and in the interest of patients, taking into account price and quality of care. This trend of 'responsibilisation' is connected with a strengthening of the patient's position, which can be summed up in the key words: 'patient empowerment', 'freedom', 'autonomy' and 'choice', but also the other side of this: The duty 'to take greater responsibility for their own present and future health, welfare and security' (Nuffield Council 2010).

18.2.3 Influence of Developments within Healthcare

A number of developments (both internal and external to healthcare) have reinforced and accelerated the trend towards more personal responsibility within healthcare.

Within healthcare, there has been a shift from more collective to more individual and personalised provision of care. The principle 'one size fits all' gets replaced by 'tailored care': 'differentiated and individualised' is the new catchphrase. Patient- or client-oriented supportive care is gaining ground. We find this development in various care sectors; just think of the fact that elderly people stay in their own homes longer. This already means that care cannot be provided everywhere in the same way, but needs to be geared to the home base where the main emphasis of care lies.

Developments in the fields of medicine, science and genetics reinforce the trend towards a more individual, personalised approach to health. Thanks to greater genetic knowledge, personally specified health risks can be recognised earlier; depending on the size of the risk, this may increase the responsibility of the carrier of the genetic deviation (usually a heightened susceptibility to a certain risk factor) to adapt his or her lifestyle to the susceptibility detected (Health Council 2010).

In the sphere of health promotion and prevention we see a comparable development towards more personal responsibility and lifestyle advice tailored to the individual person. An example: in the autumn of 2011 various patients associations in collaboration with doctors' associations launched the so-called Lifestyle Test in the Netherlands. Its aim is to spur people on to take action themselves to prevent diabetes, heart and vascular diseases and renal damage. After answering a list of questions, people are given personal advice on how to change their lifestyle, taking into account their habits and preferences.

In the focus on prevention most of us think that it is good to recognise health risks in an early stage: 'early detection'. People are increasingly focused on preventing or limiting disease. It seems to be becoming a standard norm not to accept limitations, but to anticipate them as early as possible. 'Healthy aging = prevention' (Pomp 2011). We see this trend for instance in the proposal of the Dutch College of General Practitioners (NHG) to ask people over 45 to do a regular health check—the so-called Prevention Consultation (see also later in this chapter).

18.2.4 Influence of External Developments

The trend of responsibilisation within healthcare was also reinforced by developments external to healthcare. The rapid advance of information technology, including the internet, is perhaps the most influential development at present. Citizens are gaining more direct access to their personal health data and to general information about health. As a result, they have more control over their own health. This increases the emphasis on citizens' own responsibility. Communication of citizens with doctors and between doctors themselves has been facilitated and informalised by electronic exchange of data.

Internet use in healthcare—shortly: E-health—also provides other options for prevention, treatment and care and allows citizens themselves to take more control. It produces new forms of relations and communities between doctor and patients, but also between patients with a particular disorder. Successful examples are Parkinsonnet and E-mental health, in which professionals, patients and their carers can exchange or share data and knowledge. Today's patient is not fully dependent on the care provider for information about or help for his health. He is better informed and more in control of his own health and care. In this way more and more responsibility is shifting to citizens/patients and the people around them and the role of the care provider is changing "From God to Guide" (Bloem, TedXMaastricht 2011).

As well as more medical possibilities, there are also social developments, such as the economic crisis, the prevailing (neo-)liberalism and the ever-mounting costs of healthcare, which push government policy towards more personal responsibility of citizens.

18.2.5 Responsibilisation in Health Policy for the Elderly

We will now explore how the trend of personal responsibility is present in health policy for the elderly.

We mentioned before some examples of retrospective responsibility in healthcare policy in general (for instance the proposal that people who did not live healthily should pay higher insurance premiums or higher out-of-pocket-payments). Examples of retrospective responsibility concerning specifically the elderly are less easy to find. One explanation for this is that the choices involved in healthy aging concern lifestyle choices in general, and, as a consequence, retributive measures often target the whole population instead of specifically the elderly (for instance, higher premiums for people who smoke instead of higher premiums for older people who smoke). Another explanation is that retrospective responsibility for lifestyle raises strong ethical debate in liberal societies, and therefore there are more measures that concentrate on prospective responsibility.

Prospective responsibility, on the other hand, is frequently apparent in policy that focuses on healthy aging and preventing diseases (Van Campen 2011; Health Council 2009). On municipal and regional levels there are many prevention projects for elderly people, mainly aimed at exercise, healthy nutrition, and prevention of falling, depression, loneliness and decubitus. The Prevention Consultation mentioned earlier is a new national development in the Netherlands, intended to offer a scientifically reliable risk test to people over 45 with a view to detecting and preventing a heightened risk of heart and vascular diseases, diabetes mellitus type 2 and renal diseases (Zantinge et al. 2011). A good example of focussing on self-care is the project 'Smart Care for the Turkish Elderly', an E-health project in which elderly Turkish immigrants are connected to a centralised exercise project via webcam. This project is part of a cooperation of seven European regions, focussed on innovation in changing demographic contexts (www.peopleproject.eu).

18.3 Ethical Debate about Responsibilisation in Health Policy

In this paragraph we analyse what arguments pro and contra responsibilisation are mentioned in literature, and how these general arguments apply to the specific group of the elderly.

18.3.1 Arguments in Favour of Emphasizing Personal Responsibility in Health Policy for the Elderly

There are two arguments for considering a focus on personal responsibility in health policy from an ethical point of view. We will first present these arguments in favour of personal responsibility, and then describe the ethical debate they evoke.

The first argument for considering the implementation of personal responsibility holds that it serves the personal interest. More attention for what people can do themselves to stay healthy, helps to make them less dependent on professional care. Many persons are capable of making their own decisions and taking care of their own health. After all, people today have more control over their healthcare, are better informed, better educated, financially stronger and emancipated. Empowering individuals to make their own choices is a valuable aim. In order to take responsibility for their health, people need information and sometimes a nudge (Thaler and Sunstein 2008).

One could argue that this argument equally applies to health policy for the elderly, because being older (over 70) is not different or decisive in this. What is more, if we were to spare older citizens in this respect, it could be seen as improper age discrimination or paternalism. Older citizens are usually perfectly capable of making their own decisions and looking after themselves. They may even be better at this than younger citizens, who often lead hurried lives without regard for health risks. In this sense both older and younger people need to assume responsibility.

The second argument for considering an appeal to personal responsibility holds that personal responsibility is necessary for relieving the pressure on collective means. Within the healthcare system there is pressure on financial means and the availability of professionals. A system exclusively based on collective financial means (solidarity) is simply not feasible anymore. This argument has to do with the principle 'one good turn deserves another': The right to collectively financed care is balanced by an individual duty of effort. This is all the more the case as health costs rise, partly through the availability of more (expensive) treatments. However, the question is whether the principle 'one good turn deserves another' should be enforced by equally stringent measures regarding the elderly. Would it make a difference to demand a change of lifestyle from somebody who is over 73 or is 38 years old?

18.3.2 Arguments Against Personal Responsibility in Health Policy

Arguments against personal responsibility can be divided in arguments focussed on the retrospective type and on the prospective personal responsibility. In this paragraph we focus on arguments regarding the retrospective type, because most ethical objections regarding personal responsibility for health regard interventions that concentrate on retrospective responsibility. Prospective responsibility raises fewer objections, because it mainly involves encouraging people to live a healthy lifestyle. However, interventions that encourage people to live a healthy lifestyle do raise the argument that it is paternalistic to expect people to make choices regarding their own health in a way that the government or professionals expect from people.

First of all, retrospective responsibility may be rejected because it can be a form of blaming the victim. The relationship between ill-health and a self-chosen lifestyle is not easy to prove, because unhealthiness is also related to factors beyond the individual's control.

Secondly, requiring persons to live a health-conscious lifestyle and holding them accountable for the consequences of an unhealthy lifestyle may interfere with the values of autonomy and freedom of choice.

Thirdly, it is difficult to apply retrospective responsibility in a consistent way. The types of unhealthy and risky choices are infinite (smoking, parachute jumping, work stress, playing sports with the risk of injuries, and so on), and it would be unfair to single out just one group (smokers for instance).

Fourthly and lastly, demonstrating the relationship between ill-health and a self-chosen lifestyle requires intrusions upon people's privacy, for instance, it would demand a form of 'smoking police'.

18.3.3 Arguments Against Retrospective Responsibility in Health Policy for the Elderly

The before mentioned "blaming-the-victim- argument" against retrospective responsibility may even stronger apply to health policy for the elderly. There are various reasons why older people could be less capable of taking responsibility than younger people. One of the reasons is that during the life of older people there was much less awareness and knowledge of high-risk behaviour, so that they cannot be retrospectively held accountable for its consequences. Holding older people responsible for current unhealthy (or healthy) behaviour, requires that they receive sound advice on effective prevention and a healthy lifestyle, tailored to this age group. Being held personally accountable for your health requires that you have to know what you yourself can do about it and where you can find expert support.

Another reason why older people could be regarded as less capable of taking responsibility is that behaviour change may be more difficult for older people than it is to younger people. Responsibility for the health effects of behaviour sometimes means that people need to change their behaviour. For the young this may be easier than for the old; habits may be harder to break if formed over fifty years and embedded in a certain lifestyle. Furthermore, older people sometimes have several health problems at once, which can make healthy living very complicated. Responsibility for your own health becomes a somewhat abstract and artificial notion if you take 15 pills a day for e.g. diabetes, heart failure, COPD and arthrosis.

Finally: Personal responsibility presupposes the ability to make choices, which requires competence to act. In people suffering from dementia this competence to act diminishes. However, competence to act is a criterion, but is not necessarily connected with age: Not all older people have dementia though dementia often occurs in older people, it sometimes manifests itself at a younger age too. So, this argument cannot be generally applied to all older people.

In short, it is clear that the appeal to personal responsibility in government policy for older people raises arguments on both sides which lead to the conclusion that in some respects a more flexible policy is justified.

18.4 A Health Check for People over 45?

As an example of stimulating and facilitating healthy aging and personal responsibility we will discuss here the proposal of the Dutch NHG to introduce the regular health check—the Prevention Consultation—for the over 45s. This Prevention Consultation is an example of how the government, health insurance companies or professionals can stimulate healthy behaviour by an appeal on personal responsibility. Similar initiatives are recommended in the UK (NHS Midlife Check for people over 40) and in other countries.

Prevention Consultation for the over 45s

The reason for the NHG to recommend the Prevention Consultation is the growing unhealthiness of the Dutch population and the need for one reliable, validated test as opposed to all the commercially offered tests whose reliability is unknown to the citizen. The increase in the number of chronic disorders resulting from an unhealthy lifestyle, such as smoking, lack of exercise and obesity, will continue. More and more people will spend years in an unhealthy condition. The test may stimulate people to do something about this.

Another important reason is that growing unhealthiness of citizens also leads to higher healthcare costs.

The risk test developed by the NHG for this purpose is meant for all Dutch people over 45 in order to discover diseases like diabetes and heart and vascular diseases at an early stage. Those concerned fill out an (online) questionnaire, which gives them an initial assessment of their risk of heart and vascular diseases, diabetes or renal damage. In the case of a slightly heightened risk someone is given a specific lifestyle advice. In the case of a heightened risk the person is advised to make an appointment with his GP or company doctor for a consultation. He or she will conduct an additional examination and give appropriate lifestyle advice or may start a treatment or refer to other care providers.

The idea is for the Prevention Consultation to be included in the standard health insurance package, thus making it available and affordable for all. This decision has not yet been finalised. For now, GPs themselves have to enter into agreements with health insurers in their region. The NHG Standard 'The Prevention Consultation Module Cardiometabolic Risk' was published in the periodical Huisarts & Wetenschap [GPs and Science] in March 2011. From April 2011 GPs can invite patients within their practice to fill out the questionnaire of the Prevention Consultation, on condition that there are agreements with the regional health insurers and the care providers and bodies to which referral may be made (NHG 2011).

The Prevention Consultation is an aid for exercising control over one's own health. Does this initiative developed by the NHG help older people to take responsibility for their own health? What are the potential arguments in favour of a health check?

It is a way of becoming personally aware of your state of health. The age of 45 seems an early starting-point, but if you want to prevent or reduce individual disorders and collective health costs in the future, you will have to start early. Therefore, it is important to anticipate health risks, have personal control over the process and then organise your life accordingly.

It promotes integral care for patients with multiple disorders and thus contributes to a good and suitable quality of care. Often this care is now fragmented, insufficiently coordinated and too general. The Prevention Consultation helps to take responsibility for a healthy lifestyle in a situation of multiple chronic disorders in a manner suited to the person. It is a—professional guided—way of taking your responsibility in healthy aging.

What potential objections can be articulated to a regular health check?

A health check may act as a carte blanche and lead to the following idea: 'the results are favourable, I don't have to change my lifestyle, so I'll just keep on overeating.' In that case it is counterproductive. Even though an unhealthy lifestyle may have no negative effects at present, it cannot be excluded that these effects will appear in the future.

Early detection may overshoot the mark and result in medicalization. Nobody is healthy, everybody has 'something', preventing (health) risks is something you do all your life. This may become an obsession and cause feelings of guilt.

It may also produce certain side-effects which we as a society do not really want. It may stimulate unnecessary care consumption (visiting the GP sooner and more often) and therefore stronger rises in care costs.

Making a preventive health check compulsory will probably raise the objection that people don't like to be told by the government how they should organise their lives (paternalism).

Current government policy is not based on compulsion, but it does want to enable health checks by including them in the basic package. A stumbling-block here is funding: Care insurers are not yet prepared to pay for it. Care costs are running high as it is, and when it is necessary to decide what is paid from collective resources, prevention tends to come last on the list. The fact is, the effects of prevention do not pay themselves back straightaway, but in the long term. And even that is not easy to prove.

As we argued earlier, we should not judge older people too strictly 'with today's knowledge' (retrospectively) on the (prospective) upbringing, lifestyle and habits of those days. At the same time older people can be held accountable for the consequences of current unhealthy (or healthy) behaviour. The Prevention Consultation may serve as a nudge to put them on the right track.

18.5 Concluding Remarks

Let us return to the opening question. Which ethical arguments arise when demanding personal responsibility for lifestyle from the elderly in health policy?

There are sound arguments in favour of appealing to personal responsibility for healthy behaviour, that hold for older people just as well as for younger people. It is important to show older people that they themselves can contribute to their state of health, that decline is not always necessary and that prevention in old age, too, can be worthwhile.

However, there are also objections against focussing on personal responsibility in health policy. We have argued that measures that stress prospective responsibility are least contestable, as long as they concentrate on enabling and empowering older people to take good care of their health. Measures that stress retrospective responsibility, on the other hand, evoke various objections. They can blame the victim, interfere with the values of autonomy and freedom of choice, be difficult to apply in a consistent way and require intrusions on people's privacy.

Health policy for the elderly that stresses retrospective responsibility (such as demanding higher out-of-pocket-premiums from older people who did not comply with lifestyle advices), may in particular be vulnerable for the 'blaming the victim-argument'. The reason for this is that older people may be often less capable for taking responsibility for their health compared to younger people.

Focussing on prospective responsibility by enabling and empowering older people to take care of their health as long as possible, is therefore more acceptable from an ethical point of view. The government and professionals should provide the options for healthy behaviour, advise older people on health risks and where necessary support them in preventing and reducing these risks. Rather than enforcing and compelling, policy should facilitate healthy aging, for instance by promoting opportunities for and knowledge about healthy behaviour. It is important to offer good general advice that is suited and tailored to the person and that is aimed at more exercise, fall prevention, healthy nutrition and a sensible use of alcohol, in order to make people aware that in old age, too, prevention can be greatly beneficial to health.

It is expected that the trend of a 'double' rise in the aging population will continue. A higher number of elderly people will live longer, but will also have to contend longer with ill health and limitations. The occurrence of multiple disorders will increase. Due to increased medical possibilities, healthcare costs will rise further. If care is to remain affordable and accessible for all citizens, there will be less room to pay for care from collective resources. And it is likely that choices will be made that demand more personal accountability for health. From an ethical viewpoint it is important to follow such choices critically.

References

de Beaufort, I. 1999. Individual responsibility for health. In: *HIV and AIDS, testing, screening, and confidentiality*, eds. Rebecca Bennett and Charles A. Erin. 107–124. Oxford: Oxford University Press.

Bloem, B. 2011. From God to guide. TedxMaastricht- The future of health.

Campen, C. van. 2011. *Kwetsbare ouderen.* Den Haag: Sociaal Cultureel Planbureau.

Council for Public Health and Health Care. 2002. Gezondheid en gedrag. Advies. Zoetermeer: Council for Public Health and Health Care.

Dekker, J.M., M. Alssema, P.G.H. Janssen, M. Van Der Paardt, C.C.S. Festen, M.J.W. Van Oosterhout, J.L. Van Dijk, T. Van Der Weijden, R.T. Gansevoort, J.J. Dasselaar, F. Van Zoest, T. Drenthen, E.P. Walma, A.N. Goudswaard. 2011. NHG-Standaard preventieconsult module cardiometabool risico. *Huisarts Wet* 54(3): 138–55.

Department of Health. 2010. Prevention Package for older people. http://www.dh.gov.uk

Have, H.A.M.J. ten, R.H.J.ter Meulen, and van Leeuwen E. 1998. *Medische ethiek.* Houten/Diegem: Bohn Stafleu Van Loghum.

Health Council of the Netherlands. 2009. *Preventie bij ouderen: Focus op zelfredzaamheid.* Den Haag: Gezondheidsraad.

Health Council of the Netherlands. 2010. The 'thousand-dollar genome': An ethical exploration. Den Haag: Centrum voor ethiek en gezondheid.

Nuffield Council on Bioethics. 2010. Medical profiling and online medicine: The ethics of 'personalised healthcare' in a consumer age. London: Nuffield Press.

Pomp, M. 2011 Gezond en actief ouder worden: De maatschappelijke baten van healthy aging onderzoek. Een scenario analyse tot 2050. Onderzoek in opdracht van UMCG.

Thaler, R. H. and C.R. Sunstein. 2008. *Nudge: Improving decisions about health, wealth and happiness*, Yale University Press.

Verweij, M. 1998. *Preventive medicine between obligation and aspiration.* Utrecht: Zeno, the Leiden-Utrecht Research Institute of Philosophy.

Zantinge E.M., E.A. Van Der Wilk, S. van Wieren, C.G. Schoemaker. 2011. *Gezond ouder worden in nederland. [Healthy aging in the Netherlands].* RIVM report

Chapter 19
Biogerontology: A Promising Route to Cost Containment in Healthcare?

Laura Capitaine and Guido Pennings

19.1 Introduction

Population aging is now a global phenomenon. This shift in society's age structure has been a gradual process in developed countries, spanning over more than a century. In more developing regions, however, population aging has only recently begun and is proceeding at a much faster pace than it did in developed countries (Kinsella and Phillips 2005).

The current demographic situation is the result of both improvements in life expectancy and declining fertility rates. The baby boomers will soon accelerate the process of population aging as they enter old age en masse.

Greying populations are a human success story in that they represent the culmination of social and technological progress. Nevertheless, population aging is generally viewed as a burden, rather than a blessing. 'The coming entitlement tsunami' and the 'demographic earthquake' are just a few of the expressions that are frequently used to characterise this phenomenon (Beard and Williamson 2010). One of the concerns is that an aging population will cause healthcare costs to spiral out of control. For example, according to projections made by the trustees of Medicare—the US government's health insurance programme for the elderly—the programme will go bankrupt in 2018 (Callahan and Prager 2008). Such dire predictions have initiated a widespread search for effective 'remedies'.

Age-based rationing of healthcare is one of the proposed measures for constraining the feared escalation of costs related to population aging. Various models have been put forward for prioritising young patients over elderly ones, each offering a different justification for doing so. Callahan (1987) introduces an age-limit (around late 70s—early 80s) beyond which individuals ought to be denied lifesaving treatments. From that age onward, healthcare services should be primarily aimed at alleviating patients'

L. Capitaine (✉) · G. Pennings
Bioethics Institute Ghent, Ghent University, Belgium
e-mail: Laura.Capitaine@UGent.be

G. Pennings
e-mail: guido.pennings@ugent.be

M. Schermer, W. Pinxten (eds.), *Ethics, Health Policy and (Anti-) Aging: Mixed Blessings*, Ethics and Health Policy 1, DOI 10.1007/978-94-007-3870-6_19,

suffering. The rationale behind this age-limit is that, by this age, people have lived out a 'natural lifespan', implying that they have achieved most of life's possibilities.

The fair innings argument (Harris 1985) equates a certain amount of life years (usually the traditional three score and ten) with a 'fair innings'. The argument dictates that we ought to prioritise those who have not yet had a 'fair innings' when allocating lifesaving resources. It is argued that this allocation scheme adheres to the demands of distributive justice in healthcare in that it aims to equalise people's length of life (Giordano 2005).

Daniels' (1988) starting point is the simple fact that we all age. In view of this, he urges us to conceive of the problem of justice between age groups (between the old and the young) as one of prudently allocating resources across the different stages of our lives. A prudent allocation is one in which the early stages of our lives receive more resources than the later ones as this maximises the probability of reaching an advanced age.

Proposals for rationing healthcare on the basis of age have sparked fierce criticism. A frequently cited criticism is that the elderly have an equal right to life as the nonelderly (Fleck 2010). Another type of criticism pertains to the efficacy of the proposed measures. Levinsky (1990), for instance, argues that a shift from cure to care for elderly patients would do little to curb healthcare expenditures.

Another type of cost containment proposal consists in reforming Medicare in the US. The first cries for reform can be traced back to the early years after Medicare's enactment in 1965, and ultimately resulted in a partial privatisation of the programme. Since the 1990s, attempts to further privatise Medicare have been ongoing (Geyman 2004). These reform proposals distinguish themselves from the earlier ones in that they are framed as a much needed answer to the challenges of population aging. The underlying idea is that competition between private insurers will reduce healthcare costs (Wiener and Tilly 2002). Congressman Paul Ryan has recently proposed that Medicare move toward a system wherein the government gives seniors a fixed payment to purchase a private plan of their choice (Cannon 2011). Critics fear that such a system will burden seniors with high out-of-pocket expenditures, rendering many of them unable to receive the needed care. In addition, critics point to existing data which suggest that, in the area of cost containment, private plans perform worse than traditional Medicare (Geyman 2004).

Proposals pertaining to age-based rationing and Medicare reform have been around for quite some time. A more recent proposal (see, for example, Micans 2005; Dorshkind et al. 2009; Olshansky et al. 2006) is to invest more in biogerontology—research into the biology of aging. The idea is that such research will enable us to tackle age-related diseases simultaneously, thereby ensuring that the elderly enjoy an increased healthspan (i.e. that they enjoy an increase in the number of years spent in a disease-free state). This, in turn, it is believed, will reduce the pressure on the healthcare system. To date, this argument has received no attention, which is surprising given the highly recognised need for cost containment in healthcare. The aim of this chapter is to evaluate this argument, which we will refer to as 'the cost containment argument', by critically examining its most fundamental presuppositions.

19.2 The Biogerontological Approach

Research aimed at tackling age-related diseases is primarily focused on developing methods for treating or preventing these pathologies individually. This approach, however, has only very limited potential for prolonging the healthspan of the elderly. The incidence of most age-related diseases increases exponentially during the last stage of life so that comorbidity is an inescapable fact for many elderly (Butler et al. 2008). Consequently, even if we succeeded in eradicating any one of the major age-related diseases, its place would immediately be taken by yet another.

It is only by tackling age-related diseases simultaneously that one is guaranteed a substantial impact on the overall length of healthy life. This approach amounts to intervening in the aging process as aging is the common, underlying cause of all age-related diseases. Depending on the extent to which the healthspan is prolonged, significant increases in either average life expectancy or maximum lifespan are expected (Vincent et al. 2008). The endeavour of intervening in aging is, therefore, commonly referred to as 'lifespan extension'.[1] Given the combined benefit of increased healthspan and lifespan, many biogerontologists deplore that less than 0.1 % of the NIH budget goes to anti-aging research (Olshansky et al. 2006).

Biogerontology was long viewed as a fringe science (Fishman et al. 2008). Recent developments within the field have caused it to gain scientific legitimacy.

A large part of the research efforts have been devoted to caloric restriction—an experimental setting wherein caloric intake is reduced to about 40 % below ad libitum levels. Contrary to malnutrition, the intake of important nutrients, such as vitamins and minerals, is still guaranteed. Studies on laboratory animals demonstrate that caloric restriction results in a 30–50 % increase in lifespan (Hackler 2004). Age-related diseases are postponed and their incidence is reduced (Ingram et al. 2004).

Most humans would probably have difficulty adhering to such a drastic dietary regimen. This recognition has initiated the search for substances that are able to mimic the effects of caloric restriction in the absence of a reduced caloric intake. Resveratrol, a chemical found in the skin of red grapes, offers promising prospects in this respect (Baur 2010).

Researchers are also investigating the role of gene mutations in longevity. Human centenarians and supercentenarians constitute an interesting object of scientific study in that they epitomise successful aging. Not only have they reached an advanced age, they have escaped the major age-related diseases on the way (Franceschi et al. 1995). Several gene mutations that appear to contribute to healthy aging have already been identified (Gonos 2000; Atzmon et al. 2008).

There is much disagreement concerning the expected outcomes of biogerontological research. The different opinions on this matter can be translated into four 'life extension scenarios': prolonged senescence, compression of morbidity, decelerated aging, and arrested aging.

[1] We use the terms 'lifespan extension' and 'anti-aging' interchangeably throughout this chapter.

Prolonged senescence amounts to the failure of the anti-aging research enterprise: life itself is prolonged, while the healthspan is not. For instance, decrepitude would start to set in at the age of 55 and one would die at the age of about 95 (Derkx 2009).

The prospect of a prolonged senescence is sometimes invoked as an argument against anti-aging research. However, this argument is not compelling. As previously noted, most age-related diseases show an exponential increase from a certain age onward. Each age-related disease increases our risk of death. It is, thus, unlikely that we could live to experience a period of decrepitude of the length envisaged by the prolonged senescence scenario (de Grey 2005).

Those who anticipate a compression of morbidity believe that interventions in the aging process would have but a marginal effect on the length of our lives. Individuals would live long, healthy lives and then die rather quickly after experiencing a negligible period of decrepitude (Juengst et al. 2003).

The fact that the absence of age-related diseases is extremely beneficial in terms of mortality risk explains the implausibility of the compressed morbidity scenario. The considerably extended healthspan, envisaged under this scenario, amounts to a substantial postponement of the onset of age-related diseases. It is, therefore, highly unlikely that no similarly meaningful extension of life would occur (Gems 2009; de Grey 2006).

Proponents of decelerated aging argue that we can postpone age-related diseases to such an extent that both average life expectancy and maximum lifespan are increased. Under this scenario, 90-year-olds would, for example, enjoy the health and vigour of today's 50-year-olds (Juengst et al. 2003). Miller (2002) envisages an average life expectancy of around 112 and a maximum lifespan of around 140.

We see no reason to question 'decelerated aging' as an outcome of anti-aging research. Most biogerontologists seem to share this view (Gems 2009).

Arrested aging undoubtedly constitutes the most radical scenario. It is tantamount to achieving total mastery of the aging process in that its harmful effects would be prevented. This approach involves the continuous repair of the molecular and cellular damage responsible for the onset of age-related frailty (de Grey et al. 2002). The aim is to repair the damage before it reaches a level at which it induces age-related pathologies. As these pathologies would no longer occur, a state of 'virtual immortality' would be attained in the sense that death would only result from accidents, suicide, wars, and so forth (Binstock 2004).

Whether or not the scenario of 'arrested aging' is plausible, depends upon the feasibility of the SENS-project (Strategies for Engineered Negligible Senescence), the only proposed project for achieving this state of virtual immortality. Aubrey de Grey, the man behind this project, has identified several forms of damage responsible for age-related pathologies and degeneration. For each type of damage, he has formulated a strategy targeted towards its repair (de Grey et al. 2002).

The SENS-proposal lacks persuasiveness for several reasons. Firstly, each of the therapies proposed by de Grey is unlikely to be realised any time soon. This renders the prospect of all of the proposed strategies being implemented very remote (Warner et al. 2005). Secondly, even if we were able to accomplish this feat, there is no guarantee that we would hereby have arrested the aging process. There appear to be other types of damage, besides those identified by de Grey, contributing to age-related decline (Estep III et al. 2006). Moreover, still other important forms of age-related damage could well be discovered in the future. In sum, we would probably end up having to repair an insurmountable amount of damage in order to arrest the aging process.

19.3 Presuppositions Underlying the Cost Containment Argument

The authors who advance the cost containment argument rely on a number of presuppositions. We will scrutinise the four main assumptions.

19.3.1 Life Extension will Decrease the Frailspan

Proponents of the cost containment argument support their reasoning by reference to the ability of anti-aging interventions to prolong healthspan. However, what is required for financial gains to be conceivable is not so much increases in healthspan as absolute reductions in frailspan (i.e. the period of age-related frailty). Thus, although most of them do not explicitly state this, all proponents of the cost containment argument must presuppose that any increase in healthspan will be accompanied by a decreased frailspan.

Most proponents of the cost containment argument seem to think that various life extending scenarios are plausible. Holliday, for example, advocates "measures to prevent or delay the onset of these [age-associated] diseases" (Holliday 1996, p. 90). Along the same lines, Micans speaks of the possibility to "slow or prevent the signs of aging from occurring" (Micans 2005, p. 550). None of them rule out the possibility of decelerated aging. The latter is, as we have argued, the most plausible scenario. Thus, we need to analyse its implications for the frailspan in order to assess the above presupposition.

It is often thought that decelerated aging will be accompanied by a curtailed frailspan. This prediction is based on the observed reduction in frailspan in rodents whose aging process has been slowed down through caloric restriction. Extrapolations of this kind are, however, unwarranted (Gems 2011). Moreover, even if extrapolations from rodents to humans were somehow justified, there would be little point in employing studies on caloric restriction as a reference point. For reasons previously cited, most humans are unlikely to engage in this dietary regimen as a method for decelerating aging.

We are currently unable to decelerate the human aging process. Rather than enticing us into making uneducated guesses, this fact should encourage us to refrain from any judgment concerning the effect of decelerated aging on the human frailspan. Thus, contrary to what proponents of the cost containment argument presuppose, we cannot exclude the possibility of the frailspan retaining its current length or even increasing in length.

One might argue that a curtailed frailspan is not required in order for healthcare savings to occur. Harris (2004, Chap. 14, this volume), for instance, attempts to show that life extension, even when accompanied by an increased frailspan or a frailspan of the current length, still makes economic good sense. His argument relies on economic discounting, a technique used to determine the present value of a financial cost that will be incurred at some point in the future. By enabling us to translate future costs into their present value, economic discounting provides us with a sound way for comparing costs incurred at different moments in time. Economic discounting is not to be confused with an adjustment for inflation. Future costs need to be discounted in order to account for the time value of money.

Harris reasons that we gain financially from life extension because the latter amounts to postponing the moment in time when we start incurring frailty-related healthcare costs. He uses the example of a newborn. Under present circumstances, this newborn will reach the period of frailty and its associated costs at around 70. In Harris' example, the newborn will reach this period only after 1000 years[2] in the case of life extension. According to Harris, the present-day discounted cost of treating that person in 70 years will be substantially higher than the present-day cost of treating that same person in 1000 years. There is, however, no reason why this should necessarily be the case. Harris does not seem to take into account that, over time, healthcare costs can increase considerably in real terms. Thus, healthcare costs could, between year 70 and year 1000, increase to such an extent in real terms that the discounted cost of treating that person in 1000 years is higher than that of treating the same person in 70 years. In sum, there is no guarantee that an increased frailspan or even a frailspan of the current length would be financially beneficial. Proponents of the cost containment argument must, therefore, presuppose the occurrence of a reduction in frailspan.

19.3.2 Life Extension will Enjoy a Considerable uptake Rate

Let us accept for the sake of argument that the deceleration of the aging process will be accompanied by a decrease in frailspan. The reduction in healthcare costs, envisaged by proponents of the cost containment argument, is substantial. The prospect of considerable savings presupposes a sizeable amount of people using life extending, anti-aging technologies. However, this is, as we argue below, a problematic presupposition.

The little available research concerning community attitudes towards life extending technologies points towards a rather low uptake rate. In a recent study (Partridge et al. 2011), for instance, only 35 % of the respondents answered affirmatively when

[2] Although we have previously argued that a lifespan of 1000 years is highly unrealistic, we have chosen to stick to Harris' example in order to offer an accurate representation of his argument.

asked whether they would use a life extending technology. Another study (Underwood et al. 2009) found just over half of the participants willing to consider lifespan extension. Despite appearances to the contrary, these outcomes need not be more reassuring than the ones from the former study. After all, whether or not these respondents would actually opt for lifespan extension, as opposed to merely considering it, depended on a number of conditions being fulfilled. Some participants stated that they would only use life-extending technologies if their loved ones were to do so. Others referred to the absence of any negative impact on society as a prerequisite. Thus, if life extension technologies become available, far fewer than 50 % of these respondents would actually end up using them.

Interesting results have also emerged from a study conducted by Lang et al. (2007). Participants were randomly assigned to one of 3 conditions. Whereas those in the first condition were informed that research on aging offered hopeful prospects with respect to the physical, mental, and psychological fitness in old age, participants in the second condition were told the opposite. The control group received no information concerning research on aging. Participants in each group were asked which age they would like to reach. Surprisingly, the answers did not vary significantly across the 3 conditions. In each of the 3 conditions, the average desired lifetime was approximately 86 years—well below the current maximum lifespan. Fewer than 10 % of the respondents wanted to live to 120 or beyond.

There is another reason why it is problematic to presuppose a substantial uptake rate of life extending technologies. These technologies will, like any other new medical technology, be very expensive. Thus, very few people will have access to them. In fact, relative to many other types of new technology, the cost of lifespan extension can be expected to be of an even higher order as it will most probably involve higher research and development costs. Firstly, human aging is a highly complex biological process, which suggests that any intervention designed to tackle it would need to be equally complex. Secondly, many aging mechanisms seem to act over the entire lifespan. The effectiveness of anti-aging interventions would, therefore, probably be inversely related to the age at which they are begun (Hadley et al. 2005). The early administration age, combined with the (current) lack of valid biomarkers of aging, suggests that clinical trials would probably span the entire lifetime of the enrolled subjects (Sprott 2010). Proponents of the cost containment argument might respond that the cost of life extending technology will diminish after a while. Although this is likely to happen, this fact does not necessarily do much to further their cause. Given the complexity of aging, one will most likely have to undergo numerous, different types of interventions in order to achieve the desired effect. For example, a combination of stem cell treatments, pharmaceuticals and genetic consultations could be required (Ehni and Marckmann 2009, cf Chap. 17 this volume). Thus, even if each of the needed interventions became cheaper over time, the 'whole package' would probably still not be affordable for a significant part of the population.

Another response might be that public coverage of the needed interventions will be provided in order to guarantee wide access to life extension. Mackey (2003), for instance, puts forward this argument. This line of reasoning is problematic.

Various considerations are involved in deciding whether or not a drug or intervention qualifies for public coverage. The financial cost of the drug/intervention is obviously an important consideration. As noted above, life extension involves the application of various, very expensive interventions. Thus, public coverage might not be feasible in budgetary terms. However, even if budgetary feasibility were not an issue, there would, from a purely financial perspective, probably be little incentive to provide public coverage. After all, as we will argue further on, life extension is likely to be more expensive than the current approach of treating or preventing age-related diseases individually.

Another important consideration is the extent to which the drug/intervention is medically necessary. In the case of life extension, the question of medical necessity tends to be framed in terms of whether or not aging is a disease (Caplan 2005). The latter issue is currently highly debated (Butler et al. 2004). The controversy surrounding this issue makes it difficult to predict the final outcome of the debate.

The level of public support for coverage of a drug/intervention is also taken into account in coverage decisions. The widespread reservations about using life extension among the public will need to subside for it to score well on this criterion. Once again, it is difficult to predict the chances of this happening.

We have discussed only a few of the important criteria involved in coverage decisions. Nevertheless, our discussion suffices to show that it is premature to posit public coverage of life extending interventions as a solution to the problem of access.

19.3.3 Population Aging is an Important Driver of Rising Healthcare Costs

Even if we accept, for the sake of the argument, the correctness of the two previous presuppositions, then the cost containment argument still lacks persuasiveness. After all, it encompasses a questionable presupposition concerning the problem in response to which anti-aging interventions are put forward. The argument has as its starting point the claim that population aging will cause healthcare expenditures to rise to an unsustainable level. Thus, it presupposes that the aging of the population is an important, if not the most important, contributor to rising cost pressures in the healthcare sector.

Past spending trends, however, suggest that population aging, by itself, has been only a minor driver of the annual growth in healthcare expenditures. For example, analysis of healthcare expenditure in British Columbia between 1975 and 2005 indicates that population aging increased health spending by only 0.7 % per year (Lee 2006). British Columbia is an interesting object of study given that it has a higher than average proportion of elderly relative to other provinces. In Australia, population aging has been responsible for only 10 % of increases in federal government healthcare costs over the last decade (Coory 2004). Moreover, spending patterns between 1960 and 1990 across OECD countries show that there is no relationship between population aging and increases in healthcare costs (Marmor and Oberlander 1998).

Analysis of past spending patterns is only meaningful to a certain extent. After all, population aging has yet to reach its peak, which is expected to occur around 2031 (Lee 2006). However, projections suggest that population aging will remain a minor driver of increases in health spending during this period. Richardson and Robertson (1999) present projections for Australia for 1995–2051. Regarding the effect of population aging on health expenditure, they conclude: "if aging were the only source of expenditure growth the relative size of the health sector would significantly decline as GDP would be expected to rise more rapidly than health expenditures" (p. 14). Data for other developed countries also point towards a small effect of population aging on future increases in health spending (Coory 2004).

Despite these data, the tendency to overemphasise the role of population aging in the growth of healthcare expenditures is widespread, among researchers and laymen alike. The claim that population aging is an important driver of healthcare costs is generally based on the observed positive relationship between age and average healthcare expenditure (Zweifel et al. 2004). This inference, however, overlooks the fact that this positive relationship, in part, reflects the high costs of the last year of life and the high mortality in old age. In other words, the age-related increase in healthcare costs is not only a function of age per se, but also of proximity to death (Wickstrøm et al. 2002). Thus, contrary to what is often the case, one must take the high cost of dying into account if one is to accurately predict the effect of aging on healthcare expenditures.

In exaggerating the role of population aging in (future) healthcare cost increases, proponents of the cost containment argument overestimate the cost saving potential of life extending interventions. The problems facing the cost containment argument, however, potentially go much deeper. In the following, we argue that life extending interventions not only save less than proponents of this argument claim, but could actually increase healthcare expenditures.

Medical technology is the prime determinant of the increase in healthcare costs. New technologies and the intensified use of old ones are responsible for about 50 % of the annual growth in health spending (Callahan 2009). Thus, it seems that life extension, by relying on technology, would be part of the problem, rather than the solution, when it comes to keeping healthcare costs in check. Proponents of the cost containment argument might respond that life extension distinguishes itself from the average new technology in that it would save more resources than it costs. In other words, they might claim that the savings achieved through the reduction of the frailspan would outweigh the costs of the technology needed to bring about this reduction. However, the expected characteristics of life extension render this claim dubious. Life extension will, most probably involve various types of new technologies being periodically applied from an early age until the final stages of one's significantly extended lifespan.

There is yet another way in which life extension would contribute to an intensified use of medical technology. As noted above, we are here assuming that life extension will both increase the healthspan and reduce the frailspan. Obviously, a person incurs many different types of medical costs other than those related to old age. Thus, an increase in healthspan amounts to an increase in the number of years during which such other medical costs are incurred. A part of these 'extra' medical costs will inevitably be related to the use of medical technology. In sum, life extension is likely to increase healthcare costs as both the increase in healthspan and the reduction in frailspan imply an intensified use of medical technology.

19.3.4 Any Negative Effects of Life Extension are Outweighed by the Achieved Healthcare Savings (and other Perceived Benefits)

Given the problematic nature of the above presuppositions, we currently have no reason to believe that anti-aging interventions constitute an effective means of containing healthcare costs. However, let us suppose for a moment that such interventions do have a (substantial) cost containing potential. If this potential is to be a compelling reason to increase funding for aging research, one must presuppose that these healthcare savings outweigh any negative effects of anti-aging interventions.

It is obviously extremely difficult to predict how anti-aging will affect our lives. However, as we are here assuming a considerable uptake rate of life extending interventions, we can reasonably expect a substantial population increase to occur. Projections of the US Census Bureau illustrate how profound the effects of increased longevity can be:

"Each 10-year prolongation of life expectancy will increase the eventual population of Earth at stability by 1.3 billion persons [...]. If world longevity follows the patterns that will be achieved first in the more developed countries and reaches 115 years, 5 decades longer than the current worldwide longevity, that would mean a further increase of 6.5 billion persons. Instead of the current estimated final population at stability of about 10 billion persons, there would be almost three people for every one now living worldwide" (Louria 2005, p. 317).

Marked population increases would have several detrimental effects. Biodiversity loss, deforestation, global warming, and depletion of resources (energy resources, food, water, and open space) are just a few of the expected problems.

The severity of each of the above problems is undeniable. It is, therefore, not obvious that the positive effect of cost containment outweighs these negative effects. Proponents of the cost containment argument must argue why this is so. They can follow one of two strategies in making this argument.

The first strategy consists in showing that the probability of overpopulation occurring is negligible. In this case, the obvious argument is that societies which adopt lifespan extension will most probably restrict the number of offspring people are allowed to have (Bostrom and Roache 2008). Such a policy is, however, problematic for several reasons. To begin with, ensuring compliance with any population control programme will prove challenging. For example, fining people in case of non-compliance is likely to have not much of a deterrent effect. The very poor will rely on the fact that they are unable to pay the fine, while the very rich will gladly pay it. It seems that only the use of unethical means (e.g. forced sterilisation) guarantees compliance. Furthermore, the question arises whether people who do not opt for lifespan extension should also be subjected to reproductive restrictions. If so, one would probably have a hard time justifying this. Finally, the introduction of reproductive restrictions could, by further lowering fertility rates, induce a further aging of the population. China's one-child-policy, having contributed to the dramatic aging of its population (Zhang and Goza 2006), is illustrative in this respect. Population aging could prove to be equally challenging in a world of extended lifespans, even if we assume a reduction in frailspan. The sustainability of pension systems, for instance, might still be an issue as we cannot simply assume that people will be willing to work longer.

A second strategy is to acknowledge the occurrence of overpopulation, while arguing that its negative effects can be remediated. For example, one might, following Mackey (2003), claim that societies facing food shortages would find methods for genetically engineering more nutritionally efficient food. However, it remains to be seen whether such methods will actually be developed. Nevertheless, even if we could rely on remedies being developed for some problems, there would be little reason for optimism. After all, other problems, such as the loss of biodiversity, are amenable at the most to mitigation, not remediation.

Given the problems with both of the outlined strategies, proponents of the cost containment argument will most probably be unable to successfully argue that healthcare savings outweigh the identified negative effects.

Most proponents of the cost containment argument clearly posit the cost containing potential of anti-aging interventions as a sufficient reason for investing more in aging research. Other proponents (Farrelly 2008; Butler et al. 2008), however, mention several other benefits[3] attached to anti-aging interventions, besides their cost containing potential. One of the additional perceived benefits, for example, is that longer lives contribute to a substantial growth of the national economy. The inclusion of these additional benefits could imply that, for this group of proponents, the cost containing potential of anti-aging interventions in conjunction with these other benefits constitute a sufficient reason for investing more in aging research. If this is so, they must presuppose that all of these benefits together outweigh the bad effects of overpopulation. Once again, however, an argument will need to be put forward in support of this presupposition—an arduous task, to say the least.

19.4 Concluding Remarks

Healthcare costs are rising at an unsustainable rate. Proposed measures for tackling this problem include age-based rationing and Medicare reform. A more recent proposal is to rely on life extension as a means of containing healthcare costs. We have identified four presuppositions underlying this cost containment argument. Each of

[3] The term 'longevity dividend' is generally used to refer to these other benefits as well as to the expected benefit of healthcare savings.

these presuppositions is problematic. They raise serious questions concerning both the morality (the last presupposition) and the efficacy (the other presuppositions) of life extension as a cost containment measure. Thus, life extension fares no better than 'older proposals'. The failure of all these proposals is mainly due to their misconstruing the problem of rising healthcare costs as one rooted in population aging. Society's heavy reliance on medical technology is the main driver of healthcare cost growth. Cost containment policy should, therefore, redirect its focus away from population aging towards medical technology.

There are many challenges involved in achieving a more responsible use of medical technology. A first challenge consists in specifying what constitutes the appropriate attitude towards medical technology. This attitude does not imply putting a stop to all technological innovation. Neither does it imply a reluctance to say 'no' to efficacious but overly expensive technologies. In sum, society stands for the difficult task of striking a balance between both extremes; between allowing too little and too much.

A second challenge consists in overcoming the widespread opposition to the proposed shift towards a more limited use of medical technology. We can expect strong resistance from the many industries involved in the production and distribution of medical technologies. However, the public at large will also be reluctant to embrace the required changes. In fact, relative to age-based rationing and Medicare privatisation proposals, our proposal will likely elicit even more public criticism. The former proposals 'merely' jeopardise the interests of the elderly population. However, our proposal jeopardises an interest shared by everyone as both young and old can benefit from medical technology.

A final challenge relates to the deeply ingrained nature of the attitude which needs turning around. Society's attachment to medical technology has its roots in the Enlightenment idea of infinite progress. As such, it is part of our cultural heritage. It will prove difficult to change such a deeply rooted mindset.

Despite the many challenges involved, we will need to find a way of putting the issue of medical technology at the top of the agenda. It is the only way out of the problem.

References

Atzmon, G., T.I. Pollin, J. Crandall, K. Tanner, C.B. Schechter, P.E. Scherer et al. 2008. Adiponectin levels and genotype: A potential regulator of lifespan in humans. *The Journals of Gerontology, Series A: Biological Sciences and Medical Sciences* 63(5): 447–453.

Baur, J.A. 2010. Resveratrol, sirtuins, and the promise of a DR mimetic. *Mechanisms of Aging and Development* 131(4): 261–269.

Beard, R.L., and J.B. Williamson. 2010. Social policy and the internal dynamics of the senior rights movement. *Journal of Aging Studies* 25(1): 22–33.

Binstock, R.H. 2004. Anti-aging medicine and research: A realm of conflict and profound societal implications. *The Journals of Gerontology Series A: Biological Sciences and Medical Sciences* 59(6): B523–533.

Bostrom, N., and R. Roache. 2008. Ethical Issues in Human Enhancement. In *New Waves in Applied Ethics*, eds. T. Petersen, J. Ryberg, and C. Wolf 120–152. Hampshire, UK: Palgrave Macmillan.

Butler, R.N., R. Sprott, H. Warner, J. Bland, R. Feuers, M. Forster et al. 2004. Aging: The reality. Biomarkers of aging: From primitive organisms to humans. *The Journals of Gerontology, Series A: Biological Sciences and Medical Sciences* 59(6): B560–567.

Butler, R.N., R.A. Miller, D. Perry, B.A. Carnes, T.F. Williams, C. Cassel et al. 2008. New model of health promotion and disease prevention for the 21st century. *British Medical Journal* 337(7662): 149–150.

Callahan, D. 1987. *Setting Limits. Medical goals in an aging society.* Washington, DC: Georgetown University Press.

Callahan, D. 2009. Taming the beloved beast: How medical technology costs are destroying our healthcare system. Princeton: Princeton University Press.

Callahan, D., and K. Prager. 2008. Medical care for the elderly: Should limits be set? Virtual Mentor. *American Medical Association Journal of Ethics* 10(6): 404–410.

Cannon, M. 2011. Ryan budget: A huge opportunity to improve healthcare. Kaiser health news. http://www.kaiserhealthnews.org/Columns/2011/April/040411cannon.aspx. Accessed 10 May 2011.

Caplan, A.L. 2005. Death as an unnatural process. Why is it wrong to seek a cure for aging? *EMBO Reports* 6(special issue 1): S72–75.

Coory, M.D. 2004. Aging and healthcare costs in Australia: A case of policy-based evidence? *Medical Journal of Australia* 180(11): 581–583.

Daniels, N. 1988. Am I my parents' keeper? An essay on justice between the young and the old. New York: Oxford University Press.

de Grey, A.D. 2005. Life extension, human rights, and the rational refinement of repugnance. *Journal of Medical Ethics* 31(11): 659–663.

de Grey, A. D. 2006. Compression of morbidity: The hype and the reality, Part 1. *Rejuvenation Research* 9(1): 1–2.

de Grey, A.D., B.N. Ames, J.K. Andersen, A. Bartke, J. Campisi, C.B. Heward et al. 2002. Time to talk SENS: Critiquing the immutability of human aging. *Annals of the New York Academy of Sciences* 959: 452–462.

Derkx, P. 2009. Life Extension and Meanings of Life. In *Technology, Trust, and Religion: Roles of Religions in Controversies on Ecology and the Modification of Life*, ed. W. Drees 197–220. Amsterdam: Leiden University Press.

Dorshkind, K., E. Montecino-Rodriguez., and R.A Signer. 2009. The aging immune system: Is it ever too old to become young again? *Nature Reviews Immunology* 9(1): 57–62.

Ehni, H.J., and G. Marckmann. 2009. Social justice, health inequities, and Access to new age-related interventions. *Medicine Studies* 1(3): 281–295.

Estep III, P., M. Kaeberlein, P. Kapahi, B. Kennedy, G. Lithgow, G. Martin, S. Melov et al. 2006. Life extension pseudoscience and the SENS Plan. *Technology Review* (MIT). http://www.technologyreview.com/sens/docs/estepetal.pdf. Accessed 05 May 2011.

Farrelly, C. 2008. Has the time come to take on time itself? *British Medical Journal* 337(7662): 147–148.

Fishman, J.R., R.H. Binstock., and M.A. Lambrix. 2008. Anti-aging science: The emergence, maintenance, and enhancement of a discipline. *Journal of Aging Studies* 22(4): 295–303.

Fleck, L.M. 2010. Just caring: In defense of limited age-based healthcare rationing. *Cambridge Quarterly of Healthcare Ethics* 19(1): 27–37.

Franceschi, C., D. Monti, P. Sansoni., and A Cossarizza. 1995. The immunology of exceptional individuals: the lesson of centenarians. *Immunology Today* 16(1): 12–16.

Gems, D. 2009. Eine revolution des alterns. Die neue biogerontologie und ihre implikationen. In *Länger leben? Philosophische und biowissenschaftliche Perspektiven*, eds. S. Knell, and M. Weber 25–45. Frankfurt am Main: Surhkamp Verlag.

Gems, D. 2011. Tragedy and delight: The ethics of decelerated aging. Philosophical transactions of the royal society B: *Biological Sciences* 366(1561): 108–112.

Geyman, J.P. 2004. Privatization of medicare: Toward disentitlement and betrayal of a social contract. *International Journal of Health Services* 34(4): 573–594.

Giordano, S. 2005. Respect for equality and the treatment of the elderly: Declarations of human rights and age-based rationing. *Cambridge Quarterly of Healthcare Ethics* 14(1): 83–92.

Gonos, E.S. 2000. Genetics of aging: Lessons from centenarians. *Experimental Gerontology* 35(1): 15–21.

Hackler, C. 2004. Extending the lifespan: Mythic desires and modern dangers. *HEC Forum* 16(3): 182–196.

Hadley, E.C., E.G. Lakatta, M. Morrison-Bogorad, H.R. Warner., and R.J. Hodes. 2005. The future of aging therapies. *Cell* 120(4): 557–567.

Harris, J. 1985. The value of life: An introduction to medical ethics. London: Routledge and Kegan Paul.

Harris, J. 2004. Immortal ethics. *Annals of the New York Academy of Sciences* 1019: 527–534.

Holliday, R. 1996. The urgency of research on aging. *BioEssays* 18(2): 89–90.

Ingram, D.K., R.M. Anson, R. de Cabo, J. Mamczarz, M. Zhu, J. Mattison et al. 2004. Development of calorie restriction mimetics as a prolongevity strategy. *Annals of the New York Academy of Sciences* 1019: 412–423.

Juengst, E.T., R.H. Binstock, M. Mehlman, S.G. Post., and P. Whitehouse. 2003. Biogerontology, "anti-aging medicine," and the challenges of human enhancement. *Hastings Center Report* 33(4): 21–30.

Kinsella, K., and D. Phillips. 2005. Global aging: The challenge of success. Population Reference Bureau. http://prb.org/pdf05/60.1GlobalAging.pdf. Accessed 14 Apr 2011.

Lang, F.R., P.B. Baltes., and G.G. Wagner. 2007. Desired lifetime and end-of-life desires across adulthood from 20–90: A dual-source information model. *The Journals of Gerontology, Series B: Psychological Sciences and Social Sciences* 62(5): 268–276.

Lee, M. 2006. Is BC's healthcare system sustainable? A closer look at the costs of aging and technology. Canadian Centre for Policy Alternatives. http://www.policyalternatives.ca/sites/default/files/uploads/publications/BC_Office_Pubs/bc_2006/aging_healthcare.pdf. Accessed 03 June 2011.

Levinsky, N.G. 1990. Age as a criterion for rationing healthcare. *The New England Journal of Medicine* 322(25): 1813–1816.

Louria, D.B. 2005. Extraordinary longevity: individual and societal issues. *Journal of the American Geriatrics Society* 53(9 Suppl): S317–319.

Mackey, T. 2003. An ethical assessment of anti-aging medicine. *Journal of Anti-Aging Medicine* 6(3): 187–204.

Marmor, T., and J. Oberlander. 1998. Rethinking medicare reform. *Health Affairs* 17(1): 52–68.

Micans, P. 2005. The need for anti-aging medicine: The challenges faced to incorporate preventative medicine into the clinic and into society. *Annals of the New York Academy of Sciences* 1057(1): 545–562.

Miller, R.A. 2002. Extending life: scientific prospects and political obstacles. *Milbank Quarterly* 80(1): 155–174.

Olshansky, S.J., D. Perry, R.A. Miller., and R.N. Butler. 2006. In Pursuit of the longevity dividend: What should we be doing to prepare for the unprecedented aging of humanity? *The Scientist* 20(3): 28–33.

Partridge, B., J. Lucke, H. Bartlett., and W. Hall. 2011. Public attitudes towards human life extension by intervening in aging. *Journal of Aging Studies* 25(2): 73–83.

Richardson, J., and I. Robertson. 1999. Aging and the cost of health services. Working Paper 90. Centre for Health Program Evaluation.http://www.buseco.monash.edu.au/centres/che/pubs/wp90.pdf. Accessed 15 June 2011.

Sprott, R.L. 2010. Biomarkers of aging and disease: Introduction and definitions. *Experimental Gerontology* 45(1): 2–4.

Underwood, M., H.P. Bartlett, B. Partridge, J. Lucke., and W.D. Hall. 2009. Community perceptions on the significant extension of life: An exploratory study among urban adults in Brisbane, Australia. *Social Science & Medicine* 68(3): 496–503.

Vincent, J.A., E. Tulle., and J. Bond. 2008. The anti-aging enterprise: Science, knowledge, expertise, rhetoric and values. *Journal of Aging Studies* 22(4): 291–294.

Warner, H., J. Anderson, S. Austad, E. Bergamini, D. Bredesen, R. Butler et al. 2005. Science fact and the SENS agenda. *EMBO reports* 6(11): 1006–1008.

Wickstrøm, J., N. Serup-Hansen., and I.S. Kristiansen. 2002. Future healthcare costs–do healthcare costs during the last year of life matter? *Health Policy* 62(2): 161–172.

Wiener, J.M., and J. Tilly. 2002. Population aging in the United States of America: Implications for public programmes. *International Journal of Epidemiology* 31(4): 776–781.

Zhang, Y., and F. Goza. 2006. Who will care for the elderly in China? A review of the problems caused by China's one child policy and their potential solutions. *Journal of Aging Studies* 20(2): 151–164.

Zweifel, P., S. Felder., and A. Werblow. 2004. Population aging and healthcare expenditure: New evidence on the "red herring". *Geneva Papers on Risk and Insurance—Issues and Practice* 29(4): 652–666.

Chapter 20
Aging Under the Looking-Glass

Policy Implications of Views Expressed in this Volume

Göran Hermerén

20.1 Introduction

"Age is an issue of mind over matter. If you don't mind, it doesn't matter," according to Mark Twain. Individuals can do something about this by changing their attitudes to aging. But what can the state do? To indicate the precise policy implications of all the contributions in this book, and to do this well, is a formidable task. There are several reasons for this.

First of all, this requires a very careful reading of all the chapters of this book. Moreover, it would require knowledge of the regulatory landscape of the world of a sort that few people possess. Not only would the policy implications differ in various parts of the world, but also within Europe, due to our different historical, political and religious traditions, as well as to the changing economic and technological development of the member states.

Furthermore, there are still knowledge gaps and uncertainties; the evidence available is sometimes contested or unclear. This holds, for instance, for the prospects of cures of age-related ailments and changes in appearances, in particular therapies based on developments in stem cell research and biogerontology research. Also the interests within and between many stakeholder groups vary. The variations of wishes among the elderly should warn us against treating them as a homogeneous group. Clear policy recommendations cannot always be expected.

Finally, there are also differences between some of the contributors concerning the values to be promoted and protected. Different ethical frameworks, such as the utilitarian one of John Harris and the more dignitarian or equal-value inspired one of Anders Schinkel, will not always support the same practical conclusions. Moreover, there are, as is well known, important value differences within and between societies in Europe, concerning e g the moral status of the embryo, the use of animals for research, the place of women in society, concerning collective and individual rights,

Göran Hermerén (✉)
Lund University, Sweden
e-mail: Goran.Hermeren@med.lu.se

M. Schermer, W. Pinxten (eds.), *Ethics, Health Policy and (Anti-) Aging: Mixed Blessings*, Ethics and Health Policy 1, DOI 10.1007/978-94-007-3870-6_20,

etc. Successful policy recommendations require agreement concerning relevant basic values.

What I propose to do is (a) to develop a model, which can be used to identify policy implications, and (b) to illustrate these implications with examples drawn from the contributions to the present book. This model will hopefully be of general interest, since it should be applicable also to other areas of concern than the ethics of aging populations.[1] I will thus reflect on the relevance of the ethical issues that are discussed in the volume for public health policy, and I will use the model to demonstrate the relevance of such issues for public policy.

When is a public health policy called for? The following conditions may be suggested:

1. *Threat.* There is a threat, and this threat concerns the health of a subset of the population like the elderly.
2. *Seriousness.* The threat is serious; and the more serious it is, the more urgent is some action (prevention, diagnosis, cure).
3. *Number.* Many people are potentially affected by this threat to their health; and the larger this group is; the more urgent is some action.
4. *Efficacy.* Something can be done to eliminate, reduce or circumvent this threat.
5. *Capability.* But it is beyond the capability or capacity of many or most individuals in the subset to deal effectively on their own with the threat.
6. *Proportionality.* The means used to deal with the threat are adequate/ reasonable in relation to the goal desired, or alternatively and more generally:
7. *Morality.* The means that can be used to deal with the threat are morally acceptable.

The avian influenza and other pandemics provide obvious examples of public health threats. Anti-smoking campaigns satisfy conditions of the sort mentioned above, whereas prostate cancer screening is controversial, as is treatment of moderate or high levels of cholesterol with lipids. But female circumcision does not satisfy these conditions. Anyway, if the conditions above are used as criteria, it seems evident that many of the ethical issues and suggestions discussed in this book provide a relevant basis for health policy recommendations.

Most of these conditions above admit of degrees: more or less people can be affected, and the threat can be more or less serious. Besides, many key terms, like 'health', can be interpreted in several ways (Boorse 1977; Nordenfelt 1995, 2007). This also holds for proportionality (Hermerén 2011). Threats can be classified in several ways. So a family of conditions can replace each condition. But taken together they indicate roughly when a public health policy is called for.

Clearly, methods to achieve a certain end can be effective but immoral. This applies also to attempts to prevent, diagnose, treat, and follow up health threats. Thus, also ethics conditions are required. The ethical framework used in assessing proposed actions may at least sometimes influence the judgment whether the morality clauses

[1] I have used a related but somewhat different approach in Hermerén (2007).

are met or not. Anyway, since there can be some tension between the conditions of efficacy and morality, they need to be balanced.

20.2 A Key Question

The dream of being able to stop and reverse the decay of the body is not a new one. It can be found in various forms in all cultures. This dream is also reflected in the arts. For example, Lucas Cranach finished in 1524 a well-known painting, showing aging people coming in from the left in the picture, descending into the basin of youth, moving towards the right side in the water, clearly becoming younger and younger, then leaving the basin on the right side, and disappearing to meet lovers or to go to parties.

A main question raised by these dreams is:

Should we—and who are we?—accept old age and try to adjust to the impairments and ailments that come with aging, or should we resist or delay them by preventing old age, by rejuvenating care, by treating age-related diseases?

This question is primarily addressed to each and every one of us as human beings, and as members of social communities. But certain answers will have implications for what others—in particular, governments, parliaments, social service agencies, research councils, researchers, and other stakeholders—should do.

Many more specific issues are discussed in this book regarding aspects related to the key question. Separate problems are raised depending on which individual or societal needs the authors address. For example, Govert den Hartogh, discussing the death wishes of elderly people and reminding us that the basis of the Dutch euthanasia law is beneficence rather than respect for self-determination, asks:

Should doctors be allowed to grant such requests? If this is not considered to be a proper medical task, should the law make it possible for old people to receive non-medical assistance?

Focusing on one particular aspect of the key question, several authors—some of whom are critical of Morbidity Comprehension paradigms and similar efforts—ask:

Can healthy aging be achieved through preventive interventions and life style changes? Who will benefit?

Against the background of descriptions of new interventions into aging, Hans-Jörg Ehni raises the morally important question of who will benefit from the limited access to certain kinds of medical interventions. This focuses on another aspect of the key question and is discussed in the context of a justice framework that he develops in his chapter.

Defining old age as pain and suffering is part of the problem, not a way forward, as John Vincent points out in his chapter. For the purpose of the present discussion, Miller's definition of aging, quoted by Wim Dekkers in his contribution to this book, will do well. Aging is then defined as "a process that converts healthy young adults into less healthy older ones with progressively increasing risks of illness and death". (Miller 2004, p. 228).

Another question focuses on the nature of the subject:

How should aging be conceived of? How should we understand and value aging? In discussing the key question above, some assumptions concerning aging need to be made explicit, clarified and defended. Is it good in some respects? Or bad? Such assumptions will influence what can be seen as reasonable policy implications. Some of these assumptions concern the nature of aging. Is it pathological or not? If it is, should aging be understood as a disease, an illness or a sickness? This is the question addressed in Schermer Maartje's contribution, and she begins by explaining why it matters.

She demonstrates that different ways of conceptualizing or understanding aging—as disease, illness or sickness—have important ethical and financial consequences. If the assumptions behind these conceptualizations are not made explicit, underlying normative problems like the key question stated above can be hidden. She writes: "One way to make aging itself a legitimate goal for intervention is to stress its association with disease. . . . if aging itself is a disease, it is a legitimate target for medical intervention. . . . Moreover, if interventions in aging can be labeled as treatment, this will have important financial consequences."

She also makes the important point that disease, illness and sickness are not static concepts. "What counts as disease, illness or sickness may change over time, due to new discoveries and insights, and the borders between the concepts are not sharp but rather blurred." Her own views are summarized as follows: "aging may be a disease and may cause illness, but it is not a sickness—aging is aging."

20.3 A Model

In this section I will sketch a model used as a basis for and justification of policy recommendations. The basic components in the model are the concepts of description, obstacle, goals/values, and concern—and the relations between them. This model will provide the structure for the rest of this chapter, and the headlines of the different sections in what follows will refer to the building blocks of the model.

A central element in the model is constituted by concerns. The concerns in their turn are based on descriptions of the current situation, including obstacles on the road ahead, as well as on goals and values. Stakeholders can express different concerns, and similar concerns in more than one way, depending on how they perceive the situation and current trends as well as on what they want to achieve and avoid. Incidentally, there is an interesting interplay here between descriptions and values, in that descriptions are the basis of proposals for changes—concerning what should be done in order to arrive at the goals quicker, more cost-effectively, in an ethically more acceptable way, or some combination of these possibilities.

On the basis of the concerns policy proposals can be made and justified. But not every concern qualifies as a basis of a public health proposal, even if it is regarded as legitimate from an ethical point of view. Seven conditions outlined earlier need to be met, at least to some extent.

The relations between the key concepts of the model can be represented graphically in the following way, where a line stands for 'of', referring to contents of descriptions, single arrows stand for causal relations like 'influenced by', and double arrows stand for 'basis for and/or justification of':

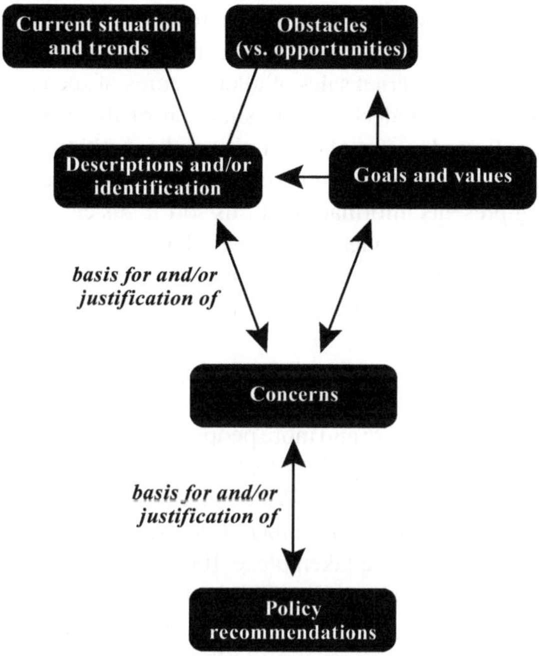

20.4 The Present Situation

The starting point for any analysis of the policy-setting landscape is a description of the present situation. But there is a complex relation between descriptions and values here, as already mentioned. Anyway, the present situation can be described in many ways, which need to be examined critically, since the descriptions can be selective, and thereby seductive, reflect a more or less hidden agenda, or be mixed with hype.

In order to describe the present situation, we need definitions, criteria and classifications of different kinds of elderly. We will also need statistics, indicating the living conditions and disease panorama of these various groups. The state of the art of relevant medical and biological research needs to be outlined.

Several chapters in this book address such issues. They present important medical, social, economic and psychological information relevant for recommending policies. For example, what can science, and in particular biogerontology, the branch of science focused on the biology of aging, provide? Will stem cell therapy have something to offer in the near future? In their chapter, Sethe and de Magalhães point out that "genetic research in aging may well pave the way to interventions other than germ line alteration, ranging from adult gene therapy, to pharmaco- and nutrigenomics".

In overviewing the claims made by the protagonists of biogerontology, transhumanists and others, we need to sort out evidence-based claims from hype. Inez de Beaufort argues convincingly in her chapter that appearance is important in our culture. Hence, it is hardly surprising that there is a growing market for products supposed to cure age-related defects in appearances and health of men and women. She describes flourishing internet sales of alleged cures of such age-related changes.

Moreover, we need to understand the situation of the elderly in our societies. Accordingly, we will need results of qualitative and quantitative investigations, indicating how the elderly perceive themselves—and how they are perceived by others. Frans Meulenberg presents information of this sort in his chapter. Such information will help us to gain a better understanding of what the present situation is like for various groups of elderly. Relevant aspects also include demographic changes and information about their disease panorama—many elderly are known to have multiple pathologies.

As several contributors have pointed out, the high prevalence of infectious disease and perinatal and infant morbidity have been combated successfully. The average lifespan has been extended, more and more people reach the age of 80, and the disease panorama of the elderly has changed. "Today we live in the area of chronic disease", Cees Hertogh underlines. Besides, as Wim Pinxten stresses "... the elimination of one disease, opens the door for the other. For example, successful reduction of cardiovascular disease may have taken place, [but] dementia is growing steadily."

Analogously, information about the goals, methods and results of scientists belong to the background. In his chapter, Wim Pinxten reports the results of interviews with Dutch scientists working in areas relevant to the understanding of problems related to aging. Information about the rapidly increasing costs for the healthcare systems—and the challenges of cost containment—is clearly important, considering of course also the possible economic benefits to society of an extended healthy lifespan.

What people want to achieve and avoid can be described, and these descriptions can be an essential part of the background information. Do people want to live forever? In his contribution John Vincent refers to an interesting survey 2011, according to which only 15 % replied that they would like to live forever (17 % for men as opposed to 13 % for women). Interestingly, Vincent points out that there was a systematic relationship to class in the sense that those with high status occupations were less likely to want to live forever.

A description of the present situation includes death wishes of the elderly. Here we need to separate as clearly as possible, as Govert den Hartogh points out, death wishes resulting from earlier psychiatric conditions and traumatic experiences and death wishes related to what in the Dutch context is referred to as a 'completed life'. In his chapter, he discusses explicitly only the latter.

Anthropological premises play an important role in this debate, a point made by several contributors to this volume. Wim Dekkers rightly points out that the "scientific work that is carried out in the field of aging, the practice of care for the elderly and ethical considerations about aging are based upon—most often implicit—presuppositions about a number of fundamental characteristics of human beings." According to Søren Holm, the standard anthropology in English language bioethics helps to explain why aging has not received the amount of attention in bioethics that reproduction and the beginning of life has, and "why old age primarily is seen through a deficit lens when it is discussed".

There are several mines in this terrain. Three simple but important points need to be remembered:

(a) Selection of information is often made on the basis of concerns the various stakeholders happen to have—and their interests are not always identical—and the same holds for the selection of what is presented by the stakeholders as important.
(b) Normative premises are often disguised as empirical premises, for instance by not making anthropological assumptions explicit. There is sometimes a very short distance between saying that "this is how human beings are" and "this is how human beings should be".
(c) Societal ideals can play a role as implicit premises in this discussion. To avoid talking at cross-purposes, this connection between bioethics and political philosophy should be acknowledged and made more explicit, a point I will return to later.

Having described where we are at present—different aspects of the present situation—let us now move to the next step, descriptions of where we want to be, and why. This will inevitably involve references to goals, values and ethical frameworks.

20.5 Goals and Values

Goals are based on values in the sense that we pursue certain goals because we consider it valuable to achieve or come closer to achieving them. As is well known, things or states of affairs can be valued as ends in themselves or as means to achieve other things or states which are valuable. Values can, philosophers remind us, be instrumental and intrinsic, even if this distinction can be drawn in more than one way.

While health and flourishing can be ends in themselves, money is clearly valuable as a means to other ends. Good social relations and a good family life may be both, as making sense of one's life. What makes life good while aging? And what, if anything, makes aging a good thing? This is how Frits de Lange puts the main questions in the opening paragraphs of his chapter.

So let us begin by considering: Which goals? Whose goals? Goals can be classified in many different ways, as is obvious from the discussion about the goals of medicine (Fleischhauer and Hermerén 2006). For one thing, it is useful to separate goals on different levels, that is, goals of individuals from the goals of states or institutions, agencies or committees. We cannot take for granted a principle of methodological individualism according to which the goals of an institution are identical to, or the sum of, or are derivable from, the goals of those individuals who at a given time happen to work in that institution.

Moreover, regulators, scientists, healthcare professionals, patients and their relatives, taxpayers and others may have somewhat different goals, both in the short term and in the long term perspective. It is also essential to distinguish between positive and negative goals—what stakeholders want to achieve and to avoid respectively. What may be a positive goal for one stakeholder may be a negative goal for another.

So: which are the goals of various stakeholders, individual or institutional? For instance, is life extension always a good thing? Under what conditions? Whose goal is it? Similar questions can be raised about immortality. In his chapter John Harris criticizes arguments against life extension. Life extension is obviously a vague notion, admitting degrees. But there is an important conceptual difference between longevity and immortality. The latter is ambiguous but not vague; it is either—or.

Suppose that combating age-related diseases successfully promotes healthy longevity, which is not self-evident for reasons indicated by Wim Pinxten and others. Suppose also that we have, as Harris suggests and I agree, a moral duty to combat age-related diseases, at least when this will promote healthy longevity. But does it follow that we have a moral duty to promote immortality? Promoting longevity and trying to achieve immortality are two distinct goals.

Many goals and values are described or hinted at in this book, somewhat different in different chapters. Sometimes, even if the same goal is advocated, like justice in the treatment of the elderly, this goal may be interpreted and conceptualized differently. It is interesting to compare, for instance, the contributions of Hans-Jörg Ehni and Anders Schinkel in this respect. Both advocate justice for the elderly, but their conceptions of justice are not identical.

Even so, there seems to be widespread agreement among the authors that important goals and values to be promoted include health, personal safety and security, being able to look after oneself, autonomy and respect for self-determination, to have a realistic self-image and maintain one's self esteem or worth, fair access to healthcare, and improvement in the quality of life and quality of care for the elderly. These are all goals of individuals, but some of them have policy implications for stakeholders at other levels—which I will return to in the last section of this chapter.

Other more general goals and values, not specifically focused on the aging population, include to eradicate poverty, to improve the level of education, to promote legal certainty and trust in social institutions, avoiding discrimination and stigmatization of the elderly as well as of other more or less vulnerable groups. Some of these goals are at least indirectly related to the goals of improving health and increasing life expectancy, since it has been demonstrated by Michael Marmot and his collaborators (Marmot and Wilkinson 2006) that poverty, smoking and alcohol has considerable impact on health and quality of life, also of the elderly.

But there is a tension between some of these goals that needs to be taken seriously. Which possible conflicts between the goals and interests of these various stakeholders can we anticipate? Is there a procedure or ethical framework with some legitimacy that can be used to deal with such conflicts? The more vulnerable and frail you are, the less likely you are to be able to take care of yourself. Wim Dekkers writes about "the tension between vulnerability and dependency on the one hand and the ideal of autonomy and independence on the other." Analogously, there is a potential conflict between the goal to prolong life, to extend the lifespan indefinitely, and the goal to improve the quality of life and care for the elderly, since sometimes one of these goals can be achieved only at the expense of the other.

Before concluding this section, just a brief comment on Anders Schinkel's general goal of justice as recognition. He argues persuasively in his contribution that the philosophical perspective on justice related to the elderly should be widened considerably. More than distributive justice needs to be discussed, and more than fair distribution of financial resources. We should ask: "what would a just society look like in its elderly-related aspects?" A problem is that the elderly is not a homogeneous group; differentiation is necessary, which means that questions concerning fair treatment need to be replaced by a family of questions, relating to different sub-categories of elderly people.

Schinkel's general position, which can form the basis of many policy recommendations, is: "what is always owed, to everyone, is recognition". He sees recognition "as the primary act of justice, and its basic form. A just society gives the elderly the recognition they are due—not collectively, not as a group, but simply as individual people." The challenge, of course, is to decide what the elderly are due, and on what ground. He continues: "Justice as recognition urges attention to the specific needs of particular individuals"—which, of course, presupposes that we know what these needs are.

Anyway, he defines recognition of every individual as a source of legitimate claims, of people's equal inherent value, as the primary act of justice. This could be interpreted as an example of a dignitarian approach in the sense explained by Roger Brownsword (Brownsword 2008). If not, an approach based on human dignity is an alternative, 'human dignity' then understood as the basis of all human rights, of what we owe to each other, also to the elderly.

Hans-Jörg Ehni approaches these problems from a different angle than Schinkel. His focus is not on justice as recognition, or a quasi-dignitarian approach, but rather theories of justice of the sort advocated by Rawls (1991) and Daniels (2008a, b) and in particular on developing a general frame of how evaluation of interventions into aging might be carried out in relation to theories concerning just healthcare.

20.6 Obstacles

The distinction between obstacles and promises is not ethically neutral. What to some stakeholders are obstacles, something that impedes progress, for instance, absence of legal regulation, may to others be the opposite, providing room for freedom of action, something that furthers progress—if their values are different enough.

Obstacles and threats on the road from where the stakeholders are at present to where they want to be in the future can be of several kinds. The importance of distinguishing between them is simply due to the fact that the obstacles need to be tackled in different ways. On a general level we may distinguish between three main approaches: circumvent the obstacle, eliminate it, or reduce it. But if we look into particular obstacles, there is a variety of possible concrete strategies.

For example, if the obstacles are particular legal rules, what is needed to change them is lobbying in parliament. If the obstacle is lack of financial resources, what is needed is more money—which of course can be obtained in a variety of ways. If the obstacle is lack of knowledge, more and better information is needed, perhaps more research. If the obstacles have to do with the attitudes of the various stakeholders, special strategies are required to change these attitudes.

For instance, Cees Hertogh claims that "care for and research directed towards improving the quality of care for the oldest old is a neglected and under financed realm." Similar points are made by other contributors. Other obstacles are referred to, explicitly or implicitly in the contributions to this book. This is hardly surprising, since the authors address at least partly different problems. A list of obstacles might include lack of clarity as to what the state and other agencies should do (Govert den Hartogh), general dislike of elderly people in our culture (Bert Keizer), lack of knowledge about the wishes and needs of the elderly (Frans Meulenberg), understaffed nursing homes (Anders Schinkel), the rising costs of healthcare and social services for the elderly (Wim Pinxten, Alies Struijs et al), the language and thinking of economists in healthcare (Anders Schinkel), lack of justice in the access to the available resources (Hans-Jörg Ehni), the discrepancy between the earlier wishes of formerly competent people and their present wishes, "possibly less competently argued but based on the experience of the new situation" (Dorothea Touwen),[2] and counterproductive regulations.

The ranking order of values may also provide a bridge to political philosophy and societal aspects of decision-making. Compare the following two ranking orders of values: (1) freedom of research, economic growth, health: longer life expectancy for future generations, well-being for future generations, individual self-determination, safety and security, integrity, solidarity: including support for vulnerable and frail people. . . and (2) solidarity: including support for vulnerable and frail people, safety and security, individual self-determination, well-being for future generations, health: longer life expectancy for future generations, economic growth, and freedom of research.

The values are the same in both cases, but the order of normative importance is reversed.[3] The top value in one ranking order is the bottom value in the other. etc. The ranking orders also reflect different two very different societies: in one the top value is freedom of research, in the other it is solidarity. For the elderly population it makes a great deal of difference as to whether they live in one or the other of these societies—and whether the implicit premises in the discussion of the elderly in society include societal ideals closer to ranking order (1) than to ranking order (2).

Clearly, many stakeholders have different views concerning what the state should do, ranging from those supporting Robert Nozick's minimal state (Nozick 1974) or Bo Rothstein's views about the importance of just institutions (Rothstein 1998),

[2] See also Broström (2007).

[3] For some interpretations of 'ranking order of values', see Hermerén (2008).

to the adherents of a communist state of a traditional sort where the state plans and organizes everything—the Soviet Union in the past, North Korea at present. Views of society, and not only anthropological premises, are important in this context. Which are the underlying societal ideals? A minimal state? A welfare society? A planned economy? Combinations or intermediaries?[4] If such underlying assumptions are not made explicit, they could be a potential obstacle for constructive discussions about who should do what.

20.7 Concerns

A general concern is obviously that the obstacles listed in the previous section may be insurmountable, impossible to bypass, reduce or eliminate in ways that are efficient and morally acceptable. If there is an obstacle to the goal of a particular stakeholder, this should be a cause of concern to that stakeholder. But at the same time it may be a relief to stakeholders with other interests and values. Likewise, a particular trend can be both a sign of hope and a reason for concern, depending on the values taken for granted—what people want to achieve and to avoid.

The logical anatomy of concerns can be outlined as follows. Concerns presuppose, firstly, descriptions of present and/or future trends and situations, including obstacles on the road between where stakeholders are and where they want to be; secondly, a statement of the goals and values of the various stakeholders, making explicit what should be protected and promoted, as well as what should be avoided; and thirdly, reasons to believe that the current situation and near future trends will undermine or threaten these goals and values.

If a concern is to be regarded as serious in an open and democratic society, this presupposes both that the values at stake are basic to the culture, for example, being rooted in or related to human rights, and that there is substantial evidence for the belief that the values may be threatened by current or near future trends. Then public policies are called for, if the conditions outlined in the introduction are met.

A concern is that the ideal outlined by the protagonists of the Compression of Morbidity paradigm is not within reach; in the words of Cees Hertogh, "for the vast majority of older persons this ideal is far from realised" and that... "the whole literature on frailty grossly neglects the perspective of the allegedly 'frail' older persons themselves." He adds:

> It is an impressive and at the same time shameful finding that so many burdensome symptoms remain under recognized and (as a consequence) under treated in older persons with complex care problems, while at the same time they are over treated with medicines... that have barely been tested for safety and effectiveness in their age group, with all related negative consequences such as hospital admission and excess mortality.

There are interesting differences between the concerns raised in some chapters in relation to the goals of enhancement, prolongevity and extension of human lifespan.

---

[4] For a different classification of types of societies see Campbell (1985).

As Alies Struijs and others point out that "people live longer does not automatically mean, however, that people live longer in a healthy condition." Those who are optimistic in describing the future possibilities of science and play down the possibility of adverse side effects—including the boring prospects of living together for ever, the problem of overpopulation that would accelerate, as well as how to feed this growing population—are likely to have other concerns than those who are less optimistic. The former may be concerned about unwarranted political and ethical constraints, whereas the latter may be more concerned about the improvement of the quality of life and the quality of care of elderly people.

The huge needs of research on the situation of the elderly and of evaluating different ways of meeting their needs may be the basis for concern about the lack of coordination of this research. Much is done without knowing what others are doing, and clear focus on the interests of the elderly is missing, as Wim Pinxten underlines in his chapter.

Issues related to discrimination of the elderly and fair access to healthcare are obviously important and a likely target for public health policy. But some distinctions are essential. First of all, concerns regarding possible benefits of particular interventions need to be separated from concerns regarding access to these interventions. Thus we need to distinguish between the following three concerns.

First, not everyone will benefit from interventions in order to prevent, slow down or reverse the biological aging process. Ehni writes that "scepticism about whether everybody will benefit from the longevity dividend in the same way or even how widespread such benefits will in fact be seems to be appropriate". Secondly, even if the previous scepticism were unjustified, access to these interventions will be limited, especially of course in a global perspective. According to Ehni:

> In many countries, public health insurance or the public health system will probably not be able to finance many of the relevant interventions and services... Due to scarce resources and cost constraints that partly result from the demographic change and technological innovation, there is some likelihood that new expensive medical services and interventions into aging will not covered by public health insurance. The appropriateness and necessity of a further lifespan extension might be doubted, as Daniel Callahan has done in an influential way....Personal limits could hinder access to new interventions into aging as well.

Third, these interventions, if and when they are available, will improve the situation of the best off, and the gap between them and those who are worse off will increase, as Ehni stresses: "... limited access to new interventions... could increase already existing inequalities in healthy life expectancy" and "... the situation of those best off is likely to further improve", "... Compared to the situation of the best off, their [those worse off] situation will be even worse than it is now."

To sum up: somewhat different concerns are expressed by the authors in this book. This is to be expected, since the problems they focus on are not identical. But a list of concerns, based on the contributions to this book includes: (a) *Unclear responsibilities*: the concern that the division of responsibility between the state, institutions, other agencies and individuals is unclear; (b) *Lack of knowledge*: the concern that we do not know enough about the needs and desires of the elderly; (c) *Inadequate support*: the concern that those needs and wishes of the elderly which are known are not catered for adequately; (d) *Dislike of the elderly*: the concern that the elderly are disliked and treated as of lesser worth than other groups in society; (e) *Internet sales of cures of dubious value*: the concern that treatments of age-related ailments and illnesses offered on the interned sometimes are based on hype and wishful thinking; (f) *Rising costs*: the concern that the costs of healthcare and social service for the elderly are rising dramatically; (g) *Fair access to healthcare*: the concern that the elderly are discriminated against unfairly, as individuals and as a group, in particular that the elderly are not provided fair access to healthcare.

20.8 Policy Options

Who should do what, when and why? For example, what could and should the EU Commission, the national governments, parliaments and various regulatory agencies at national and international level do? What should be left to the responsibility of individuals and their families?

The central point of the model generating policy implications used here is that this is relative to what is taken to be the problem, how the present situation is described, which goals and values are to be promoted, which obstacles are identified, which concerns are expressed, as well as assumptions about efficacy and morality, as suggested by the conditions in the introduction.

In other words, if the present situation is described differently, or there is disagreement concerning how it should be described, or if views are divided concerning which goals and values should be achieved, or if there is agreement about which the underlying values are but these values are ranked differently in normative importance, it will be difficult or impossible to agree on policy implications.

In a way this is a simple and obvious point, but it is worth making in the present context, since there are some interesting differences between the contributions concerning description of the situation, concerning the goals to be achieved, concerning the ethical framework used, ranging from utilitarian, Rawlsian and capability-oriented to dignitarian, as well as concerning the efficacy and morality of proposed means.

Which are the policy options? There are several and they raise different issues, also in terms of efficiency and moral acceptability. They include: regulation, monitoring, economic inducements, guidelines, and information. Each of them will have both advantages and disadvantages, somewhat different in different contexts and societies.

Laws help to create legal certainty and to define an area in which the stakeholders can operate without getting into legal difficulties. Legal regulation may be called for if, returning to the conditions in the introduction, the threats are severe and can only be dealt with adequately by society, provided that certain borders are not violated. But a problem is that laws can soon be outdated and stifle both research and creative approaches to deal with the underlying problems.

Monitoring is flexible, cheap, and probably less likely to stifle research than legal regulation. Agreement on precisely what is to be monitored, and the criteria to be used by the monitors, in particular the precise interpretation of the goals, is obviously essential. This holds also for when these goals are to be achieved. But a problem is: who selects and monitors the monitors? And will monitoring be effective?

Economic inducements and sanctions include taxation and fees. This is a policy that can be effective if it is enforced and if the grounds for taxation and fees are stated clearly. But a problem is that their effect will vary with the economic resources of the stakeholders. To some a certain fee will be an effective obstacle, to others it will mean little or nothing. From an egalitarian point of view this means that the inducements and sanctions have to be differentiated in order to achieve the desired effects.

Guidelines and self-regulation are flexible and not very costly. These regulatory approaches are often favoured by the industry. But a problem is that they can amount to little more than window-dressing if they are voluntary, vague and not combined with sanctions. Finally, information and advice may be useful if they are relevant to the needs and interests of the stakeholders, so that the latter can realize that it would be in their own long range interest to comply with the advice suggested. If the information is vague, general and not specific, it may have little impact.

20.9 Some Assumptions

Discussions of strategic policy decisions related to societal consequences and economic effects of demographic changes, entitlement to pensions, citizenship, participation, access to and distribution of healthcare do not take place in a vacuum. They will inevitably be based on certain assumptions.

The following two assumptions form the basis of my comments on proposed policy implications in the next section.

(a) All concerns cannot be dealt with in the same way,
(b) A particular concern cannot be dealt with in the same way everywhere

Why? Because of different political, legal, ethical and religious traditions, infrastructures as well as varying technological and economic development in the EU member states.

We have to look at details and particulars, and be explicit about goals and values. Moreover, the empirical evidence available and how it is interpreted can play a role for decisions about which policy option to use in a particular case and which policy to recommend. Thus, the interpretation of the evidence should be made explicit and alternative interpretations discussed, since all concerns are based on empirical evidence interpreted in a certain way:

(c) The available evidence cannot always be interpreted in one way only.

This is an obvious assumption, but it should not be forgotten. We are reminded of this room for multiple interpretations by a remark made by Bert Keizer towards the end of his contribution, discussing Alzheimer patients:

> Quite a few people would like to forego this descent into a premature oblivion by ending their lives. In the Netherlands this is possible but not often enacted because not many doctors are convinced of the intensity of the anguish which is caused by mentally falling apart in the course of Alzheimer.

Many stakeholders have different views concerning what the state should do, as already mentioned. The important thing is that what works well and will be acceptable in one country, given its traditions and current economic state, might not work in another. For example, policies that work well in a society with focus on promoting individual liberty might not work well in Japan, where the group or the family play an important role in decision-making.

Consider the case of priority setting in healthcare: what are the elderly entitled to? Do methods of rationing healthcare based on the use of QALY's (quality-adjusted life years) discriminate against the elderly? The 'fair innings argument' has a greater chance of being accepted in liberal, individualistic countries where health economists have a strong position and are backed by dominant utilitarian thinking—which incidentally is helped by a historical connection between health economy and utilitarianism. But in countries where a dignitarian or human rights-based approach is dominating, such calculi may be seen as morally objectionable.

20.10 Specific Proposed Policy Recommendations

According to the model suggested here, the starting point of the policy recommendations should be the concerns, which in their turn are related to descriptions, goals and values, as well as to identified obstacles. The ideal is that every concern is addressed by some recommendation, which is effective and not morally objectionable; and that every recommendation addresses a concern expressed by some stakeholders. In the following discussion of proposed policies, I have had to be selective; this book is rich with suggestions, some of which are just sketches while others are worked out in more detail.

All suggestions and concerns expressed in this book cannot, and should not, serve as a basis for public health policy. Whether this is so or not will depend on the extent to which they meet the seven conditions outlined in the introduction. But we have to be prepared for the existence of various grey zones and border line areas. Outside mathematics and logic most concepts have fuzzy and blurred borders. Also the specificity of the proposals needs to be considered: to what extent do they indicate clearly who should do what when and why? A list of policy suggestions, related to the earlier expressed concerns, might include the following:

(a) *Unclear responsibilities*. Let us first consider implications or recommendations related to the concern that the division of responsibility between the state, institutions, other agencies and individuals is unclear. Let us suppose we know the preferences of many or most of the elderly. What should the state and various governmental agencies do to satisfy them? What should be left to the responsibility of individuals and their families? This is obviously an important but difficult question, relevant for public health policy issues.

Discussing weariness of life amongst elderly, Govert den Hartogh makes the following point: "But precisely for the very old and frail people there is an alternative which certainly is relevant: they can stop eating and drinking. The common belief that this leads to an inhumane death is mistaken." This is clearly a reminder to individuals, not to the state or national boards of healthcare. Similarly, it is suggested: "if you want to make the best of your aging, continue to re-invent yourself by adjusting your self-image time and again to your biographical age" (de Lange). Obviously, this advice is directed to individuals, not to states or governmental agencies. It is not meant to be the basis of a public health policy, though it can be supported indirectly in various ways, for instance by information from healthcare professionals.

But in other cases the addressee is less clear. For example, suppose a person has an ugly nose and repelling skin. This gives rise to psychological suffering of the sort discussed by Inez de Beaufort in her contribution to this book. This suffering can be graded on an interval scale as more or less severe. When is there, or when should there be, an obligation for society to alleviate this suffering, and pay for the costs?

Frits de Lange is rightly critical of a policy focused only on keeping the vital senior vital—this is ageist and too limited. I completely agree. He also makes the important point that "Aging is not only a matter of loss, but a changing balance of gains and losses, throughout a life course." This has implications for policy recommendations: "Therefore it should be a public policy affair to assist older persons explicitly in finding this balance, not only by preventing the losses of old age but also by discovering, exploring and exploiting its gains."

Cees Hertogh makes similar suggestions. According to him, a double strategy is called for:

> Instead unilaterally focusing on preventing and combating frailty, more emphasis should be placed on active anticipation and on assisting people in finding an adaptive response to the implications of frailty.... This calls for an adjustment of goals in life, but also for psychosocial support and medical care tailored to the individual needs and to the consequences of progressive disability. To assist people in finding a successful adaptive response there is a need for a more positive approach towards frailty and old age.

There are good reasons for this approach and for supporting attempts to work out these policy suggestions stepwise so that they become more concrete and detailed. But the division of responsibility does not have to be the same in every situation; a certain division can be conditional. Then the conditions need to be spelled out. When has a reasonable balance been achieved? Criteria have to be worked out and justified. If the conditions in the introduction are interpreted liberally, these suggestions might be the basis of public policy recommendations. But to be effective, they need to be worked out in more detail.

These responsibilities are not static, and there is, as several authors have noted, a shift taking place from governments and social service agencies to individuals. For instance, Cees Hertogh writes:

> According to the SCP-report, this activity of monitoring [of older persons for detection of early signs of frailty] is not only the responsibility of health professionals and municipal

> officers. . . . It is also and primarily a responsibility of the older person herself. . . . All these recommendations fit in perfectly well with a government policy that focuses strongly on personal responsibility, self management and participation. . .

Alies Struijs and Marieke ten Have take this lack of clarity as their point of departure and describe recent governmental trends in healthcare making individuals more responsible for their own health. The economic background is the concern about increasing costs, and the ethical underpinning includes increased focus on freedom of choice and personal autonomy. What is particularly important is their analysis of the ethical problems this gives rise to. In particular, they focus on ethical objections to retrospective responsibility.

In the final part of their paper they discuss an ethically sound practice of a health check. This is a challenging and very interesting proposal. If people, young or old, have some personal responsibility for their own health, they also have some responsibility to use existing ways of preventing ill health. The authors refer to a Dutch model, PreventieConsult, offered to people over 45 years of age. It is so far only developed for diabetes, cardiovascular diseases and chronic kidney problems. This preventive consultation will provide a reliable alternative to commercial general self tests and health tests offered for sale on the Internet. If it works successfully in the Netherlands, it should be of interest also in other countries.

The idea is not that this preventive consultation should replace health checks and visits to doctors. But it should give a first individualized estimation of potential risks. People at risk will be recommended to consult a doctor and they will be offered certain advice concerning their life style. As is well known, prevention raises methodological, epistemological and ethical problems. Commonly asked questions about any preventive measure include these four

(i) Will the preventive methods used have any effect?
(ii) Can we, in view of the many confounders, find out what the effects are and measure them?
(iii) Will the preventive measures used be cost effective?
(iv) Are there any objections to the use of preventive efforts on moral grounds? Do they, for instance, violate people's integrity or freedom of choice?

These questions are relevant here too. But on its website, PreventieConsult claims to be the first scientifically validated test for the combination of the three conditions mentioned above.[5] Anyway, a necessary but not sufficient condition for question (iv) to be answered by no is that the three first questions are answered by yes—and possibility or stimulation rather than obligation has to be key guideline. A challenge is to prevent that obligation sneaks in through the backdoor. "You had the opportunity, you did not use it, and now you want society to pay for your healthcare. But the state cannot afford this. Well, you have to blame yourself". In such ways, the pressure can increase on people over 45 and limit their freedom of choice. The discussion of this proposal will no doubt continue.

---

[5] Those who are interested in more details might like to check the websites www.nhg.artsennet.nl and www.lekkerlangleven.nl.

(b) *Lack of knowledge*. Let us now proceed to consider recommendations related to the concern that we do not know enough about the needs and desires of the elderly. Several authors underline that we must invest more in research aiming to improve the quality of life of the oldest old. The reason is, as Cees Hertogh puts it at the end of his chapter, "adding years to life is only acceptable and worthwhile in combination with a policy that adds life to years."

This concern suggests that we should find out what the needs and desires of the elderly are, that resources should be set aside for this, that this money also should be used for measures and interventions to meet these needs. But how much? For precisely what kind of research? Who should do this? What is the role of government, parliament, agencies, research councils, researchers and relatives here?

Having discussed the contributions of genetics to our understanding of aging, stressed that aging is a fairly individual process, and emphasized phenotype plasticity, Wim Pinxten proceeds to discuss factors relevant for how we age, in particular environmental factors and life style (stress, smoking, diet, fitness). He rightly stresses that there is no cure for aging as such, but to a certain extent for age-related diseases.

In the interviews he carried out with Dutch scientists, it was repeated over and over again that how to set priorities is crucial. He argues that the starting point should be the interests of the elderly. In discussing policy implications, Pinxten suggests that aging provides a framework in which different types of research should be reframed. One of the advantages would then be not only that researchers would focus on a huge problem, since many diseases are age-related, but that "the current approach to health in the elderly lacks coordination." The policy recommendation is thus that research funding agencies should coordinate their efforts, to achieve more and better knowledge of the aging process and the challenges it presents to contemporary societies.

Towards the end of his contribution John Harris discusses some policy implications of the scientific developments he anticipates:

> We should be slow to reject cures for terrible diseases even if the price we have to pay for those cures is increasing life expectancy and even creating immortals. Better surely to accompany the scientific race to achieve immortality with commensurate work in ethics and social policy to ensure we know how to cope with the transition to parallel populations of mortals and immortals as envisaged in mythology.

Let us assume, to use Harris' words, that "the price we have to pay" is not only "increasing life expectancy", which I may favour; but also, which I doubt, creating immortals. The interviews Wim Pinxten have carried out does not support, as far as I can see, that there is a scientific race to achieve immortality. On the contrary, clinical applications of the basic research being carried out still seem far away. Even so, I agree with Harris on a general level: better to accompany the scientific developments with commensurate work in ethics and social policy. But a problem, of course, is precisely which measures and policies are "commensurate" in this situation. Persons with different ethical frameworks are likely to disagree about this.

Harris also discusses the prospects of parallel populations and the challenge of overpopulation.

> ... the most ethical course might be to contemplate a sort of "generational cleansing". This would involve deciding collectively how long it is reasonable for people to live in each generation and trying to ensure that as many as possible live healthy lives of that length. We would then have to ensure that, having lived a "fair innings", they died at the appropriate time to make way for future generations.

But this needs to clarified in order to provide a basis for public policy. The questions are simply: 'ensure' how? By whom? When? 'Die' in what way?

(c) *Inadequate support.* Next, let us consider some recommendations related to the concern that those needs and wishes of the elderly which are known are not catered for adequately. This is stressed several times by Cees Hertogh, for instance in the following earlier quoted passage:

> This calls for an adjustment of goals in life, but also for appropriate psychosocial support and medical care tailored to the individual needs and to the consequences of progressive disability. To assist people in finding a successful adaptive response there is a need for a more positive approach towards frailty and old age.

In her chapter, Dorothea Touwen makes a very interesting policy recommendation. She writes that:

> Policy makers might consider promoting that people appoint their own favored representative (healthcare power of attorney)—that is: To write an advance directive not concerning what one wants to be decided, but who one wants to be the principal representative and promotor of one's future interests.

In my view, this is an excellent suggestion. It is specific, morally acceptable and likely to be effective—even if one cannot exclude that the elderly may change their minds not only regarding what should be decided but also regarding who should decide for them.

(d) *Dislike of the elderly.* Next let us consider some suggested implications related to the concern that the elderly are disliked and treated as of lesser worth than other groups in society. The moral basis for this is clearly that this dislike can lead to violations of human rights and suffering. In his chapter Bert Keizer provides an interesting thought experiment:

> The treatment of Alzheimer patients offers a convincing illustration of that dislike of the elderly with which I started [my chapter]. For imagine that Alzheimer was not a problem of old age, but that we had 250.000 adolescents in our midst who were affected by a similar brain disease. Do you think it likely that we would dump these youngsters in care homes on the outskirts of our community so as to be able to get on with our lives? Yet that is precisely what we do with Alzheimerpatients. Of course losing so many 18 year olds to a lethal brain disease is quite a different thing than losing the aged in that manner. The difference being that the aged are, biologically speaking, not a loss, when they die. I do not mean to applaud this, I merely point out that this is what we feel and it shows in the way we care for them.

Suppose that it is proposed that the finding that many elderly themselves have low self-esteem "may suggest anti-ageism campaigns by governments". It is easy to feel sympathy for such a proposal—who would be against it? But it is not very specific. It tells us little about who should do what and when, what specifically the campaigns are to be directed against, how they are to be organized, who should be involved etc.

One will have to think in terms of a step-wise approach, where the proposal becomes more and more specific after each step.

This also holds for John Vincent's sympathetic, but somewhat general conclusion, for which he argues powerfully by pleading that "... anti-aging, science based, immortalist technology would undermine a human cultural process—the succession of generations—and hence compassion and social solidarity".

The policy conclusion of the humanist position on old age advocated in the above discussion is that increasing power to a diversity of older people's voices is essential for progress. The extent to which ageism is embedded in the cultural and scientific, knowledge-creation institutions has to be acknowledged and countered. Commercial, medical and biological institutions have re-inforced the anti-aging mind set. If the 'Third Age' agenda of a renaissance for later life is to succeed, older people must themselves be in the lead in positively valuing themselves, which means that together they/we have to find ways to live well and die well.

In Sect. 20.4 of his contribution, Schinkel outlines some policy implications of justice as recognition. The key issue, he suggests, is that we need to reduce "the number of cases in which people are treated as of lesser importance." The question is: What will best express the recognition of all those influenced by the decision? This is illustrated by an example from the Netherlands: "as long as nursing homes are understaffed, so that residents have to wear diapers, there is no justification for building an enormously expensive stretch of road [....] that will at best only temporarily relieve the problem of traffic jams".

The example is good and persuasive. But other examples can be more controversial, since there is always a room for some disagreement as to how important certain interests are. Lobbying is a political reality, not only in Brussels.

(e) *Internet sales of cures of dubious value*. Moreover, let us consider recommendations related to the concern that treatments of age-related ailments and illnesses offered on the interned sometimes are based on hype and wishful thinking. Inez de Beaufort suggests towards the end of her chapter that "policymakers should consider how they can cater to the needs of different elderly and to influence the pressure on aging people and the ideal of youthfulness."

And she adds a more specific suggestion: "Also there should be sensible and controlled portals to provide information on what treatments rest on evidence and which are based on pure speculation and wishful thinking." Regard for the importance of safety and consumer protection are underpinning this suggestion. She continues: "There may be rather dangerous treatments around." These policy suggestions are both concrete and realistic—and likely to be effective if they are carried out.

(f) *Rising costs*. Next let us consider recommendations related to the concern that the costs of healthcare and social service for the elderly are rising dramatically. Current solutions include age-based rationing and the privatisation of Medicare. But in her chapter, Laura Capitaine argues that they are based on a misconstrual of the underlying problem and that:

> Cost containment policy should redirect its focus away from population aging towards medical technology.

All European countries are struggling to find ways of dealing with the challenges of cost containment in healthcare. The challenge is to contain the costs in ways that are effective and perceived as fair on grounds the citizens cannot reasonably reject. It certainly is an issue that calls for public health recommendations, but politicians in many countries have been reluctant to deal with suggestions implying rationing. This is a politically unpopular subject especially in election times. I will now discuss certain aspects of this problem in somewhat more detail.

(g) *Fair access to healthcare*. Several recommendations are related to the concern that the elderly are discriminated against unfairly, as individuals and as a group, in particular that the elderly are not provided fair access to healthcare. Here we need to consider not only abstract theoretical problems, but also the design of basic institutional structures and what counts and should count as primary goods, as Martha Nussbaum has reminded us. (Nussbaum 2006, p. 127)

Hans-Jörg Ehni has chosen an interesting starting point. If age is the basis of much current priority setting in healthcare, though not uncontested, and research into biological aging has indicated that interventions are possible in order to prevent, slow down or reverse the aging process, what are the consequences of this for access to healthcare? These interventions are likely to involve complex, regular and personalised sets of services. Should such interventions be made widely available? What will the impact be in the context of existing health inequalities? His approach is admittedly more limited than the one pursued by Schinkel. Ehni focuses on distributive justice. He explicitly says that "... other forms of justice such as retributive justice, corrective justice or global justice are beyond the scope of this presentation".

Fair access to healthcare for the elderly is clearly a central concern in a public policy context. The policy implications related to this concern proposed by Ehni are based on a Rawlsian approach to justice. But a capability oriented approach (Sen 1985; Nussbaum and Sen 1993) would be another possible basis, as he rightly remarks. He suggests that: "... access to longevity interventions should be equal or the worst off should have priority. Otherwise, the possible impact on existing inequalities in healthy life expectancy will also lead to an increase in social injustice."

But if the access to these interventions are equal, present inequalities will be permanented. If we want to do something about these inequalities, access should be unequal; it would seem that those worst off should have priority, at least according to Rawls' Principle of Difference (Rawls 1991).

What conclusions should we draw from this? Ehni mentions three possible strategies: prohibition of longevity interventions, guaranteeing everyone universal access to such interventions, and setting priorities on publicly funded research. He rightly criticizes the first two and opts for the last one.

The policy proposal that the principles of priority setting of research should be clarified and possibly changed is interesting.

But how this is related to the concerns above is not quite clear to me. The issues raised by priority setting in research are not the same as those raised by priority setting in healthcare (Malek 1994; Hermerén 2010). There is some relation between these two priority-setting problem, but the relation is not at all clear. The criteria used are not identical. Stem cell research has for some time been one of the top priorities in the research funding, but stem-cell based therapies have not been a top priority in healthcare.

Anyway, this is intended by Ehni only as a starting point for a discussion, and he mentions "cost-effectiveness" and "social aspects" as relevant concerns. He illustrates this with research on obesity, which is a well-chosen example for the reasons he indicates.

Personally, I think that it would be interesting to apply also Scanlon's contractualist ethical framework to the concerns discussed in this book (Scanlon 2000). At the same time such an attempt might provide an interesting test of this framework.

For Scanlon, the idea of justifiability to others is morally basic. Reasons rather than desires or interests provide the starting point. The challenge is to be able to justify a proposed action to others with principles they could not reasonably reject. According to Scanlon "... thinking about right and wrong is, at the most basic level, thinking about what could be justified to others on grounds that they, if appropriately motivated, could not reasonably reject" (Scanlon 2000, p. 5). Of course, also utilitarians hold that an act should be justifiable to others on grounds they could not reasonably reject. But, according to Scanlon, the difference is that "for utilitarians, what makes an action right is having the best consequences; justifiability is merely a consequence of this." (Scanlon 2000, p. 189)

As Scanlon points out in his comments to Rawls' Difference Principle: "... there are important differences between the subject of Rawl's theory and the one being considered here. To begin with, Rawl's principles of justice are not intended to guide every choice and policy. They are proposed only for the specific task of assessing the justice of basic social institutions" (Scanlon 2000, pp. 228–229). It is clear from the contributions to this book that the ethical issues raised by treatments of the elderly in our society concern also other issues than distributive justice. This suggests that it would be useful to have a somewhat broader ethical framework than the one proposed by Rawls.

Examples where a broader framework like the one proposed by Scanlon would be of use include, for instance, cases of unclear responsibilities, controversies over how to deal with internet sales of products of dubious value, and in particular, cases of inadequate support to vulnerable and frail segments of the population.

20.11 Concluding Remarks

I hope I have demonstrated the relevance of many of the ethical issues discussed in this volume for public health policy. Finally, some personal reflections, in part related to my experience in national and international committees recommending public policies.

There is a difference between successful policymaking and contributing to lively discussions in a philosophical seminar. They do not exclude each other, of course. But in a philosophical seminar, the goal is clarity and intellectual understanding. It is possible to try out almost anything in the critical examination of an argument: to change definitions, revise assumptions, make thought experiments ("imagine a tribe...”), and explore the consequences of various scenarios. In policy making, the goal is to change current or near future trends or states of affairs, for example to fight understaffed nursing homes, internet ads based on wishful speculative thinking, unfair access to healthcare for groups in society, and other forms of discrimination and stigmatization. To be successful here, certain pragmatic considerations are essential.

The potential conflicts between what is good for certain individuals and what is good for society at large must not be swept under the carpet. The conditions in the introduction and their role throughout this chapter illustrate that values and ethics are at the heart of public health policy, and that public policy recommendations must have an ethical underpinning. The underpinning—particularly if a global view is taken—that I personally would be inclined to work with would be a human rights based approach, though there are good arguments also for several other approaches—such as capabilities-oriented or contractualist ones.

The advantage of anchoring policy recommendations in human rights is simply that these rights are enshrined in documents, which have been accepted by political assemblies like the UN and the Council of Europe. The same holds for the UN Millennium goals and similar politically agreed on treaties, even if they are lofty and vague. They provide a reference point for policy-makers, and a basis for a dialogue, which in itself is important.

It is politically difficult to propose or defend a policy that clearly violates human rights. If one wants to argue against, for instance, understaffed nursing homes, unfair access to healthcare for the elderly, or other forms of discrimination of the elderly, such moral underpinning is helpful. The challenge is to relate the proposals to the statements in such declarations. But if this can be done in a convincing way, it adds moral and political force—as well as legitimacy—to the policy recommendations.

I am aware, of course, that the philosophical basis of human and animal rights can be called into question, and has been called into question many times since the days of Jeremy Bentham. I am also aware that declarations of rights are sometimes vague and open to several interpretations. Agreement has a price, and the price is sometimes high. But this does not mean that references to human rights are arbitrary.

Relativism, cultural diversity, and national sovereignty create challenges for advocates of human rights and demonstrate the need for robust justification of these rights. Many philosophers, usually inspired by the Kantian tradition in philosophy, have attempted to provide such justifications. One of them is Alan Gewirth (1982, 1984, 1996). He bases this justification on a moral principle, according to which every agent must act in accordance with his or her own and all other agents’ generic rights to freedom and well-being. He derives this principle from the preconditions of purposeful human action. Deryck Beyleveld (1991) continues this discussion by reformulating Gewirth’s arguments, summarizing all objections to them and giving his own clear responses to them. Other influential writers in the human rights tradition include Ronald Dworkin (1977). The important discussion of the existence, justification and application of human rights is likely to continue for a long time in both political and philosophical circles.

References

Beyleveld, D. 1991. *The Dialectical Necessity of Morality*. Chicago: University of Chicago Press.

Boorse, C. 1977. Health as a theoretical concept. *Philosophy of Science* 44(4): 442–473.

Broström, L. 2007. *The Substituted Judgment Standard*. Lund University: Faculty of Medicine.

Brownsword, R. 2008. *Rights, regulation, and the technological revolution*. Oxford: Oxford University Press.

Campbell, T. 1985. *Seven theories of human society*. Oxford: Clarendon Press.

Daniels, N. 2008a. *Just Health. Meeting healthcare needs fairly*. Cambridge: Cambridge University Press.

Daniels, N. 2008b. "Justice between adjacent generations: Further thoughts". *The Journal of Political Philosophy* 16(4): 475–494.

Dworkin, R. 1977. *Taking rights seriously*. Cambridge: Harvard University Press.

Fleischhauer, K., and G. Hermerén. 2006. *Goals of medicine in the course of history and today*. Stockholm: Kungl. Vitterhets Historie och Antikvitets Akademien (Almqvist & Wiksell distr).

Gewirth, A. 1982. *Human rights: Essays on justification and applications*. Chicago: The University of Chicago Press.

Gewirth, A. 1984. *Human rights: Essays on justification and applications*. Chicago: The University of Chicago Press.

Gewirth, A. 1996. *The community of rights*. Chicago: The University of Chicago Press.

Hermerén, G. 2007. Challenges in the evaluation of nanoscale research: Ethical aspects. *NanoEthics* 1: 223–237.

Hermerén, G. 2008. European values – and others. Europe's shared values: Towards an ever-closer Union?. *European Review* 16(3): 373–385.

Hermerén, G. 2010. Looking at the future of translational stem cell research and stem cell-based therapeutic applications: Priority setting and social justice. In *Translational stem cell research: issues beyond the debate on the moral status of the human embryo*, ed. K. Hug and G. Hermerén, 431–447. New York: Springer/ Humana Press.

Hermerén. G. 2011. The principle of proportionality revisited: Interpretations and applications. *Medicine, Healthcare and Philosophy*, published online 01 November 2011.

Malek, M. 1994. *Setting priorities in healthcare*. New York: Wiley.

Marmot, M.G., and R.G. Wilkinson. 2006. *Social determinants of health*. Oxford: Oxford University Press.

Miller, R.A. 2004. Extending life: Scientific prospects and political obstacles. In *The fountain of youth. Cultural, scientific, and ethical perspectives on a biomedical goal*, ed. S.G. Post and R.H. Binstock, 228–248. Oxford: Oxford University Press.

Nordenfelt, L. 1995. *On the nature of health*. Dordrecht: Kluwer Academic Publishers.

Nordenfelt, L. 2007. The concepts of health and illness revisited. *Medicine, Healthcare and Philosophy* 10(1): 5–10.

Nozick, R. 1974. *State, anarchy, and utopia*. New York: Barnes & Noble.

Nussbaum, M., and A. Sen,1993. Eds. *The quality of life*. Oxford: Clarendon Press.

Nussbaum, M. 2006. *Frontiers of justice: Disability, nationality, species membership*. Cambridge: The Belknap Press of Harvard University Press.

Rawls, J. 1991. *A theory of justice (Revised ed.)*. Cambridge: Harvard University Press.

Rothstein, B. 1998. *Just institutions matter: The moral and political logic of the universal welfare state*. Cambridge: Cambridge University Press.

Scanlon, T.M. 2000. *What we owe to each other*. Cambridge: The Belknap Press of Harvard University Press.

Sen, A.K. 1985. *Commodities and capabilities*. Amsterdam: North-Holland.

Index

M. Schermer, W. Pinxten (eds.), *Ethics, Health Policy and (Anti-) Aging: Mixed Blessings,* Ethics and Health Policy 1, DOI 10.1007/978-94-007-3870-6,

MIX
Papier aus verantwortungsvollen Quellen
Paper from responsible sources
FSC® C105338
FSC
www.fsc.org

If you have any concerns about our products,
you can contact us on
ProductSafety@springernature.com

In case Publisher is established outside the EU,
the EU authorized representative is:
Springer Nature Customer Service Center GmbH
Europaplatz 3, 69115 Heidelberg, Germany

Printed by Libri Plureos GmbH
in Hamburg, Germany